# John Milnor

# Collected Papers

# John Milnor

# Collected Papers

## Volume 1

## Geometry

Credits (the introductions to the articles give complete sources).

*On the total curvature of knots* is reprinted courtesy of Annals of Mathematics.

*On the total curvatures of closed space curves* is reprinted by permission of Mathematica Scandinavica.

*On the existence of a connection with curvature zero* is reprinted by permission of Commentarii Mathematici Helvetici.

*Eigenvalues of the Laplace operator on certain manifolds* is reprinted courtesy of National Academy of Sciences USA.

*A note on curvature and fundamental group* is reprinted by permission of Journal of Differential Geometry.

*Problems in Differential Geometry* is reprinted by permission of the American Mathematical Society.

*Curvatures of left invariant metrics on Lie groups* is reprinted by permission of Academic Press.

*On deciding whether a surface is parabolic or hyperbolic,* is reprinted courtesy of the Mathematical Association of America.

*On fundamental groups of complete affinely flat manifolds* is reprinted by permission of Academic Press.

*On the Betti numbers of real varieties* is reprinted by permission of the American Mathematical Society.

*Isolated singularities defined by weighted homogeneous polynomials* is reprinted from Topology, Volume 9 (1970), John Milnor and Peter Orlik, pp. 385–393, Copyright 1970, with kind permission from Pergamon Press Ltd., Headington Hill Hall, Oxford OX3 0BW, UK.

*Isolated critical points of complex functions* is reprinted by permission of the American Mathematical Society.

*A problem in cartography* is reprinted courtesy of the Mathematical Association of America.

*Hilbert's problem 18: on crystallographic groups, fundamental domains, and on sphere packing* is reprinted by permission of the American Mathematical Society.

*Analytic proofs of the "Hairy Ball Theorem" and the Brouwer Fixed Point Theorem* is reprinted courtesy of the Mathematical Association of America.

*Hyperbolic geometry: the first 150 years* is reprinted by permission of the American Mathematical Society.

*On the geometry of the Kepler Problem* is reprinted courtesy of the Mathematical Association of America.

Printed in the United States of America.

ISBN 0-914098-30-6 Publish or Perish, Inc. Houston, Texas

# CONTENTS

*Starred papers are previously unpublished

CONTENTS

**Euclidean and non-Euclidean Geometry**

# PREFACE

This is the beginning of an effort to organize my published and unpublished papers. It seemed most useful to arrange these papers by subject, rather than date. This first volume consists of papers, spanning more than forty years, which have a strongly geometric flavor. It includes four papers which have not previously been published.

I am grateful to Michael Spivak for his help in bringing this collection together.

John Milnor
Stony Brook, August 1993

## *This collection of papers is dedicated to the old Fine Hall*

It is difficult to describe the uplift which I felt in coming to Princeton in 1948. Although I was a freshman, and a very young and asocial one at that, I was quickly made to feel at home in Fine Hall. Let me first describe the building itself, a marvelous place where one could absorb mathematics and the love of mathematics by osmosis. There was a large comfortable common room, where I found the games particularly intriguing: not only chess but go, Kriegspiel, and that most topological of board games "Nash" (also called hex). There was a lovely library, with an amazing collection of books, and many quiet nooks for studying. Some rooms were decorated with stained glass windows displaying important equations of mathematics and physics. A quotation from Einstein was carved in a place of honor: "Raffiniert ist der Herr Gott, aber boshaft ist er nicht".

But of course the important thing was not the building but the inhabitants. I soon learned to know not only fellow undergraduates in mathematics, but also a fascinating collection of graduate students and faculty from all over the world. At first I was particularly influenced by Al Tucker and Ralph Fox. Perhaps the most decisive experience for me was a course in General Topology taught by Fox, using the R. L. Moore method in which the students do most of the work. Every meeting was a fresh challenge in solving concrete mathematical problems. Another influence in early years was Solomon Lefschetz, a grandfatherly figure, friendly but perhaps also intimidating. Lefschetz, who had lost both hands in an industrial accident many years earlier, was difficult for me to understand, both mathematically and personally, but was an inspiring figure.

*Opposite page*: The Fine Hall faculty in May 1958.
*Front row* (left to right): Eugene Wigner, Al Tucker, Valya Bargmann.
*Second Row:* Fritz Hirzebruch, Ralph Fox, John Moore.
*Third Row:* Don Spencer, Alonzo Church, Emil Artin, Sam Wilks, John Milnor, John Tukey, Norman Steenrod, Will Feller.

Another powerful influence and powerful personality was Emil Artin, whose elegantly crafted lectures were always works of art. I am afraid that the art sometimes obscured the underlying ideas for me. Thus I attended an entire course in Algebraic Number Theory, learning many new and exciting ideas, but never finding out what an algebraic number is. My education was spotty in other ways. Some years later, when I had to teach a second year calculus course, I was chagrined to discover that I didn't know much of the material.

The most important influence on my work in subsequent years was Norman Steenrod, a friend and teacher whose care and precision in developing the ideas of algebraic topology provided an extremely valuable role model.

I am very sorry that I never got to know other members of the Princeton faculty nearly so well. Unfortunately I had an aversion to analysis and probability at this time, and therefore failed to take advantage of the presence of Salomon Bochner and Will Feller. Bochner in particular was an enigmatic figure to me, and I didn't get to know him at all well until many years later. I never did learn the reason for the permanent feud which kept Lefschetz and Bochner from speaking to each other.

Let me close by mentioning three other major influences on my mathematical development. John Moore's influence in algebraic topology is clear. A year in Zürich attending lectures by Heinz Hopf helped to nurture my feeling for geometry. Finally, I owe a clear debt to Henry Whitehead, whom I first met at a conference for young topologists which he organized in Oxford. Although I could never match his endurance in the pub, his love of topology was a constant inspiration.

jwm, Stony Brook, August 1993

# Differential Geometry and Curvature

# ON THE TOTAL CURVATURE OF KNOTS

*Reprinted from*

Annals of Mathematics

Volume 52 (1950), pp. 248–257

This paper arose from a course in Differential Geometry, taught by Al Tucker in 1948–49. Tucker described the ***total curvature***

$$\int_C \kappa\, ds = \int_C \|d^2\mathbf{x}/ds^2\|\, ds$$

of a space curve $C$, parameterized by its arclength $s$. He described Fenchel's Theorem that $\int_C \kappa\, ds \geq 2\pi$ for any closed space curve, as well as Borsuk's conjecture that $\int_C \kappa\, ds > 4\pi$ whenever the closed curve $C$ is knotted. I produced a rough sketch of a proof, which developed into the paper which follows. For more readable expositions of this proof, see Chern[1] or Spivak.[2] An independent proof by I. Fary was published at about the same time.

The argument actually shows an inequality of the form

$$\int_C \kappa\, ds > 2\pi\mu(C)$$

for any knotted curve, where $\mu(C) \geq 2$ is an integer valued topological invariant which I called the ***crookedness*** of the knot. A follow-up paper by Fox showed that $\mu(C)$ can take on arbitrarily large values. This same number $\mu(C)$ has been studied by Schubert[3] and others under the name "***bridge number***". (Compare Jones.[4])

[1] S.-S. Chern, *Curves and surfaces in Euclidean space,* pp. 16–56 of "Studies in Global Geometry and Analysis", S.-S. Chern, ed., M. A. A. 1967.

[2] M. Spivak, *A brief report on John Milnor's brief excursions into differential geometry,* pp. 31–43 of "Topological Methods in Modern Mathematics", Goldberg and Phillips eds., Publish or Perish 1993.

[3] H. Schubert, *Knoten mit zwei Brücken,* Math. Zeits. **65** (1956) 133–170.

[4] V. F. R. Jones, *Milnor's work and knot polynomials,* pp. 195–202 of "Topological Methods in Modern Mathematics", Goldberg and Phillips, eds., Publish or Perish 1993.

# ON THE TOTAL CURVATURE OF KNOTS

By J. W. Milnor

(Received October 5, 1949)

## Introduction

The *total curvature* $\int_C |\mathfrak{x}''(s)|\, ds$ of a closed curve $C$ of class $C''$, a quantity which measures the total turning of the tangent vector, was studied by W. Fenchel, who proved, in 1929, that, in three dimensional space, $\int_C |\mathfrak{x}''(s)|\, ds \geqq 2\pi$, equality holding only for plane convex curves. K. Borsuk, in 1947, extended this result to $n$ dimensional space, and, in the same paper, conjectured that the total curvature of a knot in three dimensional space must exceed $4\pi$. A proof of this conjecture is presented below.[1]

In proving this proposition, use will be made of a definition, suggested by R. H. Fox, of total curvature which is applicable to *any* closed curve. This general definition is validated by showing that the generalized total curvature $\kappa(C)$ is equal to $\int_C |\mathfrak{x}''(s)|\, ds$ for any closed curve $C$ of class $C''$. Furthermore, the theorem of Fenchel and Borsuk is true for any closed curve, if the new definition of total curvature is used.

Closely related to the concept of total curvature is a new invarient $\mu(\mathfrak{C})$, the *crookedness* of the isotopy type $\mathfrak{C}$ of closed curves. This is either a positive integer or $\infty$, according as the type $\mathfrak{C}$ is or is not represented by a polygon. In terms of the concept of crookedness it is possible to provide an alternative formulation of the generalized total curvature as a Lebesgue integral over an $(n - 1)$ dimensional sphere. The crookedness $\mu(\mathfrak{C})$ of a type $\mathfrak{C}$ of simple closed curves is connected with the total curvatures of its representative curves $C$ by the fundamental relation $2\pi\mu(\mathfrak{C}) = \text{g.l.b. } \kappa(C)$. Generally speaking this lower bound is not attained.

In the course of the paper several interesting incidental results are obtained: if the total curvature of a simple closed curve is finite, then there is an inscribed polygon equivalent to it by isotopy, and also if the curve is knotted there must be a plane which intersects it in at least six points.

I am indebted to R. H. Fox for substantial assistance in the preparation of this paper.

## 1. The Total Curvature of a Closed Polygon

By a *closed polygon* $P$ in Euclidean $n$-space $H^n$, $n \geqq 1$, will be meant a finite sequence of points $\mathfrak{a}_0, \mathfrak{a}_1, \cdots, \mathfrak{a}_{m-1}, \mathfrak{a}_m = \mathfrak{a}_0$, of which is required only that

[1] Since the completion of this paper, there has appeared an independant proof, by I. Fáry, that the relation $\int_C |\mathfrak{x}''(s)|\, ds \geqq 4\pi$ holds for all knots [6].

$\mathfrak{a}_i \neq \mathfrak{a}_{i+1}$, and the line segments $\mathfrak{a}_i\mathfrak{a}_{i+1}$ for $i = 0, 1, \cdots, m-1$. For convenience let $\mathfrak{a}_i$, where $i$ is any integer, signify $\mathfrak{a}_{(i)}$, where $(i)$ is the least positive residue mod $m$. The terms point and vector will be used synonymously, with every vector referred to a common origin, so that $\mathfrak{a}_{i+1} - \mathfrak{a}_i$ means the vector equal in length and parallel to the line segment $\mathfrak{a}_i\mathfrak{a}_{i+1}$. Denote by $\alpha_i$ the angle between the vectors $\mathfrak{a}_{i+1} - \mathfrak{a}_i$ and $\mathfrak{a}_i - \mathfrak{a}_{i-1}$ satisfying $0 \leqq \alpha_i \leqq \pi$. By *the total curvature* $\kappa(P)$ *of a closed polygon* $P$ is meant the angle sum $\sum_{i=1}^{m} \alpha_i$.

1.1 LEMMA.[2] *The adjunction of a new vertex to a closed polygon cannot decrease its total curvature. The curvature may remain constant if either the new vertex* $\mathfrak{a}_j$ *and the two adjacent vertices* $\mathfrak{a}_{j-1}$ *and* $\mathfrak{a}_{j+1}$ *are collinear or* $\mathfrak{a}_{j-2}, \mathfrak{a}_{j-1}, \mathfrak{a}_j, \mathfrak{a}_{j+1}$, *and* $\mathfrak{a}_{j+2}$ *are coplanar. Otherwise it must definitely increase.*

Let $P'$ be the closed polygon with vertices $\mathfrak{a}_1, \mathfrak{a}_2, \cdots, \mathfrak{a}_{j-1}, \mathfrak{a}_{j+1}, \cdots, \mathfrak{a}_m$. Let $P$ be the closed polygon with vertices $\mathfrak{a}_1, \mathfrak{a}_2, \cdots, \mathfrak{a}_{j-1}, \mathfrak{a}_j, \mathfrak{a}_{j+1}, \cdots, \mathfrak{a}_m$, obtained from $P'$ by adjoining the vertex $\mathfrak{a}_j$. Denote by $\alpha_i'$ for $i = 1, 2, \cdots,$

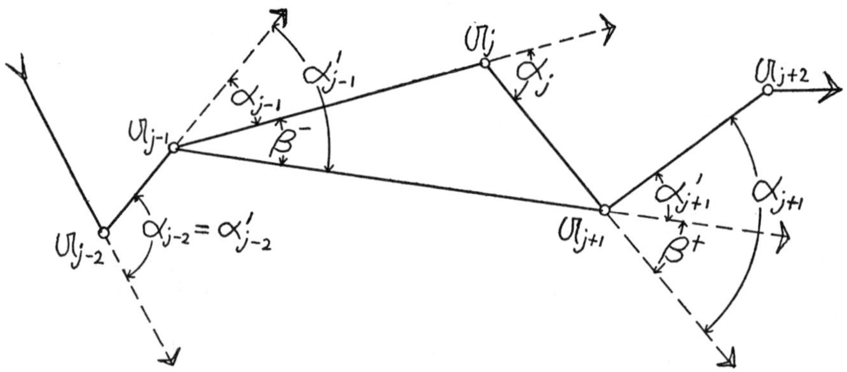


FIG. 1

$j-1, j+1, \cdots, m$ the respective exterior angles of $P'$, and by $\alpha_i$ for $i = 1, 2, \cdots, j-1, j, j+1, \cdots, m$ the respective exterior angles of $P$. Denote by $\beta^-$ the angle between $\mathfrak{a}_j - \mathfrak{a}_{j-1}$ and $\mathfrak{a}_{j+1} - \mathfrak{a}_{j-1}$, and by $\beta^+$ the angle between $\mathfrak{a}_{j+1} - \mathfrak{a}_{j-1}$ and $\mathfrak{a}_{j+1} - \mathfrak{a}_j$. (See Fig. 1.)

By the triangle inequality for spherical triangles $\alpha_{j-1} + \beta^- \geqq \alpha_{j-1}'$, where the equality can hold only if the three angles lie in a plane, that is, only if $\mathfrak{a}_{j-2}$, $\mathfrak{a}_{j-1}$, $\mathfrak{a}_j$, and $\mathfrak{a}_{j+1}$ are coplanar. Similarly $\alpha_{j+1} + \beta^+ \geqq \alpha_{j+1}'$, where the equality can hold only if $\mathfrak{a}_{j-1}, \mathfrak{a}_j, \mathfrak{a}_{j+1}$, and $\mathfrak{a}_{j+2}$ are coplanar. From the triangle with vertices $\mathfrak{a}_{j-1}, \mathfrak{a}_j$, and $\mathfrak{a}_{j+1}$ we have $\beta^- + \beta^+ = \alpha_j$. Therefore

$$\begin{aligned}\kappa(P) - \kappa(P') &= (\alpha_{j-1} - \alpha_{j-1}') + \alpha_j + (\alpha_{j+1} - \alpha_{j+1}') \\ &\geqq -\beta^- \qquad + \alpha_j \qquad -\beta^+ = 0.\end{aligned}$$

Hence $\kappa(P) \geqq \kappa(P')$ and the equality can hold only if either $\mathfrak{a}_{j-2}, \mathfrak{a}_{j-1}, \mathfrak{a}_j$, $\mathfrak{a}_{j+1}$, and $\mathfrak{a}_{j+2}$ are coplanar or $\mathfrak{a}_{j-1}, \mathfrak{a}_j$, and $\mathfrak{a}_{j+1}$ are collinear.

[2] This proof is essentially the same as a proof given by Borsuk [1. pp. 254–256].

1.2 Corollary. *If the vertices* $\mathfrak{a}_{j-2}$, $\mathfrak{a}_{j-1}$, $\mathfrak{a}_j$, *and* $\mathfrak{a}_{j+1}$ *of a closed polygon* $P$ *are not coplanar, and the vertex* $\mathfrak{a}_j$ *is replaced by a vertex* $\mathfrak{a}_j'$ *which lies on the line segment* $\mathfrak{a}_j\mathfrak{a}_{j+1}$, *then* $\kappa(P)$ *is decreased.*

## 2. The Total Curvature of a Curve

By a *closed curve* $C$ in Euclidean $n$-space $H^n$ will be meant a continuous vector function $\mathfrak{x}(t) = (\mathfrak{x}_1(t), \cdots, \mathfrak{x}_n(t))$ of period $l$ which is not constant in any $t$-interval. In particular any polygon can be described in this manner; it will be convenient to regard a polygon as a closed curve, ignoring the distinction between different parameterizations. A closed curve $\mathfrak{x}(t)$ is *simple* if $\mathfrak{x}(t_1) = \mathfrak{x}(t_2)$ only when $(t_1 - t_2)/l$ is an integer.

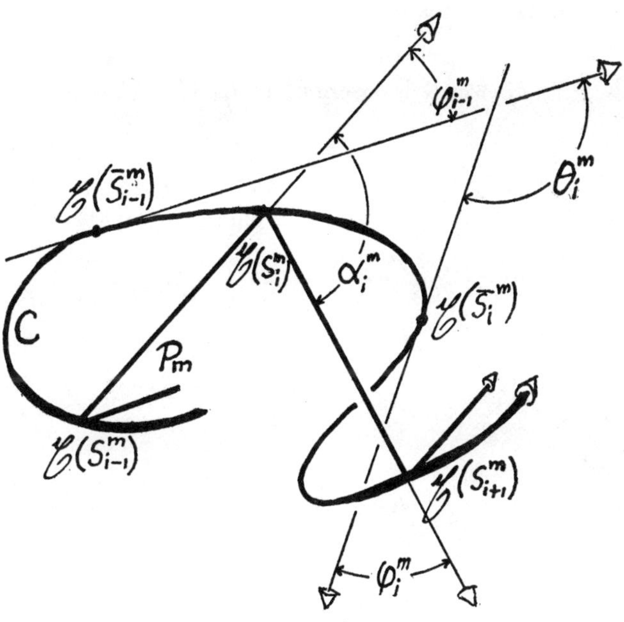


Fig. 2

A closed polygon $P$ with vertices $\mathfrak{a}_1, \cdots, \mathfrak{a}_m$ is said to be *inscribed* in a closed curve $\mathfrak{x}(t)$ if there is a set of parameter values $t_i$ such that $t_i < t_{i+1}$, $t_{i+m} = t_i + l$, and $\mathfrak{a}_i = \mathfrak{x}(t_i)$ for all integral values of $i$.

2.1 Lemma. *For any closed polygon* $P$, $\kappa(P) = \text{l.u.b. } \{\kappa(P')\}$ *where* $P'$ *ranges over all polygons inscribed in* $P$.

If $P$ is a polygon having two or more vertices coincident, it can be represented as the limit of a sequence of polygons with all vertices distinct; hence it only remains to prove the lemma for $P$ with all vertices distinct.

If $P_0'$ is a representative inscribed polygon whose vertices include all but $m$ of the vertices of $P$, we may adjoin the remaining vertices of $P$ one by one to $P_0'$, producing a sequence of polygons $P_0'$, $P_1'$, $\cdots$, $P_m'$. By 1.1, $\kappa(P_0') \leqq \kappa(P_1') \leqq, \cdots, \leqq \kappa(P_m')$; but $\kappa(P_m') = \kappa(P)$. Therefore $\kappa(P) = \text{l.u.b. } \{\kappa(P')\}$.

For each closed curve $C$ define *the total curvature* by $\kappa(C) = \text{l.u.b.}\ \{\kappa(P)\}$ where $P$ ranges over all polygons inscribed in $C$. If $C$ is itself a polygon, the preceding lemma shows that this definition is consistent with the definition of Section 1.

2.2 THEOREM. *If $C$ is a closed curve of class $C''$ parameterized by arclength $s$, then* $\kappa(C) = \int_C |\mathfrak{x}''(s)|\,ds$.

If $\mathfrak{a}_1^m = \mathfrak{x}(s_1^m), \cdots, \mathfrak{a}_m^m = \mathfrak{x}(s_m^m)$ are the vertices of a polygon $P_m$ inscribed in $C$, such that $\lim_{m\to\infty} \max_i \{(s_{i+1}^m - s_i^m)\} = 0$, it will first be shown that $\lim_{m\to\infty}\kappa(P_m) = \int_C |\mathfrak{x}''(s)|\,ds$ (Compare Fig. 2).

Define $\bar{s}_i^m = \frac{1}{2}(s_i^m + s_{i+1}^m)$ for every $i$ and $m$. Denote by $\theta_i^m$ the angle between $\mathfrak{x}'(\bar{s}_{i-1}^m)$ and $\mathfrak{x}'(\bar{s}_i^m)$. The vector $\mathfrak{x}'(s)$ describes a curve $L$ of length $\int_C |\mathfrak{x}''(s)|\,ds$ on the unit sphere $S^{n-1}$. The vectors $\mathfrak{x}'(\bar{s}_i^m)$ form the vertices of a spherical polygon of length $\sum_{i=1}^m \theta_i^m$ which is inscribed in $L$. Therefore $\lim_{m\to\infty} \sum_{i=1}^m \theta_i^m = \int_C |\mathfrak{x}''(s)|\,ds$.

Since $\mathfrak{x}''(s)$ is uniformly continuous, for each $\epsilon > 0$ there is a $\delta > 0$ such that $|\mathfrak{x}''(u) - \mathfrak{x}''(v)| < \epsilon$ for all $|u - v| < \delta$. From the identity

$$\mathfrak{x}(s_{i+1}^m) - \mathfrak{x}(s_i^m) = (s_{i+1}^m - s_i^m)\mathfrak{x}'(\bar{s}_i^m) + \int_{\bar{s}_i^m}^{s_{i+1}^m} \int_{\bar{s}_i^m}^{v} [\mathfrak{x}''(u) - \mathfrak{x}''(\bar{s}_i^m)]\,du\,dv$$

$$+ \int_{s_i^m}^{\bar{s}_i^m} \int_{v}^{\bar{s}_i^m} [\mathfrak{x}''(\bar{s}_i^m) - \mathfrak{x}''(u)]\,du\,dv$$

we have

$$\left| \frac{\mathfrak{x}(s_{i+1}^m) - \mathfrak{x}(s_i^m)}{s_{i+1}^m - s_i^m} - \mathfrak{x}'(\bar{s}_i^m) \right| < (s_{i+1}^m - s_i^m)\,\frac{\epsilon}{4} \text{ whenever } \max_i \{(s_{i+1}^m - s_i^m)\} < \delta.$$

If $\varphi_i^m$ is the angle between $\mathfrak{x}(s_{i+1}^m) - \mathfrak{x}(s_i^m)$ and $\mathfrak{x}'(\bar{s}_i^m)$, then $\sin\varphi_i^m < (s_{i+1}^m - s_i^m)\,\epsilon/4$, since the end point of the vector $(\mathfrak{x}(s_{i+1}^m) - \mathfrak{x}(s_i^m))/(s_{i+1}^m - s_i^m)$ lies within a sphere of radius $(s_{i+1}^m - s_i^m)\,\epsilon/4$ about the end point of the unit vector $\mathfrak{x}'(\bar{s}_i^m)$. Hence for sufficiently small $\epsilon$ we have $\varphi_i^m < 2\sin\varphi_i^m < (s_{i+1}^m - s_i^m)\,\epsilon/2$. The angle between $\mathfrak{x}(s_{i+1}^m) - \mathfrak{x}(s_i^m)$ and $\mathfrak{x}(s_i^m) - \mathfrak{x}(s_{i-1}^m)$ is $\alpha_i^m$, while that between $\mathfrak{x}'(\bar{s}_i^m)$ and $\mathfrak{x}'(\bar{s}_{i-1}^m)$ is $\theta_i^m$. Therefore $\alpha_i^m \leqq \theta_i^m + \varphi_i^m + \varphi_{i-1}^m$ and $\theta_i^m \leqq \alpha_i^m + \varphi_i^m + \varphi_{i-1}^m$, so that $|\alpha_i^m - \theta_i^m| < (s_{i+1}^m - s_{i-1}^m)\,\epsilon/2$, and $|\sum_{i=1}^m \alpha_i^m - \sum_{i=1}^m \theta_i^m| < l\epsilon$, where $l$ is the length of $C$. Therefore

$$\lim_{m\to\infty} \kappa(P_m) = \lim_{m\to\infty} \sum_{i=1}^m \alpha_i^m = \lim_{m\to\infty} \sum_{i=1}^m \theta_i^m = \int_C |\mathfrak{x}''(s)|\,ds.$$

In order to show that $\int_C |\mathfrak{x}''(s)|\,ds = \text{l.u.b.}\ \{\kappa(P)\}$ for $P$ inscribed in $C$, it only remains to show that $\kappa(P) \leqq \int_C |\mathfrak{x}''(s)|\,ds$.

Given any polygon $P_k$ inscribed in $C$, we may form a sequence of polygons $P_m$ for $m = k, k+1, \cdots$ by adjoining vertices to $P_k$ so that

$$\lim_{m\to\infty} \max_i \{(s_{i+1} - s_i)\} = 0.$$

By 1.1 $\kappa(P_k) \leqq \kappa(P_{k+1}) \leqq \cdots$, but $\lim_{m\to\infty} \kappa(P_m) = \int_C |\mathfrak{x}''(s)|\, ds$, and therefore $\kappa(P_k) \leqq \int_C |\mathfrak{x}''(s)|\, ds$.

### 3. The Crookedness of a Closed Curve

For each closed curve $C$ and each unit vector $\mathfrak{b}$, define $\mu(C, \mathfrak{b})$ to be the number of maxima of the function $\mathfrak{b}\cdot\mathfrak{x}(t)$ (i.e. the number of parameter values $t_0$ for which $\mathfrak{b}\cdot\mathfrak{x}(t_0) \geqq \mathfrak{b}\cdot\mathfrak{x}(t)$ for $t$ within some neighborhood of $t_0$) in a fundamental period. For each closed curve $C$ define $\mu(C) = \min_{\mathfrak{b}} \{\mu(C, \mathfrak{b})\}$. We may call $\mu(C)$ the *crookedness* of $C$.

For every vector $\mathfrak{a}_{i+1} - \mathfrak{a}_i$ in the space $H^n$ define $\mathfrak{b}_i = (\mathfrak{a}_{i+1} - \mathfrak{a}_i)/|\mathfrak{a}_{i+1} - \mathfrak{a}_i|$. According to the convention introduced earlier, $\mathfrak{b}_i$ also denotes a point on the unit sphere $S^{n-1}$, the *spherical image* of $\mathfrak{a}_{i+1} - \mathfrak{a}_i$. Given a polygon $P$ with vertices $\mathfrak{a}_1, \mathfrak{a}_2, \cdots, \mathfrak{a}_m$, a spherical polygon $Q$ is formed on $S^{n-1}$ by joining each $\mathfrak{b}_{i-1}$ to $\mathfrak{b}_i$ by a great circle arc of length $\alpha_i$. This spherical polygon $Q$ is called a *spherical image* of $P$, and is unique unless for some $j$ the vector $\mathfrak{b}_j = -\mathfrak{b}_{j+1}$. Note that it may happen that $\mathfrak{b}_j = \mathfrak{b}_{j+1}$.

3.1 THEOREM.[3] *For any closed curve $C$ in $H^n$, $n \geqq 2$, the Lebesgue integral $\int_{S^{n-1}} \mu(C, \mathfrak{b}) dS$, where $\mathfrak{b}$ ranges over the unit sphere, exists and is equal to $(M_{n-1}\kappa(C))/2\pi$, where $M_{n-1} = (2\pi^{n/2})/\Gamma(n/2)$ is the measure of $S^{n-1}$.*

We will first consider the case in which the curve is a polygon $P$. For every point $\mathfrak{b}$ of $S^{n-1}$, let $S_{\mathfrak{b}}^{n-2}$ denote the great sphere of $S^{n-1}$ which has a pole at $\mathfrak{b}$. An edge $\mathfrak{b}_{j-1}\mathfrak{b}_j$ of $Q$ crosses $S_{\mathfrak{b}}^{n-2}$ if and only if $\mathfrak{b}\cdot(\mathfrak{a}_{j+1} - \mathfrak{a}_j)$ and $\mathfrak{b}\cdot(\mathfrak{a}_j - \mathfrak{a}_{j-1})$ have opposite sign, so that $\mathfrak{b}\cdot\mathfrak{a}_j$ is a maximum or minimum of $\mathfrak{b}\cdot\mathfrak{x}(t)$. Therefore, if $S_{\mathfrak{b}}^{n-2}$ contains no vertex of $Q$, (i.e. no edge of $P$ is perpendiuclar to $\mathfrak{b}$) the number of intersections of $Q$ with $S_{\mathfrak{b}}^{n-2}$ is $2\mu(P, \mathfrak{b})$. The set of points $\mathfrak{b}$ for which $S_{\mathfrak{b}}^{n-2}$ contains some vertex of $Q$ is the union of the finite collection of great spheres $S_{\mathfrak{b}_i}^{n-2}$; in each component of the complement with respect to $S^{n-1}$ of $\bigcup_i S_{\mathfrak{b}_i}^{n-2}$ the function $2\mu(P, \mathfrak{b})$ is constant. The integral $\int_{S^{n-1}} 2\mu(P, \mathfrak{b}) dS$, where $dS$ is the element of surface on $S^{n-1}$, is therefore defined. The set of points $\mathfrak{b}$ for which $S_{\mathfrak{b}}^{n-2}$ meets a given segment $\mathfrak{b}_{i-1}\mathfrak{b}_i$ of length $0 \leqq \alpha_i \leqq \pi$ is a "double lune" bounded by the great spheres $S_{\mathfrak{b}_{i-1}}^{n-2}$ and $S_{\mathfrak{b}_i}^{n-2}$. Thus the contribution of $\mathfrak{b}_{i-1}\mathfrak{b}_i$ to $2\mu(P, \mathfrak{b})$ is 1 if $\mathfrak{b}$ is an interior point of this lune and 0 if $\mathfrak{b}$ is an exterior point. The measure of this lune is $(\alpha_i M_{n-1})/\pi$ where $M_{n-1}$ is the measure of the entire sphere. Consequently

$$\int_{S^{n-1}} 2\mu(P, \mathfrak{b})\, dS = \frac{M_{n-1}}{\pi} \sum_{i=1}^{m} \alpha_i = \frac{M_{n-1}}{\pi} \kappa(P).$$

[3] This theorem is related to Crofton's formula. Cf. [3. p. 81].

If $C$ is an arbitrary closed curve $\mathfrak{x}(t)$, let $P_m$ be a set of inscribed polygons $\mathfrak{x}_m(t)$ with vertices $\mathfrak{a}_1^m = \mathfrak{x}(t_1^m), \cdots, \mathfrak{a}_m^m = \mathfrak{x}(t_m^m)$ such that each $P_m$ contains all the vertices of $P_{m-1}$ and satisfying $\lim_{m\to\infty} \kappa(P_m) = \kappa(C)$ and $\lim_{m\to\infty} \max_i \{(t_{i+1}^m - t_i^m)\} = 0$. The values of $\mathfrak{b}$ for which $\mathfrak{b}\cdot\mathfrak{x}(t)$ or any $\mathfrak{b}\cdot\mathfrak{x}_m(t)$ has an interval of constancy form a set of measure zero, and therefore have no effect on the integral. Such values will be ignored for the remainder of the proof.

We first show that $\mu(C, \mathfrak{b}) = \lim_{m\to\infty} \mu(P_m, \mathfrak{b})$. It is certainly true that $\mu(C, \mathfrak{b}) \geqq \mu(P_m, \mathfrak{b}) \geqq \mu(P_{m-1}, \mathfrak{b})$. If $\mu(C, \mathfrak{b}) < \infty$, it is possible to select a neighborhood of each of the $\mu(C, \mathfrak{b})$ maxima of $\mathfrak{b}\cdot\mathfrak{x}(t)$ and of each of its minima sufficiently small so that a polygon with a vertex in each of these neighborhoods must have at least $\mu(C, \mathfrak{b})$ maxima; which is certainly true of $P_m$ for $m$ sufficiently large. If $\mu(C, \mathfrak{b}) = \infty$, the set of values of $t$ for which $\mathfrak{b}\cdot\mathfrak{x}(t)$ is a maximum must contain a denumerable subset $\{t_{2i}\}$ such that either $t_0 < t_2 < \cdots < \lim_{i\to\infty} t_{2i} < t_0 + l$ or $t_0 > t_2 > \cdots > \lim_{i\to\infty} t_{2i} > t_0 - l$. In either case we may select a series of intermediate values $t_{2i+1}$ such that each $\mathfrak{b}\cdot\mathfrak{x}(t_{2i}) > \mathfrak{b}\cdot\mathfrak{x}(t_{2i+1})$ and $\mathfrak{b}\cdot\mathfrak{x}(t_{2i}) > \mathfrak{b}\cdot\mathfrak{x}(t_{2i-1})$. Given any $2j < \infty$ we may select neighborhoods of the $\mathfrak{x}(t_i)$, for $i < 2j$, so small that any polygon with at least one vertex in each neighborhood has at least $j - 1$ maxima; which is true of $P_m$ for $m$ sufficiently large. Therefore $\mu(P_m, \mathfrak{b})$ increases without finite bound as $m \to \infty$. Each of the integrals $\int_{S^{n-1}} \mu(P_m, \mathfrak{b})\, dS$ exists; and the nondecreasing sequence of positive functions $\mu(P_m, \mathfrak{b})$ approaches $\mu(C, \mathfrak{b})$. Therefore[4] the integral $\int_{S^{n-1}} \mu(C, \mathfrak{b})\, dS$ exists and equals $\lim_{m\to\infty} \int_{S^{n-1}} \mu(P_m, \mathfrak{b})\, dS = \lim_{m\to\infty} \frac{(M_{n-1})}{2\pi} \kappa(P_m) = \frac{(M_{n-1})}{2\pi} \kappa(C)$.

3.2 Corollary. $\kappa(C) \geqq 2\pi\mu(C)$.

Since $$\frac{M_{n-1}}{2\pi} \kappa(C) = \int_{S^{n-1}} \mu(C, \mathfrak{b})\, dS \geqq \int_{S^{n-1}} \mu(C)\, ds = M_{n-1}\mu(C).$$

By a *convex curve* will be meant a closed plane curve described by $\mathfrak{x}(t)$ such that any line contains $\mathfrak{x}(t)$ either for not more than two values of $t$ within a fundamental period or for all values of $t$ within some interval.

3.3 Lemma. *The necessary and sufficient condition that a closed polygon $P$ in $H^2$ be convex is that for every $\mathfrak{b}$ either $\mu(P, \mathfrak{b}) = 1$ or $\mu(P, \mathfrak{b}) = \infty$.*

It is clear that this condition is necessary. Suppose that $P$ is a closed plane polygon such that either $\mu(P, \mathfrak{b}) = 1$ or $\mu(P, \mathfrak{b}) = \infty$ for each $\mathfrak{b}$ in the plane. For any $\mathfrak{b}$ such that $\mu(P, \mathfrak{b}) = 1$, any line perpendicular to $\mathfrak{b}$ will intersect $P$ in at most two points. If $\mathfrak{b}$ is a vector for which $\mu(P, \mathfrak{b}) = \infty$, and if $H^1$ is a line perpendicular to $\mathfrak{b}$ which intersects $P$ a finite number, say $r$, of times, then it is always possible to rotate $H^1$ about one of its points of intersection with $P$ in the proper direction so that $r$ is not decreased. Hence there is a $\bar{\mathfrak{b}}$ and a $\bar{H}^1$ perpendicular to $\bar{\mathfrak{b}}$ such that $\mu(P, \bar{\mathfrak{b}}) < \infty$ and the number of intersections of $\bar{H}^1$ with $P$ is $\bar{r} \geqq r$. But by the first case, $\bar{r} \leqq 2$, and therefore $r \leqq 2$.

[4] [4. Theorem 12.6 p. 28].

3.4 THEOREM. *For any closed curve $C$, $\kappa(C) \geqq 2\pi$. The equality holds if and only if $C$ is convex.*

Since $\mu(C) \geqq 1$ for every curve, or since every curve has an inscribed polygon $P$ for which $\kappa(P) = 2\pi$, we have $\kappa(C) \geqq 2\pi$. Since any curve which is not convex has an inscribed polygon which is not convex, and since a polygon inscribed in a convex curve must be convex, it only remains to prove the second portion of the theorem for polygons.

It is proved in plane geometry that the sum of the exterior angles of a convex polygon is $2\pi$. If there were a non-planar polygon $P$ for which $\kappa(P) = 2\pi$, we could select four consecutive non-coplanar vertices (neglecting vertices for which $\alpha_i = 0$). By 1.2 there would be a new polygon $P'$ such that $\kappa(P') < \kappa(P) = 2\pi$, which is impossible. If there were a non-convex plane polygon $P$ for which $\kappa(P) = 2\pi$, then 3.3 states that there would be a direction $\bar{\mathfrak{b}}$ for which $1 < \mu(P, \bar{\mathfrak{b}}) < \infty$, but there is a neighborhood of any such $\bar{\mathfrak{b}}$ within which $\mu(P, \mathfrak{b})$ is constant. This means that $\kappa(P) = \int_{S^1} \mu(P, \mathfrak{b})\, dS > \int_{S^1} dS = 2\pi$.

## 4. The Curvature and Crookedness of Isotopy Types of Curves

In $H^n$ the closed curve described by $\mathfrak{x}(t)$ of period $l$, and the closed curve described by $\bar{\mathfrak{x}}(t)$ of period $\bar{l}$ are said to be *equivalent by isotopy* if there is some isotopy of $H^n$ onto itself, which transforms $\mathfrak{x}(ul)$ into $\bar{\mathfrak{x}}(u\bar{l})$ for all $u$.

By a *curve type* $\mathfrak{C}$ in $H^n$ is meant an equivalence class of closed curves under isotopy. A curve type is *simple* if the representative closed curves are simple.

A simple curve type $\mathfrak{C}$ and its members are said to be *unknotted* if $\mathfrak{C}$ is that type which contains all circles. If a simple curve type contains no circles then the type and its members are said to be *knotted.*

A curve type $\mathfrak{C}$ and its members are said to be *tame*[5] if $\mathfrak{C}$ contains a polygon. Otherwise they are said to be *wild.*[5] It is well known that in $H^2$ every simple closed curve is unknotted. In $H^n$ for $n > 3$, every simple tame curve is unknotted.

For each curve type $\mathfrak{C}$, define $\kappa(\mathfrak{C})$ = g.l.b. $\kappa(C)$ and $\mu(\mathfrak{C}) = \min \mu(C)$, where $C$ ranges over all members of $\mathfrak{C}$.

4.1 LEMMA. *For each $\mathfrak{c}$ and $\mathfrak{p}$ in $H^{n-1}$ such that $|\mathfrak{c} - \mathfrak{p}| < r$, there is an isotopy, $f_u^{\mathfrak{c}\mathfrak{p}}(\mathfrak{x})$, $0 \leqq u \leqq 1$, of $H^{n-1}$ onto itself which transforms $\mathfrak{c}$ into $\mathfrak{p}$ and leaves fixed all points of $H^{n-1}$ outside the $(n-2)$-sphere of radius $r$ and center $\mathfrak{c}$, such that $f_u^{\mathfrak{c}\mathfrak{p}}(\mathfrak{x})$ is a continuous function of $u$, $\mathfrak{x}$, $\mathfrak{c}$, and $\mathfrak{p}$.*

For example:

$$f_u^{\mathfrak{c}\mathfrak{p}}(\mathfrak{x}) = \begin{cases} \mathfrak{x} - u\left[1 - \dfrac{|\mathfrak{x} - \mathfrak{c}|}{r}\right](\mathfrak{p} - \mathfrak{c}) & \text{for } |\mathfrak{x} - \mathfrak{c}| \leqq r. \\ \mathfrak{x} & \text{for } |\mathfrak{x} - \mathfrak{c}| \geqq r. \end{cases}$$

4.2 THEOREM. *For any simple closed curve $C$, such that $\mu(C) < \infty$, there is a polygon $P$ inscribed in $C$ and equivalent to $C$ by isotopy.*

---

[5] This definition was given by Fox and Artin [5].

If $\mathfrak{b}$ is a unit vector for which $\mu(C, \mathfrak{b}) < \infty$, there are a finite set of values $t_1 < t_2 < \cdots < t_{2\mu(C,\mathfrak{b})} < t_1 + l$, for which $\mathfrak{b} \cdot \mathfrak{x}(t)$ has a maximum or minimum. About each point $\mathfrak{x}(t_i)$ construct a cylinder $Z_i^{n-1}(0)$ with generators parallel to $\mathfrak{b}$ which intersects $C$ in exactly two points $\mathfrak{x}(t_i^-)$ and $\mathfrak{x}(t_i^+)$ such that both lie on a base of the cylinder and such that $\mathfrak{x}(t_i^+)$ is the center of this base. It will first be shown that there is an isotopy of the closed $n$-cell bounded by $Z_i^{n-1}(0)$ onto itself which leaves $Z_i^{n-1}(0)$ fixed and transforms the curve segment $\mathfrak{x}(t)$ for $t_i^- \leqq t \leqq t_i^+$ onto the polygonal line $\mathfrak{x}(t_i^-)$, $\mathfrak{x}(t_i)$, $\mathfrak{x}(t_i^+)$.

Each hyperplane $H^{n-1}$ perpendicular to $\mathfrak{b}$ which intersects $Z_i^{n-1}(0)$ intersects it in a sphere $S^{n-2}$. Perform the isotopy of each $H^{n-1}$ onto itself which transforms the curve segment $\mathfrak{x}(t)$ for $t_i \leqq t \leqq t_i^+$ into the axis of the cylinder, and which leaves all points outside of the $n$-cell bounded by the cylinder fixed, as defined by 4.1. Select a continuous sequence of coaxial cylinders $Z_i^{n-1}(v)$, $0 \leqq v < \infty$, such that any $Z_i^{n-1}(\bar{v})$ is contained within all $Z_i^{n-1}(v)$ with $v < \bar{v}$, such that each cylinder intersects $C$ only in the center of one base and in one other point of that base, and such that $Z_i^{n-1}(v)$ tends to the point $x(t_i)$ as $v \to \infty$. Rotate each $Z_i^{n-1}(v)$ about its axis so that each point $\mathfrak{x}(t)$ for $t_i^- \leqq t < t_i$ is transformed into the plane determined by $\mathfrak{x}(t_i^-)$ and the axis of the cylinders. Since we have transformed $\mathfrak{x}(t)$ for $t_i^- \leqq t \leqq t_i^+$ onto a plane curve within $Z_i^{n-1}(0)$, it is certainly possible to transform it onto the polygonal line $\mathfrak{x}(t_i^-)$, $\mathfrak{x}(t_i)$, $\mathfrak{x}(t_i^+)$, still within $Z_i^{n-1}(0)$, producing an equivalent curve $\bar{C}$ described by $\bar{\mathfrak{x}}(t)$. This curve is divided into $4\mu(C, b) = 4\mu(\bar{C}, b)$ distinct segments by the points $\mathfrak{x}(t_i^+)$ and $\mathfrak{x}(t_i^-)$. If $g > 0$ is the g.l.b. of the distances between distinct and nonconsecutive curve segments, then for each point of $\bar{C}$ which is not within any of the $Z_i^{n-1}$ we may construct the sphere $S^{n-2}$ which has its center at the point, lies in a hyperplane perpendicular to $b$, and has radius $g/3$. Since no two of these spheres can intersect, we have in effect, constructed a tube around each curve segment outside the cylinders, with no two tubes intersecting. It is now possible to inscribe a polygonal line, lying completely within the tube, in each segment of $\bar{C}$. Perform the isotopy of each $H^{n-1}$ perpendicular to $\mathfrak{b}$ onto itself which transforms each of these curve segments onto the corresponding polygonal line and which leaves fixed all points outside of the $S^{n-2}$. We have thus transformed $\bar{C}$, and therefore $C$, into an inscribed polygon $P$, equivalent by isotopy. (Note for future reference that $\mu(C, b) = \mu(P, b)$.)

4.3 Corollary. *The necessary and sufficient condition that a simple curve type $\mathfrak{C}$ be tame is that $\mu(\mathfrak{C}) < \infty$.*

4.4 Corollary. *The total curvature of a tame knot cannot equal the curvature of its type.*

Assume that $C$ is a tame knot of type $\mathfrak{C}$ with $\kappa(C) = \kappa(\mathfrak{C})$. Let $P$ be a polygon of type $\mathfrak{C}$ inscribed in $C$. Then $\kappa(P) \leqq \kappa(C)$. Since $P$ cannot lie in any plane, we may select four consecutive non-coplanar vertices (if we ignore vertices for which $\alpha_i = 0$). By 1.2 we may select a new polygon $\bar{P}$, still a member of $\mathfrak{C}$, and having $\kappa(\bar{P}) < \kappa(P) \leqq \kappa(C) = \kappa(\mathfrak{C})$; which is impossible.

4.5 Corollary. *The crookedness of any knot is greater than or equal to* 2.

If $C$ is a curve with $\mu(C) = 1$, then, in the proof of 4.2 we can select the two cylinders with a common base. The first isotopy will then transform $C$ into a plane quadrilateral, which is certainly unknotted.

4.6 COROLLARY. *The total curvature of any knot is greater than* $4\pi$.

4.7. THEOREM. *If* $\mathfrak{C}$ *is a simple curve type, then* $\kappa(\mathfrak{C}) = 2\pi\mu(\mathfrak{C})$.

It has already been shown that $\kappa(C) \geqq 2\pi\mu(C)$ for any $C \in \mathfrak{C}$ and therefore that $\kappa(\mathfrak{C}) \geqq 2\pi\mu(\mathfrak{C})$. If $\mu(\mathfrak{C}) = \infty$, this proves the proposition. If $\mu(\mathfrak{C}) < \infty$, we may select a curve $C$ of $\mathfrak{C}$ and a direction $\mathfrak{b}$ such that $\mu(C, \mathfrak{b}) = \mu(\mathfrak{C})$. By 4.2 there is a polygon $P$ which is a member of $\mathfrak{C}$ such that $\mu(P, \mathfrak{b}) = \mu(C, \mathfrak{b})$. For convenience we will select a new coordinate system so that $\mathfrak{b}$ is parallel to the $x_1$ axis. We may then define the isotopy $F_u(x_1, x_2, x_3, \cdots, x_n) = (x_1, ux_2, ux_3, \cdots, ux_n)$ for $0 < u \leqq 1$. This evidently transforms $P$ into an

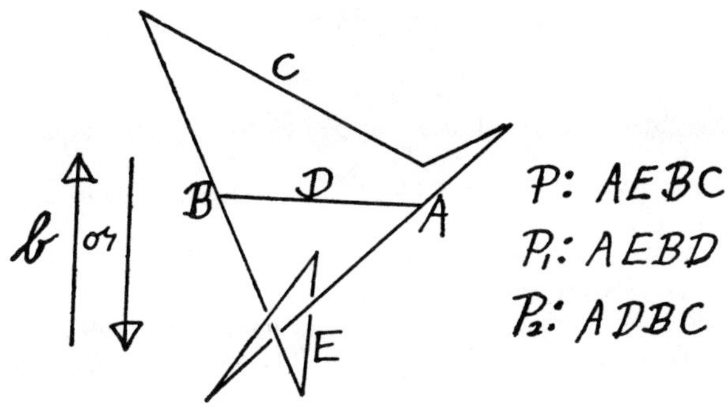


FIG. 3

equivalent polygon $P_u$, such that $\mu(P_u, \mathfrak{b}) = \mu(P, \mathfrak{b})$. If $\mathfrak{a}_i^u$, $1 \leqq i \leqq m$, is the set of vertices of $P_u$, we may divide it into four subsets:

(a) vertices $\mathfrak{a}_i^u$ such that $\mathfrak{b}\cdot\mathfrak{a}_{i-1}^u < \mathfrak{b}\cdot\mathfrak{a}_i^u < \mathfrak{b}\cdot\mathfrak{a}_{i+1}^u$,
(b) vertices $\mathfrak{a}_i^u$ such that $\mathfrak{b}\cdot\mathfrak{a}_{i-1}^u > \mathfrak{b}\cdot\mathfrak{a}_i^u > \mathfrak{b}\cdot\mathfrak{a}_{i+1}^u$,
(c) vertices $\mathfrak{a}_i^u$ such that $\mathfrak{b}\cdot\mathfrak{a}_{i-1}^u < \mathfrak{b}\cdot\mathfrak{a}_i^u > \mathfrak{b}\cdot\mathfrak{a}_{i+1}^u$,
(d) vertices $\mathfrak{a}_i^u$ such that $\mathfrak{b}\cdot\mathfrak{a}_{i-1}^u > \mathfrak{b}\cdot\mathfrak{a}_i^u < \mathfrak{b}\cdot\mathfrak{a}_{i+1}^u$.

(If an equality were to hold, we would have $\mu(P_u, \mathfrak{b}) = \infty$.) Evidently the number of vertices in (c) equals the number in (d) equals $\mu(\mathfrak{C})$. However for members of (c) and (d), $\lim_{u\to 0} (\alpha_i^u) = \pi$; whereas for (a) and (b), $\lim_{u\to 0} (\alpha_i^u) = 0$. Therefore $\lim_{u\to 0} \kappa(P_u) = 2\pi\mu(\mathfrak{C})$.

As another interesting consequence of Theorem 4.2, we have the following.

4.8 THEOREM. *Given a knot* $C$ *in* $H^3$ *for which* $\mu(C) < \infty$, *there is a plane whose intersection with* $C$ *consists of at least six components.*

Since every such $C$ has an inscribed polygon which is knotted, and since a plane intersects a curve at least as many times as it intersects an inscribed polygon, it only remains to prove the theorem for knotted polygons. If there is a polygon which does not satisfy the theorem, there must be one having a minimum number of sides. If $P$ is such a polygon, we have: $4\pi < \kappa(P) = \frac{1}{2}\int_{S^2} \mu(P, \mathfrak{b})\, dS$. Therefore there must be some unit vector $\mathfrak{b}$ such that $2 < \mu(P, \mathfrak{b})$ and also

$\mathfrak{b} \cdot \mathfrak{a}_j \neq \mathfrak{b} \cdot \mathfrak{a}_k$ for every pair of distinct vertices $\mathfrak{a}_j$ and $\mathfrak{a}_k$ of $P$. If we select a plane perpendicular to $\mathfrak{b}$ and move it parallel to itself in the direction of $\mathfrak{b}$ until it intersects $P$, it must first intersect $P$ in a minimum (i.e. an $\mathfrak{a}_j$ such that $\mathfrak{b} \cdot \mathfrak{a}_j < \mathfrak{b} \cdot \mathfrak{a}_{j\pm 1}$). After this it will intersect $P$ in two points, until it intersects another minimum, after which it will have four intersections. If it next reaches another minimum, the theorem is proved. If it next reaches a maximum, there will then be only two intersections. Join these two points by a line segment, so that two new polygons are formed by this segment and the sides of $P$. (See Fig. 3.) At least one of these new polygons, $P_1$, must be knotted. Since $\mu(P) \geqq 3$, each of the new polygons $P_1$ and $P_2$ must have at least five sides. Since $P_2$ has five or more sides, $P_1$ must have fewer sides than $P$; and therefore there must be some plane intersecting $P_1$ in six or more components. It is clear that this plane must intersect the original polygon $P$ itself in six or more components.

## Bibliography

[1] Borsuk, K., *Sur la courbure totale des courbes fermées*. Annales de la Soc. Polonaise 20 (1947), 251–265.

[2] Fenchel, W., *Uber Krummung und Windung geschlossener Raumkurven*. Math. Ann. 101 (1929), 238–252.

[3] Blaschke, W., Vorlesungen uber Integralgeometrie, zweites heft. Hamburger Mathematische Einselschriften, 22 Heft, 1937.

[4] Saks, S., Theory of the Integral (2nd revised edition). Warsaw 1937.

[5] Fox, R. H. and Artin, E., *Some Wild Cells and Spheres in three-dimensional space*. Ann. of Math. Vol. 49 (1948), 979–990.

[6] Fáry, M. I., *Sur la Courbure Totale d'une Courbe Gauche Faisant un Nœud*. Bulletin de la Soc. Math. de France LXXVII (1949), 128–138.

# ON A RELATIONSHIP BETWEEN THE BETTI NUMBERS OF A HYPERSURFACE AND AN INTEGRAL OF ITS GAUSSIAN CURVATURE

The following manuscript was submitted to Princeton University as a Junior Paper in 1950. It has not previously been published. I am grateful to the Fine Hall Library for preserving a copy.

The two main results can be restated as follows. Let $\mathcal{M}^n$ be a closed $n$-dimensional manifold, smoothly immersed in Euclidean $(n+1)$-space. For $x \in \mathcal{M}^n$ define the "***Gaussian curvature***" $\gamma(x)$ to be the product of the principal curvatures at $x$. For $n$ odd, the sign of $\gamma$ depends on a choice of transverse orientation. However, we make use only of the absolute value $|\gamma|$, which makes perfectly good sense even in the non-orientable case. Let $R_i =$ Rank $H_i(\mathcal{M}^n; \mathbf{Z}/2)$ be the $i^{\text{th}}$ Betti number of $\mathcal{M}^n$, using mod 2 coefficients. Then:

**Theorem III.** $$\int_{\mathcal{M}^n} |\gamma|\, dV_n \geq \frac{\omega_n}{2} \sum_{k=0}^{n} R_k .$$

Here $dV_n$ stands for the $n$-dimensional volume element on $\mathcal{M}^n$, and

$$\omega_n = 2\,\frac{\pi^{(n+1)/2}}{\Gamma((n+1)/2)}$$

is the $n$-dimensional volume of the unit $n$-sphere (with $\omega_1 = 2\pi$, $\omega_2 = 4\pi$, $\omega_3 = 2\pi^2, \ldots$ ). As an example, for $n = 1$ we recover Fenchel's theorem[1] that the total curvature of a closed curve parameterized by arclength $s$ satisfies

$$\int_{\mathcal{M}^1} |\gamma|\, ds \geq 2\pi,$$

at least in the plane case. Here $\gamma$ denotes the ordinary curvature of a plane curve, which is clearly not an intrinsic invariant. However:

[1] W. Fenchel, *Über Krümmung und Windung geschlossener Raumkurven,* Math. Ann. **101** (1929) 238–252.

**Theorem IV.** *For $n > 1$ the absolute value $|\gamma|$ is an intrinsic local invariant of the Riemannian manifold $\mathcal{M}^n$, independent of the choice of isometric immersion into Euclidean $(n+1)$-space.*

In fact for $n$ even it is well known[2] that $\gamma$ coincides with the intrinsic "Lipschitz-Killing" curvature of $\mathcal{M}^n$. However, for $n$ odd some proof is required.

This manuscript is mainly of historical interest. Theorem III has been quite thoroughly developed and generalized to higher dimensional immersions in later papers, for example by Chern,[3] Chern and Lashof,[4] Kuiper,[5] Wilson,[6] and by Ferus.[7] However, I have not been able to find Theorem IV in the literature, although it is quite likely known.

**Warnings.** I have not made any substantial revision, but have tried to duplicate the original typescript, including footnotes, while correcting minor errors and slightly improving the format for readability. The presentation is clumsy, relying heavily on complicated tensor computations in special coordinate systems. Global concepts are not described very well; in particular, there is some confusion between a point and its local coordinates. Also, it should be emphasized that the unit normal vector field (and hence the curvature $\gamma$ in the odd dimensional case) can only be defined locally, since the manifold $\mathcal{M}^n$ may well be non-orientable. My Riemannian geometry was self-taught from Eisenhart,[8] who considered only local differential geometry via multi-index tensor computations. The notations in this manuscript are a slightly modified version of those of Eisenhart.

---

[2] See C. Allendoerfer, *The Euler number of Riemann manifolds,* Amer. J. Math. **62** (1940) 243–248, or W. Fenchel, *On total curvatures of Riemannian manifolds* I, J. Lond. Math. Soc. **15** (1940) 15–22.

[3] S.-S. Chern, *La géométrie des sous-variétés d'un espace euclidean à plusieurs dimensions,* Enseign. Math. **40** (1955) 26–46.

[4] S.-S. Chern and R. Lashof, *On the total curvature of immersed manifolds* Amer. J. Math. **79** (1957) 306–318; and Mich. J. Math. **5** (1958) 5–12.

[5] N. Kuiper, *Immersions with minimal total absolute curvature,* Colloque Géom. Diff. Globale, Bruxelles 1958, Centre Belge Recherches Math. 1959, pp. 75–88.

[6] J. P. Wilson, *The total absolute curvature of immersed manifolds,* J. Lond. Math. Soc. **40** (1965) 362–366.

[7] D. Ferus, "Totale Absolutkrümmung in Differentialgeometrie und -topologie", Lect. Notes Math. **66**, Springer 1968.

[8] See for example L. P. Eisenhart "An Introduction to Differential Geometry with use of the Tensor Calculus", Princeton U. Press 1940.

# ON A RELATIONSHIP BETWEEN THE BETTI NUMBERS OF A HYPERSURFACE AND AN INTEGRAL OF ITS GAUSSIAN CURVATURE

Junior Paper, May 1950

John W. Milnor

By a hypersurface is meant an $n$-dimensional manifold embedded in euclidean $(n+1)$-dimensional space. By its ***Gaussian curvature*** is meant the product of its principal curvatures. The purpose of this paper is to establish, in terms of the topology of a surface, a lower bound for the "total curvature" of the surface—that is the integral over the surface of the absolute value of the Gaussian curvature. The result is based on Morse's inequalities between the numbers of critical points of a function defined on a manifold and the Betti numbers of the manifold. It is also established that the Gaussian curvature can be characterized intrinsically, so that given any Riemannian manifold, the theorem provides a necessary condition for the possibility of isometrically embedding the manifold in the euclidean space of next highest dimension.

Let $\mathcal{M}^n$ be a closed Riemannian manifold parameterized by $u^1, \ldots, u^n$, which is embedded* in euclidean $(n+1)$-space $\mathcal{R}^{n+1}$, parameterized by $x^1, \ldots, x^{n+1}$ in such a way that $x^r(u^1, \ldots, u^n) = x^r(u)$, $\dfrac{\partial x^r}{\partial u^\mu}$, and $\dfrac{\partial^2 x^r}{\partial u^\mu \partial u^\nu}$ are continuous functions, and so that the $n$ tangent vectors $\dfrac{\partial x^r}{\partial u^1}, \ldots, \dfrac{\partial x^r}{\partial u^n}$ are linearly independent. Greek indices will range from one to $n$; Latin indices from one to $n+1$. Letters from the beginning of the alphabet will be summed. The metric tensors of $\mathcal{R}^{n+1}$ and $\mathcal{M}^n$ will be denoted by $d_{rs}$ and $g_{\mu\nu}$ respectively. $d$ and $g$ will denote the determinants $[d_{rs}]$ and $[g_{\mu\nu}]$. The Riemannian volume element $\sqrt{g}\, du^1 \cdots du^n$ will always be taken as positive.

For any point of $\mathcal{M}^n$ it is possible to choose orthogonal coordinates $y^1, \ldots, y^{n+1}$ for $\mathcal{R}^{n+1}$ so that the given point lies at the origin, and so that the $y^{n+1}$ coordinate axis is orthogonal to the surface. $\mathcal{M}^n$ may then be represented by equations of the form:

$$y^{n+1} = f(y^1, \ldots, y^n) = \tfrac{1}{2} a_{\alpha\beta} y^\alpha y^\beta + K(y^1, \ldots, y^n)$$

* Self penetration being permitted.

where $K(0,\ldots,0) = \left.\frac{\partial K}{\partial y^\mu}\right|_{(0,\ldots,0)} = \left.\frac{\partial^2 K}{\partial y^\mu \partial y^\nu}\right|_{(0,\ldots,0)} = 0$. New orthogonal coordinates $x^\mu = c^\mu_\alpha y^\alpha$, $x^{n+1} = y^{n+1}$ may now be chosen for $\mathcal{R}^{n+1}$ in such a way that $a_{\alpha\beta}$ is diagonalized. Using these coordinates $x^1, \ldots, x^n$ as parameters for $\mathcal{M}^n$ in a neighborhood of the point in question, the equations of the embedding become:

$$x^\mu = u^\mu \quad [\mu = 1, \ldots, n], \qquad x^{n+1} = \tfrac{1}{2} a_{\alpha\beta} u^\alpha u^\beta + K(u)$$

where the $a_{\mu\mu}$ are principal curvatures and the $a_{\mu\nu}$ with $\mu \neq \nu$ are zero. These parameterizations will be referred to as the ***standard coordinate systems*** for $\mathcal{R}^{n+1}$ and $\mathcal{M}^n$ with respect to the point. We can now define the ***Gaussian curvature*** $\gamma(u) = \prod_{\kappa=1}^{n} a_{\kappa\kappa}$. Note that for $n$ odd the sign of $\gamma(u)$ depends on the choice of direction for the positive $x^{n+1}$ axis.

For each point of $\mathcal{M}^n$ there are two unit normal vectors, one of which is given by

$$z_r = -\frac{\sqrt{d}}{n!\sqrt{g}} \epsilon^{\alpha_1\cdots\alpha_n} \epsilon_{a_1\cdots a_n r} \frac{\partial x^{a_1}}{\partial u^{\alpha_1}} \cdots \frac{\partial x^{a_n}}{\partial u^{\alpha_n}}$$

where $\epsilon^{\kappa_1\cdots\kappa_p} = \epsilon_{\kappa_1\cdots\kappa_p}$ is plus or minus one according as $\kappa_1, \ldots, \kappa_p$ is an even or odd permutation of $1, \ldots, p$. Due to the invariant nature of this equation, it is only necessary to verify it in one coordinate system in order to establish it in all coordinate systems. (Since $\frac{1}{\sqrt{g}} \epsilon^{\alpha_1\cdots\alpha_n}$, $\sqrt{d}\, \epsilon_{a_1\cdots a_n r}$ and the $\frac{\partial x^{a_i}}{\partial u^{\alpha_i}}$ are absolute tensors, their product is an absolute tensor. Since after contraction both sides of the equation are absolute tensors with one covariant index, they transform in exactly the same way with respect to changes in coordinates of $\mathcal{R}^{n+1}$ or $\mathcal{M}^n$.) At any point, in the standard coordinate system associated with the point, this equation reduces to

$$z_r = -\frac{1}{n!} \epsilon^{\alpha_1\cdots\alpha_n} \epsilon_{a_1\cdots a_n r} \delta^{a_1}_{\alpha_1} \cdots \delta^{a_n}_{\alpha_n}$$

where the Kronecker deltas $\delta^r_\mu$ are one or zero according as $r = \mu$ or $r \neq \mu$. Any non-zero term of this sum must have each $\alpha_i$ equal to $a_i$, so that the deltas do not vanish, and must have $\alpha_1, \ldots, \alpha_n$ and $a_1, \ldots, a_n, r$ permutations of $1, 2, \ldots, n$ and $1, \ldots, n+1$ respectively so that the epsilons do not vanish. This is possible only if $r = n+1$. In this case the expression is equal to the sum of $n!$ terms, corresponding to the $n!$ permutations of $n$ elements, each of which is equal to $-1/n!$. Therefore the equation states that $z_r = 0$ for

$r = 1, \ldots, n$, and $z_{n+1} = -1$, which certain gives a unit normal vector in the standard coordinate system.

Select parameters $v^1, \ldots, v^n$ for the unit sphere* $\mathcal{S}^n$ in $\mathcal{R}^{n+1}$. Let the equation of its embedding be $z^r = z^r(v)$ and let its metric tensor be $h_{\mu\nu}$. The normal vector to $\mathcal{M}^n$ may be considered as a point on $\mathcal{S}^n$ represented by the equation $v^\mu = v^\mu(u)$ which satisfies $z_r = d_{ra} z^a(v(u))$. The image $v(\mathcal{A})$ of the set $\mathcal{A} \subset \mathcal{M}^n$ under the mapping $v\colon \mathcal{M}^n \to \mathcal{S}^n$ will be called the ***spherical image*** of $\mathcal{A}$.

**Theorem I.** *If $\{\mathcal{A}_i\}$ is a sequence of subsets of $\mathcal{M}^n$ containing the point $u$, with diameters approaching zero, such that no $\mathcal{A}_i$ has measure zero, then the ratio of the measure of the spherical image of $\mathcal{A}_i$ to the measure of $\mathcal{A}_i$ approaches $|\gamma(u)|$ as $i$ approaches infinity.*

**Proof.** $$\lim_{i\to\infty} \frac{\text{measure of } v(\mathcal{A}_i)}{\text{measure of } \mathcal{A}_i} = \lim_{i\to\infty} \frac{\displaystyle\int_{v(\mathcal{A}_i)} \sqrt{h}\, dv^1 \cdots dv^n}{\displaystyle\int_{\mathcal{A}_i} \sqrt{g}\, du^1 \cdots du^n}$$

$$= \lim_{i\to\infty} \frac{\displaystyle\int_{\mathcal{A}_i} \sqrt{h} \left\| \left[\frac{\partial v^\mu}{\partial u^\nu}\right] \right\| du^1 \cdots du^n}{\displaystyle\int_{\mathcal{A}_i} \sqrt{g}\, du^1 \cdots du^n} = \frac{\sqrt{h}}{\sqrt{g}} \left\| \left[\frac{\partial v^\mu}{\partial u^\nu}\right] \right\|,$$

where $[\cdots]$ denotes 'determinant of'.

Choosing the standard coordinate system at a representative point, and letting the parameters of $\mathcal{S}^n$ equal $z^1, \ldots, z^n$ in a neighborhood of the spherical image of the point, we have for some neighborhood of the point:

$$v^\mu = z^\mu = d^{\mu\beta} z_\beta = -d^{\mu\beta} \frac{\sqrt{d}}{n!\sqrt{g}} \epsilon^{\alpha_1\cdots\alpha_n} \epsilon_{a_1\cdots a_n\beta} \frac{\partial x^{a_1}}{\partial u^{\alpha_1}} \cdots \frac{\partial x^{a_n}}{\partial u^{\alpha_n}}.$$

Each non-zero term of this sum must have $\mu = \beta \le n$ due to the special form of $d^{\mu\beta}$. In order for the epsilon symbols to be non-zero, some $a_i$ must equal $n+1$. For all $\sigma \le n$, $\dfrac{\partial x^\sigma}{\partial u^\rho}$ must equal $\delta^\sigma_\rho$, so that $a_j = \alpha_j$ for all $j \ne i$. Such a term may be written as

$$-\frac{1}{n!\sqrt{g}} \epsilon^{\alpha_1\cdots\alpha_n} \epsilon_{a_1\cdots a_{i-1}(n+1)a_{i+1}\cdots a_n\mu} \frac{\partial x^{n+1}}{\partial u^{\alpha_i}}$$

* In all discussion of $\mathcal{S}^n$, it will be assumed that the coordinates of $\mathcal{R}^{n+1}$ are normal orthogonal.

where $\alpha_i$ must equal $\mu$. The sum of the $n!$ terms corresponding to the $n!$ permutations of $\alpha_1, \ldots, \alpha_n$ is therefore:

$$v^\mu = \frac{1}{\sqrt{g}} \frac{\partial x^{n+1}}{\partial u^\mu} = \frac{1}{\sqrt{g}} \left( a_{\alpha\mu} u^\alpha + \frac{\partial K}{\partial u^\mu} \right).$$

We can now evaluate the derivative of this quantity at the origin.

$$\begin{aligned} \frac{\partial v^\mu}{\partial u^\nu} &= -\frac{1}{(g)^{3/2}} \left( a_{\alpha\mu} u^\alpha + \frac{\partial K}{\partial u^\mu} \right) \frac{\partial g}{u^\nu} + \frac{1}{\sqrt{g}} \left( a_{\nu\mu} + \frac{\partial^2 K}{\partial u^\nu \partial u^\mu} \right) \\ &= 0 + \frac{1}{\sqrt{g}} \left( a_{\nu\mu} + 0 \right) = a_{\nu\mu}. \end{aligned}$$

Taking the determinant:

$$\frac{\sqrt{h}}{\sqrt{g}} \left[ \frac{\partial v^\mu}{\partial u^\nu} \right] = \left[ \frac{\partial v^\mu}{\partial u^\nu} \right] = [a_{\nu\mu}] = \prod_{\kappa=1}^{n} a_{\kappa\kappa} = \gamma(0),$$

which completes the proof. □

Let $\mathcal{F}$ denote the set of flat points of $\mathcal{M}^n$, that is the set of points $u$ such that $\gamma(u) = 0$. Let $\mathcal{T}$ denote the set of $\mathcal{S}^n$ which represent normals to points of $\mathcal{F}$. (Thus $\mathcal{T}$ is the union of $v(\mathcal{F})$ and the set of antipodes of $v(\mathcal{F})$.)

**Lemma.** *$\mathcal{T}$ has measure zero.*

**Proof.** $\displaystyle\int_{\mathcal{T}} \sqrt{h}\, dv^1 \cdots dv^n \leq 2 \int_{v(\mathcal{F})} \sqrt{h}\, v_1 \cdots dv^n$

$$\begin{aligned} &\leq 2 \int_{\mathcal{F}} \sqrt{h} \left| \left[ \frac{\partial v^\mu}{\partial u^\nu} \right] \right| du^1 \cdots du^n = 2 \int_{\mathcal{F}} \frac{\sqrt{h}}{\sqrt{g}} \left| \left[ \frac{\partial v^\mu}{\partial u^\nu} \right] \right| \sqrt{g}\, du^1 \cdots du^n \\ &= 2 \int_{\mathcal{F}} |\gamma(u)| \sqrt{g}\, du^1 \cdots du^n = 0. \ \square \end{aligned}$$

If $v$ is a point of $\mathcal{S}^n$ and $\mathcal{A}$ is a subset of $\mathcal{M}^n$, let $2\mu(v, \mathcal{A})$ denote the number of points of $\mathcal{A}$ at which $z^r(v)$ is one of the normals to $\mathcal{M}^n$.

**Lemma.** *If $\mathcal{A}$ is compact and contains no flat points, then*

$$\int_{\mathcal{A}} |\gamma(u)| \sqrt{g}\, du^1 \cdots du^n = \int_{\mathcal{S}^n} \mu(v, \mathcal{A}) \sqrt{h}\, dv^1 \cdots dv^n.$$

**Proof.** Each of these two integrals represents a function of the set $\mathscr{A}$ which vanishes if $\mathscr{A}$ has measure zero. Let $s(\mathscr{A})$ denote one of the two integrals, and $\ell(\mathscr{A})$ the other. If the Lemma were false, then we could find a set $\mathscr{A}^0$ with say $0 \leq s(\mathscr{A}^0) < \ell(\mathscr{A}^0)$. Then a nested sequence of sets $\mathscr{A}^0 \supset \mathscr{A}^1 \supset \mathscr{A}^2 \supset \cdots$ will be defined by an inductive process, in such a way that:

1) Each $\mathscr{A}^i$ is a compact set of positive measure,
2) $\dfrac{s(\mathscr{A}^{i+1})}{\ell(\mathscr{A}^{i+1})} \leq \dfrac{s(\mathscr{A}^i)}{\ell(\mathscr{A}^i)} < 1$, and
3) $\lim\limits_{i\to\infty} \text{diameter}\, \mathscr{A}^i = 0$.

Since $\ell(\mathscr{A}^0)$ is not zero, $\mathscr{A}^0$ cannot have measure zero. Given $\mathscr{A}^i$, the set $\mathscr{A}^{i+1}$ can be constructed as follows. Since $M^n$ can be triangulated, there exist coverings of $\mathcal{M}^n$ and therefore of $\mathscr{A}^i$ by a finite number of compact sets of arbitrarily small diameter, such that the intersection of any two sets of the covering has measure zero. Choose a covering $\mathscr{A}^i = \mathscr{A}^i_1 \cup \cdots \cup \mathscr{A}^i_{n_i}$ in such a way that the diameter of each $\mathscr{A}^i_j$ is less than $1/i$ and so that the first $m_i$ of the sets have $\ell(\mathscr{A}^i_j) > 0$ while the remaining $n_i - m_i$ have $\ell(\mathscr{A}^i_j) = 0$ for some number $m_i$. Now $s(\mathscr{A}^i) = s(\mathscr{A}^i_1) + \cdots + s(\mathscr{A}^i_{n_i})$ and $\ell(\mathscr{A}^i) = \ell(\mathscr{A}^i_1) + \cdots + \ell(\mathscr{A}^i_{m_i})$. If $\dfrac{s(\mathscr{A}^i_j)}{\ell(\mathscr{A}^i_j)}$ were greater than $\dfrac{s(\mathscr{A}^i)}{\ell(\mathscr{A}^i)}$ for $j = 1, \ldots, m_i$, then $\sum\limits_{j=1}^{m_i} s(\mathscr{A}^i_j)$ would be greater than $\sum\limits_{j=1}^{m_i} \dfrac{s(\mathscr{A}^i)}{\ell(\mathscr{A}^i)} \ell(\mathscr{A}^i_j) = s(\mathscr{A}^i)$ which is impossible. Therefore there is some $j_i$ such that $\dfrac{s(\mathscr{A}^i_j)}{\ell(\mathscr{A}^i_j)} \leq \dfrac{s(\mathscr{A}^i)}{\ell(\mathscr{A}^i)}$ and we may let $\mathscr{A}^{i+1} = \mathscr{A}^i_j$ (a set of positive measure since $\ell(\mathscr{A}^i_j) > 0$).

Let $u = \bigcap\limits_{i=1}^{\infty} \mathscr{A}^i$. By Theorem I,

$$|\gamma(u)| = \lim_{i\to\infty} \frac{\displaystyle\int_{v(\mathscr{A}^i)} \sqrt{h}\, dv^1 \cdots dv^n}{\displaystyle\int_{\mathscr{A}^i} \sqrt{g}\, du^1 \cdots du^n}.$$

Since none of the principal curvatures vanish at $u$, distinct points have distinct normals in some sufficiently small neighborhood of $u$. That is, for sufficiently large $i$, $\mu(v, \mathscr{A}^i) = 1/2$ for $v$ in $v(\mathscr{A}^i)$ or its antipodal set and

$\mu(v, \mathscr{A}^i) = 0$ elsewhere, so that

$$\int_{v(\mathscr{A}^i)} \sqrt{h}\, dv^1 \cdots dv^n = \int_{\mathscr{S}^n} \mu(v, \mathscr{A}^i)\sqrt{h}\, dv^1 \cdots dv^n.$$

Therefore

$$1 = \frac{1}{|\gamma(u)|} \lim_{i\to\infty} \frac{\displaystyle\int_{v(\mathscr{A}^i)} \sqrt{h}\, dv^1 \cdots dv^n}{\displaystyle\int_{\mathscr{A}^i} \sqrt{g}\, du^1 \cdots du^n} = \lim_{i\to\infty} \frac{\displaystyle\int_{\mathscr{S}^n} \mu(v, \mathscr{A}^i)\sqrt{h}\, dv^1 \cdots dv^n}{\displaystyle\int_{\mathscr{A}^i} \sqrt{g}\, du^1 \cdots du^n}$$

and

$$1 = \lim_{i\to\infty} \frac{s(\mathscr{A}^i)}{\ell(\mathscr{A}^i)} \le \frac{s(\mathscr{A})}{\ell(\mathscr{A})} < 1,$$

which is impossible. □

**Theorem II.**

$$\int_{\mathcal{M}^n} |\gamma(u)|\sqrt{g}\, du^1 \cdots du^n = \int_{\mathscr{S}^n} \mu(v, \mathcal{M}^n)\sqrt{h}\, dv^1 \cdots dv^n.$$

Construct a sequence of compact sets $A^i$ such that $\mathscr{A}^0 \subset \mathscr{A}^1 \subset \mathscr{A}^2 \subset \cdots$ and $\mathcal{M}^n - \mathscr{F} = \mathscr{A}^0 \cup \mathscr{A}^1 \cup \mathscr{A}^2 \cup \cdots$. This is possible since $\mathscr{F}$ is closed and $\mathcal{M}^n$ is a separable regular compact space. Since $\mathscr{F}$ is of measure zero, $\int_{\mathscr{S}^n} \mu(v, \mathscr{F})\sqrt{h}\, dv^1 \cdots dv^n = 0$ and

$$\begin{aligned}
\int_{\mathcal{M}^n} |\gamma(u)|\sqrt{g}\, du^1 \cdots du^n &= \int_{\mathcal{M}^n - \mathscr{F}} |\gamma(u)|\sqrt{g}\, du^1 \cdots du^n \\
&= \lim_{i\to\infty} \int_{\mathscr{A}^i} |\gamma(u)|\sqrt{g}\, du^1 \cdots du^n \\
&= \lim_{i\to\infty} \int_{\mathscr{S}^n} \mu(v, \mathscr{A}^i)\sqrt{h}\, dv^1 \cdots dv^n \\
&= \int_{\mathscr{S}^n} \mu(v, \mathcal{M}^n - \mathscr{F})\sqrt{h}\, dv^1 \cdots dv^n \\
&= \int_{\mathscr{S}^n} \mu(v, \mathcal{M}^n)\sqrt{h}\, dv^1 \cdots dv^n. \ \square
\end{aligned}$$

Morse theory shows that for a twice differentiable function from a manifold to the real numbers, such that $\left[\dfrac{\partial^2 f}{\partial u^\mu \partial u^\nu}\right] \neq 0$ for all critical points (that is points for which $\dfrac{\partial f}{\partial u^\mu} = 0$, $\mu = 1, \ldots, n$), the number of critical points of type $k$ is greater than or equal to the $k^{\text{th}}$ mod 2 Betti number, which will be denoted by $R_k$. This result will be used to establish the principal theorem of this paper.

**Theorem III.** $\displaystyle\int_{\mathcal{M}^n} |\gamma(u)|\sqrt{g}\, du^1 \cdots du^n \geq \frac{\omega_n}{2} \sum_{k=0}^{n} R_k,$

*where*

$$\omega_n = \int_{\mathcal{S}^n} \sqrt{h}\, dv^1 \cdots dv^n = \frac{2\pi^{(n+1)/2}}{\Gamma((n+1)/2)}$$

*is the measure of a unit n-sphere.*

**Proof.** For each point $v$ of $\mathcal{S}^n$ consider the function $f_{(v)} \colon \mathcal{M}^n \to$ reals defined by the inner product $f_{(v)}(u) = d_{ab}x^a(u)z^b(v)$. The number of critical points of this function is $2\mu(v, \mathcal{M}^n)$. The Hessian $\left[\dfrac{\partial^2 f}{\partial u^\mu \partial u^\nu}\right]$ is equal to $\pm g\gamma(u)$ for critical points as can be seen by evaluation in the standard coordinate system. The condition that the Hessian be non-zero is therefore equivalent to the condition that $v \in \mathcal{S}^n - \mathcal{T}$; and for such $v$, $2\mu(v, \mathcal{M}^n) \geq \sum_{k=0}^{n} R_k$. It follows that

$$\begin{aligned}\int_{\mathcal{M}^n} |\gamma(u)|\sqrt{g}\, du^1 \cdots du^n &= \int_{\mathcal{S}^n} \mu(v, \mathcal{M}^n)\sqrt{h}\, dv^1 \cdots dv^n \\ &= \int_{\mathcal{S}^n - \mathcal{T}} \mu(v, \mathcal{M}^n)\sqrt{h}\, dv^1 \cdots dv^n \\ &\geq \int_{\mathcal{S}^n - \mathcal{T}} \frac{1}{2}\left(\sum_{k=0}^{n} R_k\right)\sqrt{h}\, dv^1 \cdots dv^n = \frac{\omega_n}{2}\sum_{k=0}^{n} R_k\end{aligned}$$

which completes the proof. □

It will now be shown that the Gaussian curvature can be defined intrinsically, that is in terms of $g_{\mu\nu}$ alone. As with the preceding two definitions of $\gamma(u)$, the sign will be indeterminate in the odd dimensional case. (In the one dimensional case, such a definition is of course impossible.) Let $R_{\mu_1\mu_2\nu_1\nu_2}$ be the Riemann tensor derived from $g_{\mu\nu}$.

**Theorem IV.** *For* $n = 2, 4, 6, \ldots$

$$\gamma(u) = \frac{\epsilon^{\alpha_1\cdots\alpha_n}\epsilon^{\beta_1\cdots\beta_n}}{2^{n/2}n!\, g} R_{\alpha_1\alpha_2\beta_1\beta_2} R_{\alpha_3\alpha_4\beta_3\beta_4} \cdots R_{\alpha_{n-1}\alpha_n\beta_{n-1}\beta_n}.$$

*For* $n = 3, 5, 7, \ldots$

$$\left(\gamma(u)\right)^2 = \frac{\epsilon^{\alpha_1\cdots\alpha_n}\,\epsilon^{\beta_1\cdots\beta_n}\,\epsilon^{\gamma_1\cdots\gamma_n}\,\epsilon^{\delta_1\cdots\delta_n}}{2^{n-1}(n-1)(n-1)!\,n!\,g^2} R_{\alpha_1\alpha_2\beta_1\beta_2}\,R_{\alpha_3\alpha_4\beta_3\beta_4}\cdots$$
$$\cdots R_{\alpha_{n-2}\alpha_{n-1}\beta_{n-2}\beta_{n-1}}\,R_{\alpha_n\gamma_1\beta_n\delta_1}\,R_{\gamma_2\gamma_3\delta_2\delta_3}\cdots R_{\gamma_{n-1}\gamma_n\delta_{n-1}\delta_n}.$$

Since this formula is already well known for the even dimensional case, only the odd dimensional case will be proved. Since the quantity on the right is obtained by the multiplication and contraction of tensors, but is itself a scalar, it will be sufficient to evaluate it in the standard coordinate system associated with a representative point:

$$x^\mu(u) = u^\mu, \qquad \frac{\partial x^\mu}{\partial u^\nu} = \delta^\mu_\nu \quad [\mu = 1, \ldots, n],$$

$$x^{n+1}(u) = \frac{1}{2}a_{\alpha\beta}u^\alpha u^\beta + K(u), \qquad \frac{\partial x^{n+1}}{\partial u^\mu} = a_{\alpha\mu}u^\alpha + \frac{\partial K}{\partial u^\mu},$$

$$g_{\mu\nu} = d_{ab}\frac{\partial x^a}{\partial u^\mu}\frac{\partial x^b}{\partial u^\nu} = d_{\mu\nu} + \left(a_{\alpha\mu}u^\alpha + \frac{\partial K}{\partial u^\mu}\right)\left(a_{\beta\nu}u^\beta + \frac{\partial K}{\partial u^\nu}\right)$$
$$= d_{\mu\nu} + a_{\alpha\mu}a_{\beta\nu}u^\alpha u^\beta + a_{\alpha\mu}u^\alpha\frac{\partial K}{\partial u^\nu} + a_{\beta\nu}u^\beta\frac{\partial K}{\partial u^\mu} + \frac{\partial K}{\partial u^\mu}\frac{\partial K}{\partial u^\nu},$$

$$\frac{\partial g_{\mu\nu}}{\partial u^\rho} = a_{\alpha\mu}a_{\rho\nu}u^\alpha + a_{\rho\mu}a_{\beta\nu}u^\beta + a_{\rho\mu}\frac{\partial K}{\partial u^\nu} + a_{\alpha\mu}u^\alpha\frac{\partial^2 K}{\partial u^\rho\partial u^\nu} + \cdots,$$

$$\frac{\partial^2 g_{\mu\nu}}{\partial u^\rho\partial u^\sigma} = a_{\sigma\mu}a_{\rho\nu} + a_{\rho\mu}a_{\sigma\nu} + a_{\rho\mu}\frac{\partial^2 K}{\partial u^\nu\partial u^\sigma} + a_{\sigma\mu}\frac{\partial^2 K}{\partial u^r ho\partial u^\nu}$$
$$+ a_{\alpha\mu}u^\alpha\frac{\partial^3 K}{\partial u^\rho\partial u^\nu\partial u^\sigma} + \cdots,$$

$$g_{\mu\nu}\big|_{(0)} = \delta_{\mu\nu}, \qquad \frac{\partial g_{\mu\nu}}{\partial u^\rho}\bigg|_{(0)} = 0, \qquad \frac{\partial^2 g_{\mu\nu}}{\partial u^\rho\partial u^\sigma}\bigg|_{(0)} = a_{\sigma\mu}a_{\rho\nu} + a_{\rho\nu} + a_{\sigma\nu},$$

$$[\mu\nu, \rho]\big|_{(0)} = \tfrac{1}{2}\left(\frac{\partial g_{\nu\rho}}{\partial u^\mu} + \frac{\partial g_{\rho\mu}}{\partial u^\nu} - \frac{\partial g_{\mu\nu}}{\partial u^\rho}\right)\bigg|_{(0)} = 0,$$

$$R_{\mu_1\mu_2\nu_1\nu_2}\big|_{(0)} = \frac{1}{2}\left(\frac{\partial^2 g_{\mu_1\nu_2}}{\partial u^{\mu_2}\partial u^{\nu_1}} + \frac{\partial^2 g_{\mu_2\nu_1}}{\partial u^{\mu_1}\partial u^{\nu_2}} - \frac{\partial^2 g_{\mu_1\nu_1}}{\partial u^{\mu_2}\partial u^{\nu_2}} - \frac{\partial^2 g_{\mu_2\nu_2}}{\partial u^{\mu_1}\partial u^{\nu_1}}\right)\bigg|_{(0)}$$
$$+ g^{\alpha\beta}\left([\mu_2\nu_1, \alpha][\mu_1\nu_2, \beta] - [\mu_2\nu_2, \alpha][\mu_1\nu_1, \beta]\right)\big|_{(0)}$$
$$= a_{\mu_1\nu_1}a_{\mu_2\nu_2} - a_{\mu_1\nu_2}a_{\mu_2\nu_1}.$$

Since $a_{\mu\nu}$ is diagonal, the only non-vanishing components of this tensor at the origin are those of the form $R_{\sigma\tau\sigma\tau} = a_{\sigma\sigma}a_{\tau\tau}$ or $R_{\sigma\tau\tau\sigma} = -a_{\sigma\sigma}a_{\tau\tau}$ with $\sigma \neq \tau$. The expression

$$R_{\mu_1\mu_2\nu_1\nu_2} \cdots R_{\mu_{n-2}\mu_{n-1}\nu_{n-2}\nu_{n-1}} R_{\mu_n\sigma_1\nu_n\tau_1} R_{\sigma_2\sigma_3\tau_2\tau_3} \cdots R_{\sigma_{n-1}\sigma_n\tau_{n-1}\tau_n}$$

where $(\mu_1, \ldots, \mu_n)$, $(\nu_1, \ldots, \nu_n)$, $(\sigma_1, \ldots, \sigma_n)$, and $(\tau_1, \ldots, \tau_n)$ are each permutations of 1, 2, ... , $n$, has three possible value: $\pm \prod_{\kappa=1}^{n} a_{\kappa\kappa}^2$ and zero. It is positive if each $\mu_i = \nu_i$, each $\sigma_i = \tau_i$, and $\mu_n = \nu_n \neq \sigma_1 = \tau_1$; or for any set of permutations obtained from these by an even number of interchanges

$$\nu_1 \leftrightarrow \nu_2,\ \nu_3 \leftrightarrow \nu_4,\ \ldots,\ \nu_{n-2} \leftrightarrow \nu_{n-1}; \quad \tau_2 \leftrightarrow \tau_3,\ \ldots,\ \tau_{n-1} \leftrightarrow \tau_n.$$

Its sign is negative for an odd number of such interchanges, and in all other cases it is zero.

In each case when this expression is not zero, it has the same sign as

$$\epsilon^{\mu_1\cdots\mu_n}\,\epsilon^{\nu_1\cdots\nu_n}\,\epsilon^{\sigma_1\cdots\sigma_n}\,\epsilon^{\tau_1\cdots\tau_n}.$$

Therefore their product can only be positive or zero. The number of terms which do not vanish is given by the product of:

the number of permutations of $\mu_1, \mu_2, \ldots, \mu_n$;

$2^{(n-1)/2}$, the number of interchanges of $\nu_1$ and $\nu_2, \ldots, \nu_{n-2}$ and $\nu_{n-1}$;

$n! - (n-1)!$, the number of permutations of $\sigma_1, \ldots, \sigma_n$ for which $\mu_n \neq \sigma_1$;

and

$2^{(n-1)/2}$, the number of possible interchanges of $\tau_2$ and $\tau_3, \ldots,$ or $\tau_{n-1}$ and $\tau_n$.

Therefore

$$\begin{aligned}&\epsilon^{\alpha_1\cdots\alpha_n}\,\epsilon^{\beta_1\cdots\beta_n}\,\epsilon^{\gamma_1\cdots\gamma_n}\,\epsilon^{\delta_1\cdots\delta_n}\,R_{\alpha_1\alpha_2\beta_1\beta_2}\,R_{\alpha_3\alpha_4\beta_3\beta_4}\cdots\\ &\qquad\cdots R_{\alpha_{n-2}\alpha_{n-1}\beta_{n-2}\beta_{n-1}}\,R_{\alpha_n\gamma_1\beta_n\delta_1}\,R_{\gamma_2\gamma_3\delta_2\delta_3}\cdots R_{\gamma_{n-1}\gamma_n\delta_{n-1}\delta_n}\\ &= n!\,2^{(n-1)/2}\big(n! - (n-1)!\big)\,2^{(n-1)/2}\prod_{\kappa=1}^{n} a_{\kappa\kappa}^2 = (n-1)\,(n-1)!\,n!\,2^{n-1}\big(\gamma(u)\big)^2\end{aligned}$$

which completes the proof. □

Taking this as the definition of $\gamma(u)$, we may restate the preceding theorem as follows.

**Theorem III′.** *Given a closed manifold, a necessary condition that it be possible to embed it isometrically and with continuous second derivatives in the euclidean space of next highest dimension is that:*

$$\int_{\mathcal{M}^n} |\gamma(u)|\sqrt{g}\, du^1 \cdots du^n \geq \frac{\omega_n}{2} \sum_{k=0}^{n} R_k.$$

This result still leaves open the question as to whether the lower bound which has been given for the total curvature of a hypersurface is actually the greatest lower bound. In the simpler cases, such as spheres or orientable two dimensional manifolds, this bound is actually attained in the standard embeddings. For a two dimensional projective plane, on the other hand, the lower bound can probably not be obtained, although arbitrarily close approximations are possible.

Another unanswered question is that of finding the greatest lower bound of the total curvature of a manifold, if no restrictions as to embedding are made. In the even dimensional case a lower bound of

$$\frac{\omega_n}{2}|\chi(\mathcal{M}^n)| = \frac{\omega_n}{2}\left|\sum_{k=0}^{n}(-1)^k R_k\right|$$

is given by the generalized Gauss-Bonnet formula*

$$\int_{\mathcal{M}^n} \gamma(u)\sqrt{g}\, du^1 \cdots du^n = \frac{\omega_n}{2}\chi(\mathcal{M}^n).$$

Again this bound seems to be the best obtainable for the simplest manifolds, but may not be the best in more complex cases.

* C. Allendoerfer and A. Weil, *The Gauss-Bonnet Theorem for Riemannian Polyhedra*; Transactions of the American Mathematical Society, Vol. 53 (1943) pp. 101–129.

# ON TOTAL CURVATURES OF CLOSED SPACE CURVES

*Reprinted from*

Mathematica Scandinavica
Volume 1 (1953), pp. 289–296

This paper relates the topology of a closed space curve to integrals which involve not only the curvature but also the torsion.[1]

**More recent developments.** Relations between the topology and geometry of space curves have been noted since the time of Gauss. For more modern work see Pohl,[2] Langer and Singer,[3] and particularly the study of knot ***energy*** by O'Hara,[4] as well as Bryson, Freedman, He and Wang.[5]

[1] Compare the discussion by M. Spivak, *A brief report on John Milnor's brief excursions into differential geometry*, pp. 31–43 of "Topological Methods in Modern Mathematics", Publish or Perish 1993.

[2] W. F. Pohl, *The self linking number of a closed space curve*, J. Math. Mech. **17** (1968) 975–985.

[3] J. Langer and D. A. Singer, *Curve straightening and a minimax argument for closed elastic curves*, Topology **24** (1985) 75–88.

[4] J. O'Hara, *Energy of a knot*, Topology **30** (1991) 241–247.

[5] S. Bryson, M. Freedman, Z. He, and Z. Wang, *Möbius invariance of knot energy*, Bull. Amer. Math. Soc., **28** (1993) 99–103. See also Freedman and He, *On the "energy" of knots and unknots*, to appear.

# ON TOTAL CURVATURES OF CLOSED SPACE CURVES

JOHN MILNOR

## 1. Introduction.

Let $C$ be a closed curve of class 3 in euclidean 3-space, having curvature $\varkappa(s) > 0$ and torsion $\tau(s)$, where $s$ denotes arc-length. Define

$$\varkappa(C) = \int_C \varkappa(s)\, ds, \qquad \tau(C) = \int_C |\tau(s)|\, ds\,,$$

$$\omega(C) = \int_C (\varkappa^2(s) + \tau^2(s))^{1/2}\, ds\,.$$

These quantities measure the lengths of the spherical indicatrices of the tangent vector, the binormal vector, and the principal normal vector, respectively. They are related by the inequalities

$$2\pi \leqq \varkappa(C) \leqq \omega(C) \leqq \varkappa(C) + \tau(C) \leqq 2^{1/2}\, \omega(C)\,.$$

(The first inequality was proved by Fenchel [4]. The remaining three are obvious.)

The quantity $\varkappa(C)$, known as the total curvature, has been studied by several authors (see references [3]–[7]). This paper will attempt a similar study of the quantities $\varkappa(C) + \tau(C)$ and $\omega(C)$.

Two simple closed curves belong to the same *isotopy type* if there exists an isotopy of euclidean space onto itself (not necessarily differentiable) which transforms one curve into the other. It will be proved that the greatest lower bound of $\varkappa(C) + \tau(C)$ over an isotopy class of closed curves is a number of the form $2\pi n$, $n$ a positive integer, where $n = 1$ only for the class of unknotted curves, and where $n = 2$ only for certain especially simple knots (figure 1). On the other hand an example is given to show that the greatest lower bound of $\omega(C)$ over an isotopy class is not necessarily $2\pi$ times an integer.

If $C$ has linking number $n$ with some straight line, it is proved that $\varkappa(C)+\tau(C) \geqq 2\pi n$.

If the torsion $\tau(s)$ does not change sign and is not identically zero, it

Received August 24, 1953.

The author was a National Science Foundation fellow during 1952–53.

is proved that $\omega(C) \geqq 4\pi$. (This result answers a question raised by Fenchel [5, p. 51]: Does the condition $\tau(s) > 0$ imply a lower bound greater than $2\pi$ for $\varkappa(C)+\tau(C)$?)

**2. The quantity $\varkappa(C)+\tau(C)$.**

For each plane $P$ through the origin let $C_P$ denote the image of $C$ in $P$ under orthogonal projection. The following proposition has been proved by I. Fáry [3].

LEMMA 1. *The total curvature $\varkappa(C)$ is equal to the average over all planes $P$ through the origin of $\varkappa(C_P)$.*

(The word average is to be interpreted as follows. To each plane $P$ there corresponds a unique pair of unit vectors which are perpendicular to $P$. Therefore the usual Lebesgue measure on the unit sphere induces a measure on the set of all planes through the origin. The average may now be interpreted as a Lebesgue integral.)

A similar assertion holds for $\tau(C)$. Let $\nu(C_P)$ denote the number of points of $C$ at which the osculating plane is perpendicular to $P$. This can also be interpreted as the number of points of the plane curve $C_P$ which are either inflection points or cusps.

LEMMA 2. *$\tau(C)$ equals the average over all planes through the origin of $\pi\nu(C_P)$. Furthermore $\nu(C_P)$ is an even integer for almost all $P$.*

The proof of lemma 2 will depend on the following proposition of integral geometry.

LEMMA 3. *The length of a curve $B$ on the unit sphere equals $\pi$ times the average over all great circles $G$ of the number of times that $B$ intersects $G$.*

This is proved in [2], p. 81.

Now let $B$ be the spherical indicatrix of the binormal vectors, with length $\tau(C)$. Let $P$ be the plane spanned by $G$. A binormal vector lies in $P$ if and only if the corresponding osculating plane is perpendicular to $P$. Therefore $\nu(C_P)$ equals the number of times that $B$ intersects $G$. Since this is an even number for almost all $G$ (all but a set of measure zero), lemma 2 follows from lemma 3.

Finally we will need a two dimensional analogue of lemma 1. Let $\mu(\boldsymbol{v}, C)$ denote the number of points of $C$ at which the tangent vector is perpendicular to the unit vector $\boldsymbol{v}$.

LEMMA 4. *The total curvature $\varkappa(C_P)$ of a plane curve equals the average* *over the unit circle of* $\pi\mu(\boldsymbol{v}, C_P)$. *Furthermore* $\mu(\boldsymbol{v}, C_P)$ *is even for almost all* $\boldsymbol{v}$.

The proof is easily supplied. (See [7].)

For each isotopy type $\mathfrak{C}$ of closed space curves, let $(\varkappa+\tau)(\mathfrak{C})$ denote the greatest lower bound of $\varkappa(C)+\tau(C)$ over all representative curves $C$, which are of class 3 and satisfy $\varkappa(s) > 0$. Similarly define $\varkappa(\mathfrak{C})$ and $\omega(\mathfrak{C})$.

THEOREM 1. *For every isotopy type* $\mathfrak{C}$ *of closed curves the quantity* $(\varkappa+\tau)(\mathfrak{C})$ *is an integral multiple of* $2\pi$.

(The corresponding assertion for $\varkappa(\mathfrak{C})$ was proved in [7]. Examples given by R. H. Fox [6] show that $\varkappa(\mathfrak{C})$ can be *any* positive integral multiple of $2\pi$. The same examples apply to $(\varkappa+\tau)(\mathfrak{C})$.)

Suppose that $2\pi n \leqq (\varkappa+\tau)(\mathfrak{C}) < 2\pi(n+1)$. Then there exists a representative $C$ with $\varkappa(C)+\tau(C) < 2\pi(n+1)$. We will construct isotopic curves $C'$ with $\varkappa(C')+\tau(C')$ arbitrarily close to $2\pi n$. This will prove that $(\varkappa+\tau)(\mathfrak{C}) = 2\pi n$.

By lemmas 1 and 2 there exists a plane projection $C_P$ of $C$ so that $\varkappa(C_P)+\pi\nu(C_P) \leqq \varkappa(C)+\tau(C)$, where $\nu(C_P)$ is an even integer. Furthermore the plane $P$ can be chosen so that it is not perpendicular to any tangent of $C$. By lemma 4 there exists a unit vector $\boldsymbol{v}$ in $P$ so that $\pi\mu(\boldsymbol{v}, C_P) \leqq \varkappa(C_P)$, where $\mu(\boldsymbol{v}, C_P)$ is even. Since $\mu(\boldsymbol{v}, C_P) = \mu(\boldsymbol{v}, C)$ we have

$$\pi\big(\mu(\boldsymbol{v}, C)+\nu(C_P)\big) \leqq \varkappa(C_P)+\pi\nu(C_P) \leqq \varkappa(C)+\tau(C) < 2\pi(n+1)\,,$$

hence

$$\mu(\boldsymbol{v}, C)+\nu(C_P) \leqq 2n\,.$$

Choose coordinates $(x_1, x_2, x_3)$ so that $\boldsymbol{v}$ is in the direction of the $x_1$-axis, and so that $P$ is the $(x_1, x_2)$-plane. Define the isotopy $h_t$ by $h_t(x_1, x_2, x_3) = (x_1, tx_2, t^2x_3)$. As $t \to 0$, the tangent vector of the curve $C_t = h_t(C)$ approaches $\pm\boldsymbol{v}$, except at the $\mu(\boldsymbol{v}, C)$ points where the tangent to $C$ is perpendicular to $\boldsymbol{v}$. In the neighborhood of each of these points, the tangent to $C_t$ rotates through an angle which converges to $\pi$ as $t \to 0$. This proves that $\varkappa(C_t) \to \pi\mu(\boldsymbol{v}, C)$; and similarly one can prove that $\tau(C_t) \to \pi\nu(C_P)$. Consequently

$$\varkappa(C_t)+\tau(C_t) \to \pi\big(\mu(\boldsymbol{v}, C)+\nu(C_P)\big) \leqq 2\pi n\,,$$

which completes the proof.

Since $\varkappa(C) \geqq 4\pi$ for any knot (see [3], [5], or [7]), the case $(\varkappa+\tau)(\mathfrak{C}) = 2\pi$ can occur only if $\mathfrak{C}$ is the class of unknotted curves.

THEOREM 2. *If* $(\varkappa+\tau)(\mathfrak{C}) = 4\pi$, *then* $\mathfrak{C}$ *is represented by one of the curves of figure* 1.

Fig. 1.

In the language of braid theory [1], these curves can be described as closed braids with two strands. The following converse to theorem 2 is evident: If $\mathfrak{C}$ can be represented by a closed braid with $n$ strands, then $(\varkappa+\tau)(\mathfrak{C}) \leqq 2\pi n$. These assertions remain valid if knots with several components (links) are considered.

The proof follows. Choose a representative curve $C$ and a plane projection $C_P$ so that $\varkappa(C_P)+\pi\nu(C_P) < 6\pi$, where $\nu(C_P)$ is even. If $\nu(C_P) \geqq 2$, this would imply that $\varkappa(C_P) < 4\pi$, hence that $C$ were unknotted. Therefore $\nu(C_P) = 0$. Since $C_P$ is a plane curve with no inflection points, $\varkappa(C_P)$ is a multiple of $2\pi$, and therefore $\varkappa(C_P) = 4\pi$.

Fig. 2.

Without loss of generality we may assume that $C_P$ has only finitely many crossing points.

A list of the possible types of plane curves having no inflection points, having total curvature $4\pi$, and having finitely many crossing points is given in figure 2. It is not hard to show that this list is complete. Since a knot having one of these curves as plane projection is clearly isotopic to one of the curves of figure 1, this completes the proof.

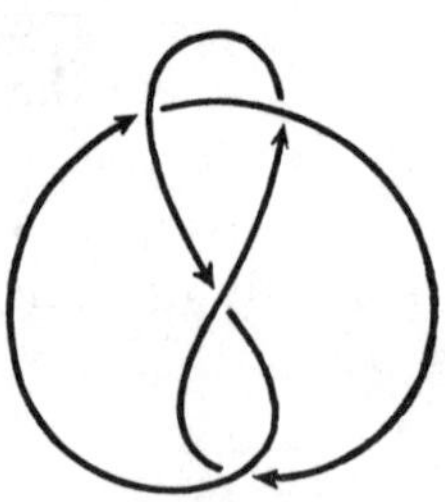

Fig. 3.

As an example to illustrate theorems 1 and 2, consider the class $\mathfrak{C}$ of the "figure eight" knot (figure 3). Since $\nu(C_P) = 2$ for the given projection $C_P$, and since $\mu(\boldsymbol{v}, C) = 4$ for a suitable vector $\boldsymbol{v}$ in $P$, it follows that $(\varkappa+\tau)(\mathfrak{C}) \leqq (2+4)\pi$, hence $(\varkappa+\tau)(\mathfrak{C}) = 2\pi, 4\pi$, or $6\pi$. But the Alexander polynomial of $\mathfrak{C}$ is $t^2-3t+1$, while the curves of figure 1 have Alexander polynomials of the form $\Sigma_{i=0}^{2k}(-t)^i$. (For definition of the Alexander polynomial, see for example Reidemeister [8, pp. 37–41].) Therefore $\mathfrak{C}$ can not be represented by any of these curves, and we must have $(\varkappa+\tau)(\mathfrak{C}) = 6\pi$. (Note that $\varkappa(\mathfrak{C}) = 4\pi$ for this example.)

As a final application of lemmas 1 and 2, the following theorem will be proved.

THEOREM 3. *If* $C$ *has linking number* $n$ *with the straight line* $L$, *then* $\varkappa(C)+\tau(C) \geqq 2\pi n$.

Let $P$ be any plane not perpendicular to $L$, and let $L_P$ and $C_P$ be the projections of $L$ and $C$ on $P$. It will be shown that $\varkappa(C_P)+\pi\nu(C_P) \geqq 2\pi n$. By lemmas 1 and 2 this will imply that $\varkappa(C)+\tau(C) \geqq 2\pi n$.

Since $C_P$ must cross $L_P$ at least $2n$ times, it follows that the tangent to $C_P$ must be parallel to $L_P$ at least $2n$ times. Consider two consecutive points of $C_P$ for which the tangents are parallel. If the tangents point in opposite directions, then the total curvature of the curve segment between the two points is at least $\pi$. If the tangents point in the same direction, then there must be some inflection point in between. In either case there is a contribution of at least $\pi$ to $\varkappa(C_P)+\pi\nu(C_P)$. Therefore $\varkappa(C_P)+\pi\nu(C_P) \geqq 2\pi n$, which completes the proof.

## 3. The quantity $\omega(C)$.

THEOREM 4. *If* $\tau(s) \geqq 0$ *for all* $s$, *but* $\tau(s)$ *is not identically* 0, *then* $\omega(C) \geqq 4\pi$.

An example given by Fenchel [5] shows that $\omega(C)$ can be arbitrarily close to $4\pi$.

It is first necessary to prove a lemma and to review some formulas of differential geometry. Suppose that two antipodal points of the unit sphere have been specified as north and south poles.

LEMMA 5. *Let* $G$ *and* $G'$ *be oriented great circles, such that* $G$ *does not pass through the north and south poles. If the tangent vector to* $G'$, *at the point where* $G'$ *crosses south of* $G$, *lies north of the equator, then* $G'$ *makes a smaller angle with the equator than* $G$.

Since the crossing point lies on the segment between the southernmost point of $G'$ and the northernmost point, it follows that the southernmost point of $G'$ lies north of $G$. Hence $G'$ makes a smaller angle with the equator than $G$.

Let $T$ denote the spherical indicatrix of the tangent vectors of $C$. The tangent, principal normal, and binormal vectors $\boldsymbol{t}$, $\boldsymbol{n}$, and $\boldsymbol{b}$ are related by the Frenet formulas:

$$\begin{aligned} d\boldsymbol{t}/ds &= \qquad \varkappa\boldsymbol{n} \\ d\boldsymbol{n}/ds &= -\varkappa\boldsymbol{t} \qquad +\tau\boldsymbol{b} \\ d\boldsymbol{b}/ds &= \qquad -\tau\boldsymbol{n}\,. \end{aligned} \tag{1}$$

In particular the tangent vector of the curve $T$ is just $\boldsymbol{n}$. Considered as a spherical curve, $T$ has geodesic curvature $\tau/\varkappa$, where a positive value indicates that $T$ curves towards the binormal vector $\boldsymbol{b}$.

The proof of theorem 4 follows. By lemma 3 it is sufficient to prove that the principal normal vector $\boldsymbol{n}(s)$ intersects every great circle at least four times. Suppose that it intersected some great circle $E$ only twice, at $\boldsymbol{n}(s_0)$ and $\boldsymbol{n}(s_1)$. Taking $E$ as equator, let $-\pi/2 \leqq \varphi(\boldsymbol{v}) \leqq \pi/2$ denote the latitude (angle north of equator) of any point $\boldsymbol{v}$. By proper choice of $s_0, s_1$, and the direction north, we may assume that

$$|\varphi(\boldsymbol{b}(s_0))| \geqq |\varphi(\boldsymbol{b}(s_1))| \tag{2}$$

and

$$\varphi(\boldsymbol{n}(s)) > 0 \qquad \text{for } s_0 < s < s_1\,. \tag{3}$$

Let $G$ be the great circle which is tangent to $T$ at $\boldsymbol{t}(s_0)$. It will be shown that $T$ cannot cross $G$; hence that $T$ cannot be the tangent indicatrix of a closed curve (see for example [5, p. 49]). By symmetry it will be sufficient to restrict our attention to the interval $s_0 \leqq s \leqq s_1$.

Formulas (1) and (3) imply that $\varphi(\boldsymbol{b}(s))$ is a monotone decreasing function of $s$; strictly decreasing whenever $\tau(s) > 0$. Together with (2) this implies

$$|\varphi(\boldsymbol{b}(s_0))| \geqq |\varphi(\boldsymbol{b}(s))|\,. \tag{4}$$

If $\varphi(\boldsymbol{b}(s_0))$ were equal to $\varphi(\boldsymbol{b}(s_1))$, it would follow that $\tau(s)$ were identically zero, not only for $s_0 \leqq s \leqq s_1$, but also, by a similar argument, for the remaining values of $s$. Since this is excluded, it follows that $\varphi(\boldsymbol{b}(s_0)) > \varphi(\boldsymbol{b}(s_1))$; hence by (2):

$$\varphi(\boldsymbol{b}(s_0)) > 0\,. \tag{5}$$

Suppose that $T$ crosses $G$ at some point $\boldsymbol{t}(s')$ with $s_0 < s' < s_1$. Let $\boldsymbol{t}(s')$ be the first such crossing, and let $G'$ be the great circle which is tangent to $T$ at $\boldsymbol{t}(s')$. It will be proved that $G$ and $G'$ satisfy the hypothesis of lemma 5. First note that $T$ cannot be tangent to $G$ at a crossing: this follows from the fact that the geodesic curvature $\tau/\varkappa$ of $T$ cannot change sign. Therefore the great circles $G$ and $G'$ are distinct. According to (5), $\boldsymbol{b}(s_0)$ lies north of the equator. Since $\boldsymbol{b}(s_0)$ is the pole of the great circle $G$, this implies that $G$ does not pass through the north and south poles. Since the geodesic curvature of $T$ is non-negative, $T$ must always either curve towards $\boldsymbol{b}(s)$ or else follow a great circle. Therefore $T$ either curves north of $G$ at $\boldsymbol{t}(s_0)$, or else follows the circle $G$ for some interval and then curves north. In either case $T$, and therefore $G'$, must cross south of $G$ at $\boldsymbol{t}(s')$. The tangent to $G'$ at $\boldsymbol{t}(s')$ is just $\boldsymbol{n}(s')$, which lies north of the equator by (3).

Now by lemma 5, $G'$ makes a smaller angle with the equator than $G$. But the poles of the great circles $G'$ and $G$ are just the points $\boldsymbol{b}(s')$ and $\boldsymbol{b}(s_0)$. Therefore

$$\left|\varphi\big(\boldsymbol{b}(s')\big)\right| > \left|\varphi\big(\boldsymbol{b}(s_0)\big)\right| ,$$

which contradicts (4), and completes the proof.

In conclusion an example will be given to show that $\omega(\mathfrak{C})$ is not necessarily an integral multiple of $2\pi$. In fact for the class $\mathfrak{C}$ of the figure eight knot (figure 3) it will be shown that

$$4\pi < \omega(\mathfrak{C}) < 6\pi .$$

For any representative curve $C$, the inequality $\varkappa(C)+\tau(C) \geqq 6\pi$ implies that

$$\omega(C) \geqq \big(\varkappa(C)+\tau(C)\big)2^{-1/2} \geqq 6\pi\, 2^{-1/2} > 4\pi .$$

(Actually the slightly better result $\omega(C) > 2\pi\, 5^{1/2}$ may be obtained by also using the inequalities $\omega(C) \geqq \big(\varkappa^2(C)+\tau^2(C)\big)^{1/2}$ and $\varkappa(C) > 4\pi$.) It is now necessary to construct a representative curve $C$ having $\omega(C) < 6\pi$.

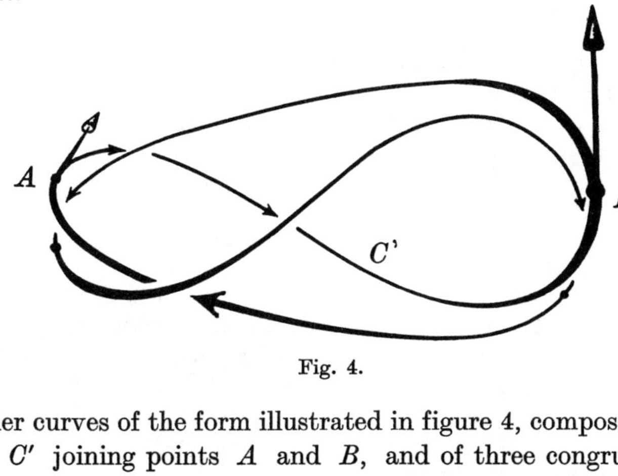


Fig. 4.

Consider curves of the form illustrated in figure 4, composed of the arc segment $C'$ joining points $A$ and $B$, and of three congruent arc segments (one properly congruent, and two congruent with reversed orientations). The tangent vectors $\boldsymbol{t}_A$, $\boldsymbol{t}_B$, and the displacement vector $\overrightarrow{AB}$ are to be perpendicular to each other, up to an $\varepsilon$ approximation. To compute $\omega(C')$, it is convenient to go over to the spherical indicatrix $T$ of the tangent vectors. Since the curve $T$ has tangent vector $\boldsymbol{n}(s)$, it follows that $\omega(C) = \varkappa(T)$ and $\omega(C') = \varkappa(T')$. Thus it is necessary to find a curve segment $T'$ joining $\boldsymbol{t}_A$ and $\boldsymbol{t}_B$ on the unit sphere, having total curvature as small as possible, subject to the condition $\int_{T'} \boldsymbol{t}(s)\, ds = \overrightarrow{AB}$.

This last condition may also be expressed as follows. The vector $\overrightarrow{AB}$ should be an interior point of the convex cone generated by the curve segment $T'$. Take $T'$ to be a circle arc with $\boldsymbol{t}_A$ and $\boldsymbol{t}_B$ as end points, and $\overrightarrow{AB}$ as interior point. Letting the mid-point of this circle arc approach $\overrightarrow{AB}$, the total curvature $\varkappa(T')$ can be brought arbitrarily close to $\frac{2}{3} \cdot 2\pi$. Now piecing the arc $T'$ together with three congruent arcs (two of which have reversed orientations) we obtain a closed spherical curve $T_1$ of class 1, with $\varkappa(T_1)$ arbitrarily close to $4 \cdot \frac{2}{3} \cdot 2\pi$. This curve $T_1$ can clearly be approximated by a curve $T$ of class 2. If $T$ is described by the vector $\boldsymbol{t}(u)$, then, for a suitably chosen weight function $w(u) > 0$, the vector $\boldsymbol{x}(u) = \int_0^u w(u)\, \boldsymbol{t}(u)\, du$ will describe a closed space curve $C$, of class 3 with $\varkappa(s) > 0$, belonging to the isotopy class $\mathfrak{C}$, and having $\omega(C) = \varkappa(T)$ arbitrarily close to $\frac{8}{3} \cdot 2\pi$. Thus it has been proved that

$$4\pi < 2 \cdot 5^{1/2}\pi \leqq \omega(\mathfrak{C}) \leqq \tfrac{8}{3} \cdot 2\pi < 6\pi \,.$$

## REFERENCES


1. E. Artin, *Theorie der Zöpfe*, Abh. Math. Sem. Hamburg. Univ. 4 (1926), 47–72. (See also Ann. of Math. 48 (1947), 101–126.)
2. W. Blaschke, *Vorlesungen über Integralgeometrie* II, Leipzig und Berlin, 1937; New York, 1950.
3. I. Fáry, *Sur la courbure totale d'une courbe gauche faisant un nœud*, Bull. Soc. Math. France 77 (1949), 128–138.
4. W. Fenchel, *Über Krümmung und Windung geschlossener Raumkurven*, Math. Ann. 101 (1929), 238–252.
5. W. Fenchel, *On the differential geometry of closed space curves*, Bull. Amer. Math. Soc. 57 (1951), 44–54.
6. R. H. Fox, *On the total curvature of some tame knots*, Ann. of Math. 52 (1950), 258–260.
7. J. Milnor, *On the total curvature of knots*, Ann. of Math. 52 (1950), 248–257.
8. K. Reidemeister, *Knotentheorie* (Ergebnisse der Mathematik 1,1), Berlin, 1932.

For further references see [5].


PRINCETON UNIVERSITY, PRINCETON, N. J., U. S. A.

# ON THE EXISTENCE OF A CONNECTION WITH CURVATURE ZERO

*Reprinted from*

Commentarii Mathematici Helvetici

Volume 32 (1958), pp. 215–223

This note characterizes those orientable 2-plane bundles over an orientable surface which possess a flat connection.

**More recent developments.** John Wood[1] has considered the corresponding question for arbitrary circle bundles over a surface. He gives a particularly neat construction of the 2-plane bundle with Euler number equal to $\chi/2 = 1 - g$ over a surface of genus $g \geq 2$. Such a surface can be described as a quotient $M^2 \cong D/\Gamma$ where $D$ is the open unit disk of complex numbers and $\Gamma \cong \pi_1(M^2)$ is a discrete torsionfree group of conformal automorphisms of $D$. The action of $\pi_1(M^2)$ on $D$ gives rise to an associated flat $D$-bundle over $M^2$. He shows that this $D$-bundle is topologically equivalent to the tangent bundle of $M^2$, with Euler number $2 - 2g$. If we replace $D$ by the upper half-plane which is conformally equivalent, then its group of conformal automorphisms can be identified with **PSL**(2, **R**). Thus we obtain an associated **PSL**(2, **R**)-bundle with Euler number $2 - 2g$. Since this number is even, we can lift the inclusion $\pi_1(M^2) \to$ **PSL**(2, **R**) to the 2-fold covering group **SL**(2, **R**). This lifting induces the required flat **SL**(2, **R**)-bundle over $M^2$ with Euler number $1 - g$.

Benzécri,[2] Sullivan,[3] and Gromov[4] have proved analogous results for higher dimensional manifolds, and Smillie[5] used these methods to construct

[1] J. Wood, *Bundles with totally disconnected structural group*, Comment. Math. Helv. **46** (1971) 257–273.

[2] J.-P. Benzécri, *Sur la classe d'Euler (ou de Stiefel-Whitney) de fibrés affins plats*, Compt. Rend. Acad. Sci. Paris **260** (1965) 5442–5444.

[3] D. Sullivan, *A generalization of Milnor's inequalities concerning affine foliations and affine manifolds*, Comment. Math. Helv. **51** (1976) 183–189.

[4] M. Gromov, *Volume and bounded cohomology*, Pub. Math. IHES **56** (1983) 5–99.

[5] J. Smillie, *Flat manifolds with non-zero Euler characteristic*, Comment. Math. Helv. **52** (1977) 453–455.

manifolds of dimension four and higher such that the tangent bundle possesses a flat connection even though the Euler characteristic is non-zero.

See also Ivanov and Turgaev.[6] A subject which is closely related to the study of bundles with flat connection is the study of Lie groups provided with the discrete topology.[7]

[6] N. V. Ivanov and V. G. Turgaev, *The canonical cocycle for the Euler class of a flat bundle,* Sov. Math. Dokl. **26** (1982) 78–81.

[7] J. Milnor, *On the homology of Lie groups made discrete,* Comment. Math. Helv. **58** (1983) 72–85.

# On the existence of a connection with curvature zero[1])

by John Milnor

## 1. Introduction

Let $M^2$ be a closed oriented $C^\infty$-surface of genus $g$ and let $GL^+(2)$ denote the group of $2\times 2$ real matrices with positive determinant. To each $GL^+(2)$-bundle[2]) over $M^2$ there corresponds the Euler (or Stiefel-Whitney) class

$$X \in H^2(M^2, Z) \ ,$$

and conversely each class $X$ comes from a unique equivalence class of bundles. Let $X[M^2]$ denote the Kronecker index of $X$ with the fundamental cycle of $M^2$.

**Theorem 1.** *If* $|X[M^2]| \geqslant g > 0$ *then the* $GL^+(2)$*-bundle over* $M^2$ *with* Euler *class* $X$ *does not possess a connection*[3]) *with curvature zero.*

By an *affine connection* on $M^2$ is meant a connection in the tangent bundle of $M^2$. Bezecri [3] has shown that $M^2$ can possess a symmetric affine connection with curvature zero if and only if $g = 1$. A somewhat stronger result follows from Theorem 1. For the tangent bundle of $M^2$ the integer $X[M^2]$ is equal to the Euler characteristic $2-2g$. Since $|2-2g| \geqslant g$ for $g \geqslant 2$ we have:

**Corollary.** *A surface with genus* $g \geqslant 2$ *does not possess any (not necessarily symmetric) affine connection with curvature zero.*

(The corresponding assertion for $g = 0$ is easy. In fact a bundle over a simply connected manifold can have a flat connection only if it is a product bundle). A study of the flat affine connections which do exist on the torus has been made by Kuiper [6].

It would be interesting to ask if every manifold with a flat affine connection has Euler characteristic zero, but only partial results have been obtained in higher dimensions. (See Auslander [8].)

The following shows that Theorem 1 was a best possible result.

**Theorem 2.** *If* $|X[M^2]| < g$ *then the* $GL^+(2)$*-bundle over* $M^2$ *with* Euler *class* $X$ *does have a connection with curvature zero.*

**Corollary.** *If* $g \geqslant 2$ *then there exist* $GL^+(2)$*-bundles over* $M^2$ *with* $X \neq 0$ *which have connections with curvature zero.*

---

1) The author holds an Alfred P. Sloan fellowship.

2) By a *$GL^+(2)$-bundle* is meant a principal fibre bundle with structural group $GL^+(2)$.

3) For the theory of connections see Chern [4] or Ambrose and Singer [1].

This behavior of the EULER class is in contrast with the known behavior of the PONTRJAGIN classes. (See § 4.)

The proofs of Theorems 1 and 2 are outlined in § 2 and completed in § 3. Some related problems are discussed in § 4.

## 2. Proof of Theorems

Let $G$ be a LIE group and $M$ a $C^\infty$-manifold. The universal covering space of $M$ can be considered as a $\pi_1(M)$-bundle over $M$. Therefore any homomorphism $\pi_1(M) \to G$ induces a $G$-bundle over $M$.

**Lemma 1.** *A $G$-bundle over $M$ possesses a connection with curvature zero if and only if it is induced from the universal covering bundle by a homomorphism $h$: $\pi_1(M) \to G$.*

(Compare STEENROD [7], § 13. The image of $h$ in $G$ is called the holonomy group. See AUSLANDER and MARKUS [2].) More generally for any cell complex $K$ a homomorphism $h: \pi_1(K) \to G$ induces a $G$-bundle over $K$. If $G$ is connected, then the first obstruction to the existence of a cross-section in this bundle is an element

$$c(h) \in H^2(K, \pi_1(G)) .$$

The next lemma gives an algorithm for computing this obstruction.

Assume that $K$ has only one vertex so that $\pi_1(K)$ has a canonical presentation with generators $\alpha_1, \ldots, \alpha_m$ corresponding to the oriented edges and relations $\varphi_1(\alpha_1, \ldots, \alpha_m) = 1, \ldots, \varphi_n(\alpha_1, \ldots, \alpha_m) = 1$ corresponding to the oriented faces. It is important that the relation $\varphi_k$ should be obtained by listing the edges incident to the face $e_k^2$ (with appropriate signs) in the cyclic order determined by the orientation of $e_k^2$. Consider the exact sequence

$$1 \to \pi_1(G) \xrightarrow{i} \tilde{G} \xrightarrow{p} G \to 1$$

where $\tilde{G}$ is the universal covering group. For each edge of $K$ define $\gamma_j = h(\alpha_j) \in G$, and choose a representative $\Gamma_j \in p^{-1}(\gamma_j)$. Then for each face we have

$$p\varphi_k(\Gamma_1, \ldots, \Gamma_m) = \varphi_k(\gamma_1, \ldots, \gamma_m) = 1$$

so that $i^{-1}\varphi_k(\Gamma_1, \ldots, \Gamma_m) \in \pi_1(G)$ is defined.

**Lemma 2.** *The cocycle in $K$ which assigns to the $k$-th face of $K$ the element $i^{-1}\varphi_k(\Gamma_1, \ldots, \Gamma_m)$ of $\pi_1(G)$ represents the negative $-c(h)$ of the obstruction cohomology class.*

The proofs of Lemmas 1 and 2 will be given in § 3.

Now let $K$ be the surface $M^2$ of genus $g > 0$. Using the cell subdivision of

$M^2$ obtained by matching the edges of a $4g$-gon, we obtain the presentation for $\pi_1(M^2)$ with generators $\alpha_1, \ldots, \alpha_{2g}$ and relation

$$\varphi(\alpha_1, \ldots, \alpha_{2g}) = \alpha_1\alpha_2\alpha_1^{-1}\alpha_2^{-1} \ldots \alpha_{2g-1}\alpha_{2g}\alpha_{2g-1}^{-1}\alpha_{2g}^{-1} = 1 .$$

The problem of determining which classes $c(h)$ can occur is therefore equivalent to the following: Which elements of $\pi_1(G)$ have the form

$$i^{-1}\varphi(\Gamma_1, \ldots, \Gamma_{2g}) \quad \text{with} \quad \Gamma_1, \ldots, \Gamma_{2g} \in \tilde{G}?$$

Every nonsingular matrix $\gamma$ can be written uniquely in the form

$$\gamma = r(\gamma)s(\gamma) \tag{1}$$

where $r(\gamma)$ is orthogonal and $s(\gamma)$ is symmetric and positive definite. This defines a retraction

$$r\colon\ GL^+(n) \to SO(n) .$$

It is easily verified that $r$ has the following properties:

$$r(\gamma^{-1}) = r(\gamma)^{-1} , \tag{2}$$

(3) if $\gamma_1$ or $\gamma_2$ is orthogonal then

$$r(\gamma_1\gamma_2) = r(\gamma_1)r(\gamma_2) .$$

Now consider the special case $n = 2$. For the rest of the section let $G$ denote $GL^+(2)$ and let $S$ denote the circle group $SO(2)$.

The universal covering group $\tilde{S}$ is isomorphic to the real numbers $R$ under the isomorphism $\operatorname{Exp}\colon R \to \tilde{S}$ which covers the homomorphism

$$\alpha \to \begin{pmatrix} \cos\alpha & \sin\alpha \\ -\sin\alpha & \cos\alpha \end{pmatrix} .$$

The retraction $r$ is covered by a retraction $\tilde{r}\colon \tilde{G} \to \tilde{S}$. Define $\theta = \operatorname{Exp}^{-1}\tilde{r}$ so that the following diagram is commutative.

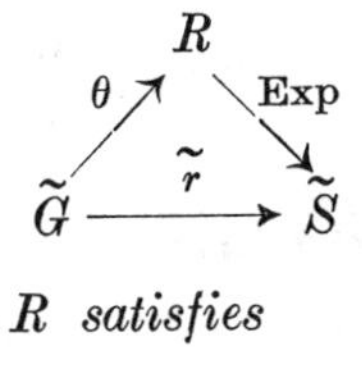


**Lemma 3.** *The map* $\theta\colon\ \tilde{G} \to R$ *satisfies*

$$|\,\theta(\Gamma_1\Gamma_2) - \theta(\Gamma_1) - \theta(\Gamma_2)| < \frac{\pi}{2}$$

*for all* $\Gamma_1, \Gamma_2 \in \tilde{G}$.

**Proof.** By direct computation we obtain the following explicit formula for $r$:

$$r\begin{pmatrix} a & b \\ c & d \end{pmatrix} = \frac{1}{\sqrt{x^2+y^2}}\begin{pmatrix} x & y \\ -y & x \end{pmatrix}$$

where $x = a + d$, $y = b - c$. This implies that

(4) if $\gamma \epsilon G$ has positive trace then $r(\gamma) \epsilon S$ also has positive trace.

If $\sigma_1$ and $\sigma_2$ are two positive definite matrices then the product $\sigma_1 \sigma_2$ has positive trace. (Proof. Choose an orthogonal matrix $\varrho$ so that $\varrho \sigma_1 \varrho^{-1}$ is diagonal. Then

$$\text{Trace}\,(\sigma_1 \sigma_2) = \text{Trace}\,((\varrho \sigma_1 \varrho^{-1})(\varrho \sigma_2 \varrho^{-1}))$$

which is clearly a sum of positive terms.) Together with (4) this implies

(5) if $\sigma_1$ and $\sigma_2$ are two positive definite elements of $G$, then $r(\sigma_1 \sigma_2)$ has positive trace.

Now let $\gamma_1$ and $\gamma_2$ be arbitrary elements of $G$. According to (1)

$$\gamma_1 \gamma_2 = r(\gamma_1) r(\gamma_2) \sigma_1 \sigma_2 \ ,$$

where $\sigma_1 = r(\gamma_2)^{-1} s(\gamma_1) r(\gamma_2)$, $\sigma_2 = s(\gamma_2)$. According to (3) this implies that

$$r(\gamma_1 \gamma_2) = r(\gamma_1) r(\gamma_2) r(\sigma_1 \sigma_2) \ .$$

Therefore according to (5) we have:

(6) for any $\gamma_1, \gamma_2 \epsilon G$ the matrix $r(\gamma_2)^{-1} r(\gamma_1)^{-1} r(\gamma_1 \gamma_2)$ has positive trace.

Given elements $\Gamma_1$, $\Gamma_2$ of $\tilde{G}$ set $p(\Gamma_i) = \gamma_i$ and define

$$\Delta(\Gamma_1, \Gamma_2) = \theta(\Gamma_1 \Gamma_2) - \theta(\Gamma_1) - \theta(\Gamma_2) \ .$$

Then $p \operatorname{Exp} \Delta(\Gamma_1, \Gamma_2) = r(\gamma_1 \gamma_2) r(\gamma_1)^{-1} r(\gamma_2)^{-1}$ has positive trace by (6). Since $p \operatorname{Exp} \Delta = \begin{pmatrix} \cos \Delta & \sin \Delta \\ -\sin \Delta & \cos \Delta \end{pmatrix}$ by definition, this implies that $\cos \Delta$ is positive. But since $\Delta$ is a continuous function of $\Gamma_1$ and $\Gamma_2$ which vanishes for $\Gamma_1 = 1$, this implies that $-\frac{\pi}{2} < \Delta < \frac{\pi}{2}$ ; which completes the proof of Lemma 3.

**Proof of Theorem 1.** The maps which are used are summarized in the following diagram.

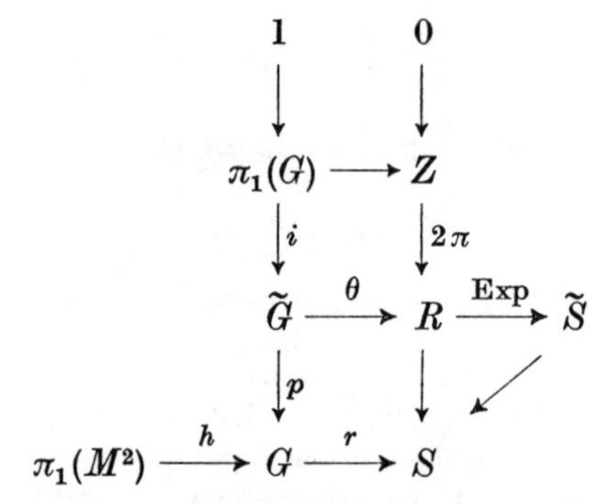


Here "$2\pi$" denotes multiplication by $2\pi$. The above maps are homomorphisms, with the exception of $\theta$ and $r$. The coefficient homomorphism $\pi_1(G) \to Z$

carries the obstruction class $c(h)\in H^2(M^2,\pi_1(G))$ into the Euler class X. The cocycle representing $-c(h)$ carries the unique 2-cell of $M^2$ into

$$i^{-1}(\Gamma_1\Gamma_2\Gamma_1^{-1}\Gamma_2^{-1}\ ..\ \Gamma_{2g}^{-1})\in\pi_1(G)\ .$$

Therefore $X[M^2]$ is equal to

$$-\frac{1}{2\pi}\theta(\Gamma_1\Gamma_2\Gamma_1^{-1}\Gamma_2^{-1}\dots)\in Z\ .$$

Applying the previous lemma $4g-1$ times we have

$$|\theta(\Gamma_1\Gamma_2\Gamma_1^{-1}\Gamma_2^{-1}\dots)-\theta(\Gamma_1)-\theta(\Gamma_2)-\theta(\Gamma_1^{-1})\dots|<(4g-1)\frac{\pi}{2}\ .$$

Cancelling the terms $-\theta(\Gamma_j)-\theta(\Gamma_j^{-1})$ and dividing by $2\pi$ this gives

$$|X[M^2]|<g-\tfrac{1}{4}<g$$

which completes the proof of Theorem 1.

Let $\gamma_0$ denote the matrix $\begin{pmatrix}2&0\\0&\frac{1}{2}\end{pmatrix}$ and let $\Gamma_0$ be the matrix in $p^{-1}(\gamma_0)$ satisfying $\theta(\Gamma_0)=0$. Let $K$ and $\tilde{K}$ denote the conjugate classes of $\gamma_0$ and $\Gamma_0$ respectively.

**Lemma 4.** *Any element of the conjugate class* $(\mathrm{Exp}\,\pi)\tilde{K}$ *can be expressed as the product of two elements in* $\tilde{K}$.

(Note that $\mathrm{Exp}\,\pi$ is in the center of the group $\tilde{G}$.) Proof. Consider the identity

$$\begin{pmatrix}2&0\\0&\frac{1}{2}\end{pmatrix}\begin{pmatrix}-2\frac{1}{2}&4\frac{1}{2}\\-3&5\end{pmatrix}=\begin{pmatrix}-5&9\\-1\frac{1}{2}&2\frac{1}{2}\end{pmatrix}\ .$$

The first two matrices have determinant 1 and trace $2\frac{1}{2}$, and therefore belong to the conjugate class $K$. Similarly the third belongs to $\begin{pmatrix}-1&0\\0&-1\end{pmatrix}K$. Let $\Gamma_1,\Gamma_2\in\tilde{K}$ be the elements corresponding to the first two matrices. Then $\Gamma_1\Gamma_2$ must belong to $(\mathrm{Exp}\,\pi n)\tilde{K}$ for some odd integer $n$. We will prove that $n=\pm 1$. (Actually $n=+1$.)

Since elements of $K$ have positive trace, assertion (4) implies that $\cos\theta(\Gamma)>0$ for $\Gamma\in\tilde{K}$. Since $\tilde{K}$ is connected this implies that $|\theta(\Gamma)|<\frac{\pi}{2}$ for $\Gamma\in\tilde{K}$. Therefore $|\theta(\Gamma)|>\frac{5}{2}\pi$ for $\Gamma\in(\mathrm{Exp}\,\pi n)\tilde{K}$, $n\geqslant 3$.

But since

$$|\theta(\Gamma_1\Gamma_2)|<|\theta(\Gamma_1)|+|\theta(\Gamma_2)|+\frac{\pi}{2}<\frac{3}{2}\pi\ ,$$

it follows that $\Gamma_1\Gamma_2$ can only belong to $(\operatorname{Exp}\pi n)\tilde{K}$ for $n = \pm 1$. If $n = +1$ then every element of $(\operatorname{Exp}\pi)\tilde{K}$ can be written in the form

$$\Gamma(\Gamma_1\Gamma_2)\Gamma^{-1} = (\Gamma\Gamma_1\Gamma^{-1})(\Gamma\Gamma_2\Gamma^{-1})\in\tilde{K}\tilde{K} .$$

If $n = -1$ then $\Gamma_2^{-1}\Gamma_1^{-1}$ can be substituted for $\Gamma_1\Gamma_2$ in this formula. This completes the proof of Lemma 4.

One consequence of this lemma is the following

(7) Every element $\Gamma$ of $(\operatorname{Exp}\pi)\tilde{K}$ has the form $\Gamma_1\Gamma_2\Gamma_1^{-1}\Gamma_2^{-1}$.

In fact choose $\Gamma_1, \Gamma_3\in\tilde{K}$ so that $\Gamma_1\Gamma_3 = \Gamma$. Since $\Gamma_1^{-1}$ also belongs to $\tilde{K}$, there exists an element $\Gamma_2\in\tilde{G}$ so that $\Gamma_2\Gamma_1^{-1}\Gamma_2^{-1} = \Gamma_3$, and hence

$$\Gamma = \Gamma_1\Gamma_2\Gamma_1^{-1}\Gamma_2^{-1} .$$

We are now ready to give the proof of Theorem 2. First recall the statement:

**Theorem 2.** *If $|X[M^2]| < g$ then the $G$-bundle over $M^2$ with* Euler *class $X$ does have a connection with curvature zero.*

According to Lemma 1 it is sufficient to prove that this bundle is induced by a homomorphism $\pi_1(M^2)\to G$. We will concentrate on the case

$$X[M^2] = 1 - g \ , \quad g > 2 \ ,$$

since the other cases can easily be obtained by the same method.

Applying the previous lemma $g-2$ times there exist elements $\bar{\Gamma}_1, \ldots, \bar{\Gamma}_{g-1}\in\tilde{K}$ so that the product $\bar{\Gamma}_1\ldots\bar{\Gamma}_{g-1}$ is equal to $(\operatorname{Exp}(g-2)\pi)\Gamma_0$. Setting $\bar{\Gamma}_g = \Gamma_0^{-1}\in\tilde{K}$ this gives

$$\bar{\Gamma}_1\ldots\bar{\Gamma}_{g-1}\bar{\Gamma}_g = \operatorname{Exp}(g-2)\pi \ .$$

Now choose elements $\Gamma_{2i-1}, \Gamma_{2i}$ in $\tilde{G}$ so that $\Gamma_{2i-1}\Gamma_{2i}\Gamma_{2i-1}^{-1}\Gamma_{2i}^{-1} = (\operatorname{Exp}\pi)\bar{\Gamma}_i$. Then

$$\begin{aligned}\Gamma_1\Gamma_2\Gamma_1^{-1}\Gamma_2^{-1}\ldots\Gamma_{2g-1}\Gamma_{2g}\Gamma_{2g-1}^{-1}\Gamma_{2g}^{-1} &= (\operatorname{Exp}\pi)\,\bar{\Gamma}_i\ldots(\operatorname{Exp}\pi)\,\bar{\Gamma}_g\\ &= (\operatorname{Exp}\pi)^g\operatorname{Exp}(g-2)\pi = \operatorname{Exp}2\pi(g-1).\end{aligned}$$

Since this is in the kernel of $p\colon\tilde{G}\to G$, we can define $h\colon\pi_1(M^2)\to G$ by $h(\alpha_i) = p(\Gamma_i)$. The characteristic number of the corresponding $G$-bundle is

$$-\frac{1}{2\pi}\theta(\operatorname{Exp}(2\pi(g-1))) = 1 - g \ ,$$

which completes the proof.

## 3. Proof of Lemmas

Suppose that $q\colon\ E \to M$ is the projection map of a $G$-bundle with a flat connection. Then each $C^\infty$-path $\lambda\colon\ [0, 1] \to M$ induces an isomorphism of $q^{-1}\lambda(1)$ onto $q^{-1}\lambda(0)$, invariant under $C^\infty$-homotopies of the path which keep the end points fixed. Identify $G$ with the fibre over a base point. If $\lambda$ is a closed path representing the element $\alpha$ of $\pi(M)$, then the corresponding isomorphism of the fiber $G$ onto itself is left-translation by an element $h(\alpha)$ of $G$. It is easily verified that

$$h\colon\ \pi_1(M) \to G$$

is a homomorphism.

Define a map $\eta\colon\ \tilde{M} \times G \to E$ as follows, where $\tilde{M}$ denotes the universal covering space. Given $(x,\gamma) \in \tilde{M} \times G$ choose a $C^\infty$-path from $x$ to the base point in $\tilde{M}$, and project onto a path $\lambda$ in $M$. Then the isomorphism along $\lambda$ carries $\gamma \in G$ into a point $\eta(x,\gamma)$ in the fibre corresponding to $x$. The identity

$$\eta(x\cdot\alpha,\, \gamma) = \eta(x,\, h(\alpha)\cdot\gamma)$$

which is easily verified, shows that $E$ is the $G$-bundle over $M$ induced by $h$. This proves the first half of Lemma 1.

It will be convenient to let $\pi_1(M)$ act on the left of $\tilde{M}$ by the rule

$$\alpha\cdot x = x\cdot\alpha^{-1}$$

for $\alpha \in \pi_1(M)$, $x \in \tilde{M}$; and on the left of $G$ by the rule

$$\alpha\cdot\gamma = h(\alpha)\gamma\ .$$

Then $\pi_1(M)$ acts on the left of $\tilde{M} \times G$, and the collapsed space $E$ is the total space of the $G$-bundle over $M$ induced by $h$. The product bundle $\tilde{M} \times G$ has a canonical flat connection. Since this connection is invariant under the action of $\pi_1(M)$, it induces a flat connection in $E$. This completes the proof of Lemma 1.

The proof of Lemma 2 will be based on the following. Since $\pi_1(K)$ acts on the left of $\tilde{K}$ without fixed points, and also acts on the left of $G$, we can ask if there exists a $\pi_1(K)$-equivariant map $\tilde{K} \to G$. Let

$$c'(h) \in H^2(K, \pi_1(G))$$

be the first obstruction to the existence of such a map. We will first compute this class $c'(h)$.

For each edge $e^1_j$ of $K$ let $\lambda_j\colon\ [0, 1] \to K$ denote a characteristic map, and let $\Lambda_j\colon\ [0, 1] \to \tilde{K}$ be the unique lifting which carries 0 into the base point $\tilde{e}^0$. Then $\Lambda_j(1)$ will be equal to $\alpha_j\cdot\tilde{e}^0$.

Let $\mu_j$: $[0, 1] \to G$ be any path from 1 to $\gamma_j$. An equivariant map $f$ from the 1-skeleton $(\tilde{K})^1$ to $G$ is defined by

$$\alpha \cdot \Lambda_j(t) \to \alpha \cdot \mu_j(t)$$

for each $\alpha \in \pi_1(K)$, $t \in [0, 1]$, and $1 \leqslant j \leqslant m$.

Consider the boundary of the $k$-th 2-cell $e_k^2$ in $K$. This closed path, which represents the element

$$\varphi_k(\alpha_1, \ldots, \alpha_m) = \alpha_{l_1}^{\varepsilon_1} \ldots \alpha_{l_r}^{\varepsilon_r}$$

of $\pi_1(K^1)$, lifts into a path in $\tilde{K}$ which goes from $\tilde{e}^0$ to $\alpha_{l_1}^{\varepsilon_1} \cdot \tilde{e}^0$ to $\alpha_{l_1}^{\varepsilon_1} \alpha_{l_2}^{\varepsilon_2} \cdot \tilde{e}^0$ to ... to $\varphi_k(\alpha_1, \ldots, \alpha_m) \cdot \tilde{e}^0 = \tilde{e}^0$. Mapping this closed path into $G$ we obtain a representative $f \partial \tilde{e}_k^2$ of the required homotopy class

$$c'(h)(e_k^2) \in \pi_1(G) \; .$$

To evaluate this homotopy class we will lift the path to $\tilde{G}$. Suppose that the path $\mu_j$, when lifted to $\tilde{G}$, goes from 1 to $\Gamma_j$. Then the path $f \partial \tilde{e}_k^2$, lifted to $\tilde{G}$, goes from 1 to $\Gamma_{l_1}^{\varepsilon_1}$ to $\Gamma_{l_1}^{\varepsilon_1} \Gamma_{l_2}^{\varepsilon_2}$ to ... to $\varphi_k(\Gamma_1, \ldots, \Gamma_m)$. Therefore this path represents the element $-i^{-1} \varphi_k(\Gamma_1, \ldots, \Gamma_m)$ of $\pi_1(G)$.

To complete the proof of Lemma 2 the following is needed.

**Lemma 5.** *The obstruction class* $c'(h)$ *is equal to the obstruction class* $c(h)$.

**Proof.** A cross section $K^1 \to E = \eta(\tilde{K} \times G)$ over the 1-skeleton is defined by the rule

$$\lambda_j(t) \to \eta\,(\Lambda_j(t), \mu_j(t)) \; ,$$

for each $t \in [0, 1]$ and each $j$. The obstruction to extending this over the 2 cell $e_k^2$ is defined as follows. Lifting $e_k^2$ to the 2-cell $\tilde{e}_k^2$, over which the bundle is a product, the given cross section on the 1-skeleton can be considered as a map $\partial \tilde{e}_k^2 \to \partial \tilde{e}_k^2 \times G$. Projecting into $G$ we obtain a representative of the required class $c(h)(e_k^2) \in \pi_1(G)$. But evidently the class obtained in this way is the same as the class $c'(h)(e_k^2)$ studied previously. This completes the proof of Lemma 2.

## 4. Further remarks

Although the following two results are topological in nature, the author only knows how to prove them by methods of differential geometry, using theorems of Chern and Weil.

**Theorem 3.** *Let* $K$ *be a finite complex.*

(a) *The* $SO(n)$*-bundle over* $K$ *induced by any homomorphism* $\pi_1(K) \to SO(n)$ *has trivial* Euler *class with rational coefficients.*

(b) *The $GL(n)$-bundle over $K$ induced by any homomorphism $\pi_1(K) \to GL(n)$ has trivial* PONTRJAGIN *classes with rational coefficients.*

These results are in contrast with Theorem 2, which showed that a homomorphism $\pi_1(K) \to GL^+(n)$ may induce a bundle with non-trivial EULER class.

Both results would be false with integer coefficients. For example consider the Lens space $K = S^5/Z_p$ and a non-trivial homomorphism $h: \pi_1(K) \to SO(2)$. The EULER class $X \in H^2(K, Z) \approx Z_p$ can be computed by Lemma 2, and turns out to be non-zero. This implies that the PONTRJAGIN class $p_1 = X^2$ is also non-zero.

**Proof of 3a.** Imbed $K$ in some euclidean space as a deformation retract of a neighborhood $U$. Then we may replace $K$ by the $C^\infty$-manifold $U$ for the rest of the proof. The following assertion follows easily from CHERN [4]. (See also [5].) For any $SO(n)$-bundle over $U$ choose a connection with curvature forms $\Omega_{ij}$. Then the form

$$\Sigma \varepsilon_{i_1 \dots i_n} \Omega_{i_1 i_2} \dots \Omega_{i_{n-1} i_n}$$

represents a cohomology class which is a non-zero multiple of the EULER class.

If this $SO(n)$-bundle is induced by a homomorphism $\pi_1(U) \to SO(n)$, then according to Lemma 1 there exists a connection with $\Omega_{ij} = 0$. Therefore the EULER class must be zero.

The proof of 3b is similar, being based on the following assertion which also follows easily from [4]. For any $GL(n)$-bundle over $U$ choose a connection with curvature forms $\Omega^i_j$. Then the form

$$\Sigma \delta^{j_1 \dots j_{2k}}_{i_1 \dots i_{2k}} \Omega^{i_1}_{j_1} \dots \Omega^{i_{2k}}_{j_{2k}}$$

represents a cohomology class which is a non-zero multiple of the $k$-th PONTRJAGIN class. The rest of the proof is clear.

REFERENCES

[1] W. AMBROSE and I. M. SINGER, *A theorem on holonomy*, Trans. Amer. Math. Soc. 75 428–443 (1953).
[2] L. AUSLANDER and L. MARKUS, *Holonomy of flat affinely connected manifolds*, Ann. Math. 62 (1955) 139–151.
[3] J. P. BENZECRI, *Variétés localement plates*, Thesis, Princeton University 1955.
[4] S. S. CHERN, *Topics in differential geometry (mimeographed)*, Institute for Advanced Study, Princeton 1951.
[5] S. S. CHERN, *On curvature and characteristic classes of a Riemann manifold*, Abhandl. Math. Sem. Univ. Hamburg 20 (1955) 117–126.
[6] N. H. KUIPER, *Sur les surfaces localement affines*, Colloque de Géometrie différentielle, Strasbourg 1953, 79–86.
[7] N. E. STEENROD, *The topology of fibre bundles*, Princeton 1951.
[8] L. AUSLANDER, *On Euler characteristics of locally affine spaces*, to appear.

(Received August 23, 1957.)

# EIGENVALUES OF THE LAPLACE OPERATOR ON CERTAIN MANIFOLDS

*Reprinted from*

Proceedings of the National Academy of Sciences USA

Volume 51 (1964), p. 542

This example of two compact Riemannian manifolds which are isospectral but not isometric was pure serendipity, the shortest and easiest paper I ever wrote. Since then, many such examples have been constructed,[1] all of the others requiring much more serious work. The manifolds can have constant positive[2] curvature, or constant negative[3] curvature. They can be bounded domains in the Euclidean plane[4] or bounded convex[5] domains in a higher dimensional Euclidean space. There can be continuous families of such examples,[6] and there can be examples which are not even locally isometric.[7]

On the other hand, there is a substantial literature going back as far as Hermann Weyl which shows that the spectrum of the Laplacian does have very substantial implications for the topology and geometry of the manifold.[8]

---

[1] See for example C. Gordon and D. DeTurck, *Isospectral metrics and finite Riemannian coverings*, pp. 79–92 of "The Legacy of Sonya Kovalevskaya", Contemp. Math. **64**, Amer. Math. Soc. 1987.

[2] A. Ikeda, *On lens spaces which are isospectral but not isometric*, Ann. Sci. Éc. Norm. Sup. Paris **13** (1980) 303–315.

[3] M.-F. Vignéras, *Variétés riemanniennes isospectrales et non isométriques*, Annals Math. **112** (1980) 21–32.

[4] C. Gordon, D. Webb and S. Wolpert, *One cannot hear the shape of a drum*, Bull. Amer. Math. Soc. **27** (1992) 134–138.

[5] C. Gordon and D. Webb, *Isospectral convex domains in Euclidean space*, preprint, MSRI 1994.

[6] C. Gordon and E. Wilson, *Isospectral deformations of compact solvmanifolds*, J. Diff. Geom. **19** (1984) 241–256.

[7] C. Gordon, *Isospectral closed Riemannian manifolds which are not locally isometric*, to appear.

[8] See for example M. Berger, P. Gauduchon and E. Mazet, *Le Spectre d'une Variété Riemannienne*, Lect. Notes Math. **194**, Springer 1971; and I. Kiyotaka, *Curvature and spectrum of Riemannian manifold*, Tôhoku Math. J. **25** (1973) 557–567.

# *EIGENVALUES OF THE LAPLACE OPERATOR ON CERTAIN MANIFOLDS*

By J. Milnor

PRINCETON UNIVERSITY

*Communicated February 6, 1964*

To every compact Riemannian manifold $M$ there corresponds the sequence $0 = \lambda_1 \leq \lambda_2 \leq \lambda_3 \leq \ldots$ of eigenvalues for the Laplace operator on $M$. It is not known just how much information about $M$ can be extracted from this sequence.[1] This note will show that the sequence does not characterize $M$ completely, by exhibiting two 16-dimensional toruses which are distinct as Riemannian manifolds but have the same sequence of eigenvalues.

By a *flat torus* is meant a Riemannian quotient manifold of the form $R^n/L$, where $L$ is a lattice (= discrete additive subgroup) of rank $n$. Let $L^*$ denote the dual lattice, consisting of all $y \in R^n$ such that $x \cdot y$ is an integer for all $x \in L$. Then each $y \in L^*$ determines an eigenfunction $f(x) = \exp(2\pi\, ix \cdot y)$ for the Laplace operator on $R^n/L$. The corresponding eigenvalue $\lambda$ is equal to $(2\pi)^2 y \cdot y$. Hence, the number of eigenvalues less than or equal to $(2\pi r)^2$ is equal to the number of points of $L^*$ lying within a ball of radius $r$ about the origin.

According to Witt[2] there exist two self-dual lattices $L_1, L_2 \subset R^{16}$ which are distinct, in the sense that no rotation of $R^{16}$ carries $L_1$ to $L_2$, such that each ball about the origin contains exactly as many points of $L_1$ as of $L_2$. It follows that the Riemannian manifolds $R^{16}/L_1$ and $R^{16}/L_2$ are not isometric, but do have the same sequence of eigenvalues.

In an attempt to distinguish $R^{16}/L_1$ from $R^{16}/L_2$ one might consider the eigenvalues of the Hodge-Laplace operator $\Delta = d\delta + \delta d$, applied to the space of differential $p$-forms. However, both manifolds are flat and parallelizable, so the identity

$$\Delta(f\, dx_{i_1} \wedge \ldots \wedge dx_{i_p}) = (\Delta f) dx_{i_1} \wedge \ldots \wedge dx_{i_p}$$

shows that one obtains simply the old eigenvalues, each repeated $\binom{16}{p}$ times.

[1] Compare Avakumović, V., "Über die Eigenfunktionen auf geschlossenen Riemannschen Mannigfaltigkeiten," *Math. Zeits.*, **65**, 327–344 (1956).

[2] Witt, E., "Eine Identität zwischen Modulformen zweiten Grades," *Abh. Math. Sem. Univ. Hamburg*, **14**, 323–337 (1941). See p. 324. I am indebted to K. Ramanathan for pointing out this reference.

# A NOTE ON CURVATURE AND FUNDAMENTAL GROUP

*Reprinted from*

Journal of Differential Geometry

Volume 2 (1968), pp. 1–7

This note relates the curvature of a closed manifold to "growth" in its fundamental group. It is closely related to earlier work by Švarc.[1]

**More recent developments.** The study of growth in finitely generated groups leads to a number of interesting questions.[2] One particularly noteworthy result in this field is Gromov's proof[3] that a group of polynomial growth is virtually nilpotent. Another is Grigorchuk's example[4] of a group which has more than polynomial but less than exponential growth. (It is not known whether such a group can have only a finite number of defining relations.) An important recent idea in the relationship between geometry and combinatorial group theory is Gromov's theory of ***hyperbolic groups***.[5] Another is the theory of ***automatic groups***, which has been developed by a number of authors.[6] Note also that there is a close relationship between growth in $\pi_1(M)$ and the entropy of the geodesic flow[7] in the unit tangent bundle of $M$.

---

[1] A. S. Švarc, *A volume invariant of coverings*, Dokl. Akad. Nauk SSSR **105** (1955) 32–34 (Russian).

[2] See for example J. Wolf, *Growth of finitely generated solvable groups and curvature of Riemannian manifolds*, J. Diff. Geom. **2** (1968) 421–446; as well as J. Milnor, *Growth of finitely generated solvable groups*, J. Diff. Geom. **2** (1968) 447–449.

[3] M. Gromov, *Groups of polynomial growth and expanding maps*, Pub. Math. IHES **53** (1981) 53–73.

[4] Compare R. Grigorchuk, *On growth in group theory*, Proc. ICM Kyoto 1990, pp. 325–338, Math. Soc. Japan 1991.

[5] See M. Gromov, *Hyperbolic groups*, pp. 75–263 of "Essays in Group Theory", MSRI Publ. **8** (Gersten, ed.) Springer 1987; as well as E. Ghys, *Les groupes hyperboliques*, Sem. Bourbaki 1989/90, Astérisque 1990 **189–190** (1991) Exp. **722**, pp. 203–238.

[6] See J. Cannon, *The combinatorial structure of cocompact discrete hyperbolic groups*, Geom. Dedic. **16** (1984) 123–148; as well as D. B. A. Epstein, *Computers, groups and hyperbolic geometry*, pp. 5, 9–29, 281 of "On the Geometry of Differentiable Manifolds", Rome 1986, Astérisque **163–164** (1988); and Epstein, Cannon, Holt, Levy, Patterson and Thurston, *Word Processing in Groups*, Jones and Bartlett 1992.

[7] A. Manning, *Topological entropy for geodesic flows*, Ann. Math. **110** (1979) 567–573.

# A NOTE ON CURVATURE AND FUNDAMENTAL GROUP

J. MILNOR

Define the *growth function* $\gamma$ associated with a finitely generated group and a specified choice of generators $\{g_1, \cdots, g_p\}$ for the group as follows (compare [9]). For each positive integer $s$ let $\gamma(s)$ be the number of distinct group elements which can be expressed as words of length $\leq s$ in the specified generators and their inverses. (For example, if the group is free abelian of rank 2 with specified generators $x$ and $y$, then $\gamma(s) = 2s^2 + 2s + 1$.) We will see that the asympotic behavior of $\gamma(s)$ as $s \to \infty$ is, to a certain extent, independent of the particular choice of generators (Lemma 1).

This note will make use of inequalities relating curvature and volume, due to R. L. Bishop [1], [2] and P. Günther [3], to prove two theorems.

**Theorem 1.** *If $M$ is a complete n-dimensional Riemannian manifold whose mean curvature tensor $R_{ij}$ is everywhere positive semidefinite, then the growth function $\gamma(s)$ associated with any finitely generated subgroup of the fundamental group $\pi_1 M$ must satisfy*

$$\gamma(s) \leq \text{constant} \cdot s^n .$$

It is conjectured that the group $\pi_1 M$ itself must be finitely generated.

The constant in this inequality will depend, of course, on the particular set of generators which is used to define $\gamma(s)$.

**Theorem 2.** *If $M$ is compact Riemannian with all sectional curvatures less than zero, then the growth function of the fundamental group $\pi_1 M$ is at least exponential:*

$$\gamma(s) \geq a^s$$

*for some constant $a > 1$.*

In both cases, any set of generators for the group may be used in defining $\gamma(s)$.

**Remarks.** Note that there is always an exponential upper bound for $\gamma(s)$. In fact the inequality

$$\gamma(s + t) \leq \gamma(s)\gamma(t)$$

Received February 12, 1968. This work was supported in part by National Science Foundation Grant # GP-7917.

implies that $\gamma(s) \leq \gamma(1)^s$. With only a little more work one can verify that the sequence $\gamma(s)^{1/s}$ necessarily converges to a finite limit as $s \to \infty$. (For each fixed $t$, setting $k = [s/t] + 1$, the inequality $\gamma(s) \leq \gamma(kt) \leq \gamma(t)^k \leq \gamma(t)^{1+s/t}$ implies that $\limsup \gamma(s)^{1/s} \leq \gamma(t)^{1/t}$. Therefore $\limsup \gamma(s)^{1/s} \leq \inf \{\gamma(t)^{1/t}\} \leq \liminf \gamma(s)^{1/s}$, and the sequence converges.) It will be convenient to say that a group has *exponential growth* whenever $\lim \gamma(s)^{1/s} > 1$.

The two theorems stated above may be contrasted with three classical theorems; namely Myers' theorem which asserts that if mean curvature is positive definite on a compact manifold then the fundamental group is finite; the Hadamard-Cartan theorem which asserts that if all sectional curvatures are $\leq 0$ on a complete manifold then the higher homotopy groups $\pi_i M$, $i \geq 2$, are zero; and Preissmann's theorem which asserts that if all sectional curvatures are $<0$ on a compact manifold then every abelian subgroup of $\pi_1 M$ is cyclic.

The proofs of Theorems 1 and 2 follow. Some examples and further discussion will be given at the end.

*Proof of Theorem 1.* Choose a base point $x_0$ in the universal covering space $\tilde{M}$, and let $V(r)$ denote the volume of the neighborhood $N_r(x_0)$ consisting of all $y \in \tilde{M}$ with $d(x_0, y) \leq r$, using the Riemannian distance function and the Riemannian $n$-dimensional volume element in $\tilde{M}$.

If the mean curvature is everywhere $\geq 0$, then according to Bishop:

$$V(r) \leq \omega_n r^n \,, \tag{1}$$

where $\omega_n$ denotes the volume of the unit disk in euclidean $n$-space.

We will identify $\pi_1 M$ with the group of all covering transformations of $\tilde{M}$ over $M$. Let $g_1, \cdots, g_p : \tilde{M} \to \tilde{M}$ be the preferred generators for some subgroup of $\pi_1 M$, and let $\mu$ denote the maximum of the numbers $d(x_0, g_i(x_0))$ for $i = 1, 2, \cdots, p$. Note that the neighborhood $N_{\mu s}(x_0)$ contains at least $\gamma(s)$ distinct points of the form $g(x_0)$ with $g \in \pi_1 M$.

Choose a number $\varepsilon > 0$ which is small enough so that the neighborhood $N_\varepsilon(x_0)$ in $\tilde{M}$ is disjoint from all of the translates $g(N_\varepsilon(x_0))$, with $g \neq 1$. Then the neighborhood $N_{\mu s+\varepsilon}(x_0)$ contains at least $\gamma(s)$ disjoint sets of the form $g(N_\varepsilon(x_0))$. This proves that

$$\gamma(s) V(\varepsilon) \leq V(\mu s + \varepsilon) \,. \tag{2}$$

Combining (1) and (2) we obtain the required inequality

$$\gamma(s) \leq c s^n \,, \qquad \text{for} \qquad s \geq 1 \,,$$

where $c = \omega_n(\mu + \varepsilon)^n / V(\varepsilon)$. This proves Theorem 1.

Before proving Theorem 2 we need to investigate the dependence of the growth function $\gamma(s)$ of a group on the particular choice of generators $\{g_1, \cdots, g_p\}$ for the group.

**Lemma 1.** *Let $\{g_1, \cdots, g_p\}$ and $\{h_1, \cdots, h_q\}$ be two different sets of generators for the same group, and let $\gamma(s)$ and $\gamma'(t)$ be the corresponding growth functions. Then there exist positive constants k and k′ so that*

$$\gamma'(t) \leq \gamma(kt)$$

*for all t and $\gamma(s) \leq \gamma'(k's)$ for all s.*

*Proof.* If $k$ is large enough so that each $h_j$ can be expressed as a word of length $\leq k$ in the generators $g_i$ and $g_i^{-1}$, then the required inequality $\gamma'(t) \leq \gamma(kt)$ is clearly satisfied.

*Proof of Theorem 2.* Let $-\alpha^2 < 0$ be an upper bound for the sectional curvature of $M$. Then according to Günther the volume $V(r)$ is greater than or equal to the volume

$$n\omega_n \int_0^r (\alpha^{-1} \sinh \alpha x)^{n-1} dx$$

of a corresponding ball in the space of constant negative curvature. This expression is asymptotically equal to $2c \exp(\lambda r)$ as $r \to \infty$ (where $c = n\omega_n/(n-1)(2\alpha)^n$ and $\lambda = (n-1)\alpha$ are positive constants). Hence

$$V(r) > c \exp(\lambda r) \tag{3}$$

for $r$ sufficiently large.

Let $\delta$ be the diameter of $M$. The projection $\tilde{M} \to M$ clearly maps the compact neighborhood $N = N_\delta(x_0)$ onto $M$. Hence the set of all translates $gN$, with $g \in \pi_1 M$, forms a covering of $\tilde{M}$. Note that this covering is locally finite. (For otherwise, for some finite $r$ the neighborhood $N_r(x_0)$ would intersect infinitely many of the $gN$, and hence $N_{r+\delta+\varepsilon}(x_0)$ would contain infinitely many disjoint sets $gN_\varepsilon(x_0)$, each of volume $V(\varepsilon) > 0$. This is impossible since $V(r + \delta + \varepsilon) < \infty$.)

Let $F$ be the finite set consisting of all elements $g$ in $\pi_1 M$ such that the translate $gN$ intersects $N$. Let $\nu > 0$ be the minimum of $d(gN, N)$ as $gN$ ranges over all translates of $N$ which do not intersect $N$.

**Lemma 2.** *If $d(x_0, gN) < \nu t + \delta$ for some positive integer t, then g can be expressed as a t-fold product, $g = f_1 f_2 \cdots f_t$, with $f_1, \cdots, f_t \in F$.*

*Proof.* Choose $y \in gN$ with $d(x_0, y) < \nu t + \delta$, and choose points $y_1, y_2, \cdots, y_{t+1} = y$ along the minimal geodesic from $x$ to $y$ so that

$$d(x_0, y_1) \leq \delta \qquad \text{and} \qquad d(y_i, y_{i+1}) < \nu\,.$$

Each $y_i$ belongs to some translate $h_i N$, where we can choose $h_1$ to be 1 and $h_{t+1}$ to be $g$. Let $f_i = h_i^{-1} h_{i+1}$ so that $f_1 f_2 \cdots f_t = g$. Since the two points $h_i^{-1} y_i$ and $h_i^{-1} y_{i+1}$ have distance $< \nu$, and belong to $N$ and to $f_i N$ respectively, it follows from the definitions of $\nu$ and $F$ that $f_i \in F$. This proves Lemma 2.

In particular it follows that the elements of $F$ generate the group $\pi_1 M$, Let $\gamma'(t)$ denote the growth function of $\pi_1 M$, computed using $F$ as the set of preferred generators.

According to Lemma 2, the interior of the neighborhood $N_{\nu t+\delta}(x_0)$ is completely covered by the translates $f_1 f_2 \cdots f_t N$ with $f_i \in F$. Since precisely $\gamma'(t)$ of these translates are distinct, this proves that

$$(4) \qquad \gamma'(t)V(\delta) \geq V(\nu t + \delta) \ .$$

By Lemma 1 there exists a positive integer $k$ so that

$$\gamma(kt) \geq \gamma'(t)$$

or in other words

$$(5) \qquad \gamma(s) \geq \gamma'([s/k]) \ , \qquad \text{for} \qquad s \geq k \ .$$

Combining (3), (4) and (5) we clearly obtain

$$(6) \qquad \gamma(s) \geq c_1 \exp(\lambda_1 s)$$

for all sufficiently large $s$, where the constants $c_1 = c \exp(\lambda\delta - \lambda\nu)/(V(\delta)$ and $\lambda_1 = \lambda\nu/k$ are both positive. Since $\gamma(s) > 1$ for all $s \geq 1$, it is now easy to choose a constant $a > 1$ so that

$$\gamma(s) \geq a^s$$

for all $s$. This completes the proof of Theorem 2.

**Remark.** It may be conjectured that some weaker hypothesis, in which some zero sectional curvatures were allowed, would still be sufficient to imply exponential growth. Perhaps the hypothesis of negative definite mean curvature would already suffice?

**Examples.** First consider the 2-dimensional case. On a sphere or projective plane or torus or Klein bottle we can impose a metric with curvature $\geq 0$, so that Theorem 1 applies. Choosing the most familiar generators for $\pi_1 M$, the growth function is given by

$\gamma(s) \equiv 1$ for the sphere,

$\gamma(s) \equiv 2$ for the projective plane $P^2$,

$\gamma(s) = 2s^2 + 2s + 1$ for the torus or Klein bottle.

Thus $5s^2$ is an upper bound in each case.

Similarly for an $n$-dimensional torus the growth function of $\pi_1$ is a polynomical of degree $n$, providing that the standard generators are used (compare [6]).

Now consider the non-orientable surface $P^2 \# P^2 \# \cdots \# P^2$ which is obtained from the 2-sphere by attaching $p$ "cross-caps". If $p \geq 3$ it is known that this surface admits a metric of (constant) negative curvature. The fundamental group has generators, say $x_1, x_2, \cdots, x_p$, subject to the single relation $x_1^2 x_2^2 \cdots x_p^2 = 1$. To estimate the growth function we will use the following.

**Lemma 3.** *For a free group, which is freely generated by the preferred set of generators* $\{g_1, \cdots, g_p\}$, *the growth function is given by*

$$\gamma(s) = (p(2p-1)^s - 1)/(p-1) \ .$$

*Hence*

$$(2p-1)^s \leq \gamma(s) \leq (2p+1)^s \ .$$

The proof is not difficult.

Now for the required group, with generators $x_1, \cdots, x_p$ and relation $x_1^2 \cdots x_p^2 = 1$, note that the elements $x_1, \cdots, x_{p-1}$ generate a free subgroup (see [7]). Thus we obtain the estimate

$$(2p-3)^s \leq \gamma(s) \leq (2p+1)^s \ ,$$

which supports Theorem 2.

In the case of an orientable surface of genus $q \geq 2$, a similar argument shows that $(4q-3)^s \leq \gamma(s) \leq (4q+1)^s$. (For examples of higher dimensional manifolds of negative curvature, see Borel, *Compact Clifford-Klein forms of symmetric spaces*, Topology **2** (1963) 111–122.)

As a final example, let $G$ denote the nilpotent Lie group consisting of all $3 \times 3$ triangular real matrices with 1's on the diagonal, and let $H$ be the subgroup consisting of all integer matrices of the same form. Then the coset space $G/H$ is a compact 3-dimensional manifold with fundamental group $H$.

**Lemma 4.** *The growth function of H is quartic:*

$$c_1 s^4 \leq \gamma(s) \leq c_2 s^4 \ , \qquad \text{with} \qquad c_1 > 0.$$

*Outline of proof.* As preferred generators take the matrices $x = I + e_{23}$ and $y = I + e_{12}$. Let $c$ denote the central element $yxy^{-1}x^{-1} = I + e_{13}$. Note that every expression of the form $x^t y^t c^k$ with $0 \leq k \leq t^2$ can be expressed as a word of length $2t$ in $x$ and $y$. Hence every $x^i y^j c^k$ with $1 \leq i \leq t$, $1 \leq j \leq t$, and $1 \leq k \leq t^2$ has length $\leq 4t$, which proves that

$$\gamma(4t) \geq t^4 \ .$$

Conversely, if $x^i y^j c^k$ can be expressed as a word of length $\leq s$ then it is easily verified that $|i| \leq s$, $|j| \leq s$, and $|k| \leq (s/2)^2$, which yields a quartic upper bound for $\gamma(s)$.

**Corollary.**[1] *No Riemannian metric on $G/H$ can satisfy either the hypothesis of Theorem 1 or the hypothesis of Theorem 2.*

For Lemma 4 shows that $\gamma(s)$ is larger than any constant times $s^3$, but smaller than any $a^s$, $a > 1$, for sufficiently large $s$.

If we impose the most natural metric, which comes from a left invariant metic on $G$, computation shows that the mean curvature tensor of $G/H$ is indefinite, with signature $(+, -, -)$. The scalar curvature $\sum R^j{}_j$ turns out to be negative. Thus: *The fundamental group of a compact manifold of negative scalar curvature need not have exponential growth.* On the other hand the product of a surface of genus 2 with a small sphere provides an example of a manifold of positive scalar curvature whose fundamental group does have exponential growth. Perhaps the scalar curvature of a manifold of dimension $\geq 3$ has no influence at all on the fundamental group?

(Here is an alternative example of a 3-manifold whose fundamental group is the nilpotent group $H$ of Lemma 4. Let $M^3$ be the intersection of the algebraic variety $z_1^2 + z_2^3 + z_3^6 = 0$ in complex 3-space with the unit sphere. Compare [4], [8, §9]. This manifold can also be described as the 6-fold cyclic branched covering of the 3-sphere $|z_1|^2 + |z_2|^2 = 1$, branched along the trefoil knot $z_1^2 + z_2^3. = 0$. I do not know whether or not this manifold $M^3$ is diffeomorphic to $G/H$.)

**A concluding problem.** It is interesting to compare the growth function $\gamma(s)$ of a group with the behavior of left-invariant, symmetric random walks on the group, as studied by[2] Kesten [5]. Each such random walk is described by a symmetric matrix $m_{gh}$ whose spectrum has largest element

$$\lambda = \limsup\, (m_{11}^{(s)})^{1/s} .$$

Here $m_{gh}^{(s)}$ denotes the probability of passing from $g$ to $h$ in $s$ steps of the random walk. We will assume that every pair $g, h$ of group elements "communicate" (that is $m_{gh}^{(s)} > 0$ for some $s$).

This number $\lambda$ is equal to 1 if the group is solvable, but is less than 1 if there exists a free non-cyclic subgroup. The distinction $\lambda = 1$ or $\lambda < 1$ depends only on the group and not on the particular probability distribution used to define the random walk.

**Lemma 5.** *If $\lambda < 1$ then the growth function satisfies* $\lim \gamma(s)^{1/s} > 1$ *(exponential growth).*

*Proof.* Choose the random walk so that only right multiplication by the preferred generators or their inverses has positive probability at each step. Applying Schwarz's inequality to the equation

$$m_{1g}^{(2t)} = \sum_h m_{1h}^{(t)} m_{hg}^{(t)}$$

[1] The second half of this corollary could equally well be derived from Preissmann's theorem.

[2] I am indebted to R. Lyndon for pointing this out.

we obtain $m_{1g}^{(2t)} \leq m_{11}^{(2t)}$. Averaging this inequality over the $\gamma(2t)$ elements $g$ represented by words of length $\leq 2t$, we obtain

$$\frac{1}{\gamma(2t)} \leq m_{11}^{(2t)} ;$$

hence

$$\frac{1}{\lim \gamma(s)^{1/s}} \leq \lim \sup (m_{11}^{(s)})^{1/s} < 1 ,$$

which completes the proof.

Surprisingly, the converse statement is false. For example the group with generators $x, y$ and defining relation $xyx^{-1} = y^2$ is solvable, so that $\lambda = 1$. But $\gamma(2t) \geq 2^t$, since the words

$$xy^{\varepsilon_1}xy^{\varepsilon_2} \cdots xy^{\varepsilon_t}$$

with $\varepsilon_i = 0$ or $1$ all represent distinct elements of this group.

**Problem.** Consider such a random walk on the fundamental group of a compact manifold of negative curvature. Is $\lambda$ necessarily less than 1?

## References

[1] R. L. Bishop, *A relation between volume, mean curvature, and diameter,* Notices Amer. Math. Soc. **10** (1963) 364.

[2] —— & R. J. Crittenden, *Geometry of manifolds,* Academic Press, New York, 1964.

[3] P. Günther, *Einige Sätze über das Volumenelement eines Riemannschen Raumes,* Publ. Math. Debrecen **7** (1960) 78–93.

[4] F. Hirzebruch, *The topology of the normal singularities of an algebraic surface* (d'après Mumford), Séminaire Bourbaki, 15e année 1962–63, N° 250.

[5] H. Kesten, *Symmetric random walks on groups,* Trans. Amer. Math. Soc. **92** (1959) 336–354.

[6] H. U. Krause, *Gruppenstruktur und Gruppenbild,* Thesis, Eidg. Techn. Hochs. Zürich 1953, 45 pp.

[7] W. Magnus, *Über diskontinuierliche Gruppen mit einer definierenden Relation* (*Der Freiheitssatz*), J. reine angew. Math. **163** (1930) 141–165.

[8] J. Milnor, *Singular points of complex hypersurfaces,* Annals of Math. Studies, Princeton University Press, Princeton, to appear.

[9] ——, *Problem,* Amer. Math. Monthly, to appear.*

[10] S. Myers, *Riemannian manifolds with positive mean curvature,* Duke Math. J. **8** (1941) 401–404.

[11] A. Preissmann, *Quelques propriétés globales des espaces de Riemann,* Comment. Math. Helv. **15** (1942–43) 175–216.

[12] J. L. Synge, *On the connectivity of spaces of positive curvature,* Quart. J. Math. Oxford, Ser. (1), **7** (1936) 316–320.

UNIVERSITY OF CALIFORNIA, LOS ANGELES

*[Updated reference] Volume **75** (1968) 685–686.

# PROBLEMS IN DIFFERENTIAL GEOMETRY

*Reprinted from*

Mathematical Developments arising from Hilbert Problems
(Conference held in De Kalb, Illinois, May 1974)
Proceedings of Symposia in Pure Mathematics
Volume XXVIII, Part 1, Felix Browder, ed.
American Mathematical Society 1976, pp. 54–57

As part of a session on the ***Problems of Present Day Mathematics*** organized by Felix Browder, I described some of the central problems of differential geometry. There has been very substantial progress on nearly all of these questions in the intervening years. Here is the original problem list, followed by a discussion of some of this later work.

**XV. Differential Geometry.**[3] **(J. Milnor).**

1. *The (self-intersecting) soap bubble problem.* According to Plateau[4] in 1873: "Very probably the sphere is the only closed surface of constant mean curvature." For the boundary of a star-shaped region this had been proved by Jellett in 1853. For immersed surfaces of genus zero it has been proved by Hopf in 1950, and for embedded surfaces of arbitrary genus by Aleksandrov in 1958. However, the case of an immersed surface of higher genus remains open. Chern remarks that the situation is the same for elliptic $W$-surfaces, that is, surfaces with a relation $W(k_1, k_2) = 0$ between principal curvatures, where the equation defines $k_2$ as a monotone decreasing function of $k_1$.

2. *Understanding the scalar curvature* $R = \Sigma g^{ik} g^{jl} R_{ijkl}$. In 1960, Yamabe claimed the following theorem: *For every smooth compact manifold of dimension at least* 3 *with Riemannian metric* $g_{ij}$ *there exists a conformally equivalent metric* $e^{2\rho} g_{ij}$ *which has constant scalar curvature.* Yamabe's proof was defective, and his theorem may well be false in some cases. However, his paper has led to important work by Aubin, Trudinger, Eliasson, Fischer and Marsden, and Kazdan and Warner. In particular Eliasson showed that every manifold of dimension $\geqslant 3$ admits a metric of constant negative scalar curvature. It is not known which manifolds admit metrics of positive scalar curvature (constant or not), or zero scalar curvature. One important contribution is Lichnerowicz's 1963 proof that a $4k$-dimensional spin manifold $M$ with $\hat{A}[M] \neq 0$ cannot admit a metric with $R \geqslant 0, R \not\equiv 0$. Furthermore, if the first Betti number is nonzero, then the case $R \equiv 0$ cannot occur either. (See Kazdan and Warner, "*Prescribing curvatures*", §5.4. For more on spin manifolds with $\hat{A}[M] \neq 0$, see Atiyah and Hirzebruch.) Hitchin has shown that an exotic sphere which does not bound a spin manifold cannot admit a metric of positive scalar curvature.

Closely related is a problem which has been posed by R. Geroch in connection with general relativity theory. Consider a Riemannian metric on the real coordinate space $\mathbf{R}^n$ which coincides with the Euclidean metric $\delta_{ij}$ outside of a bounded open set. If $R \geqslant 0$ everywhere, does it follow that this metric is flat? Can it happen that $R > 0$ throughout the open set? (The case $R < 0$ can occur. Compare Aubin.)

---

[3] I am indebted to S. S. Chern, J. Nitsche, J. O'Sullivan, N. Wallach, and S. T. Yau for useful suggestions and references.

[4] This problem was perhaps first posed by Delaunay in 1843.

3. *Understanding the Ricci tensor* $R_{ik} = \Sigma g^{jl} R_{ijkl}$. If a compact Riemannian manifold has $R_{ik}$ identically zero, does it follow that it is flat? In this connection we note Calabi's conjecture that any Kähler manifold with $c_1 = 0$ admits a Kähler metric with $R_{ik} \equiv 0$.

Which compact manifolds admit metrics of positive definite (or negative definite) Ricci curvature? A key result is Meyer's classical theorem that positive Ricci curvature implies finite fundamental group. Here are some examples of manifolds of negative Ricci curvature. Nagano and Smyth consider a nonsingular hyperplane section of an abelian variety of complex dimension $n \geqslant 3$. The resulting manifold has negative definite Ricci tensor almost everywhere, hence by a theorem of Ehrlich the metric can be deformed so that the Ricci tensor will be negative definite everywhere. Yet the fundamental group is free abelian of rank $2n$, and the universal covering space is not contractible, so this manifold cannot admit a metric of nonpositive sectional curvature. Yau has shown that the connected sum of two manifolds of constant nonpositive curvature admits a metric of negative Ricci curvature. Again the universal covering space is not contractible provided that the dimension is 3 or more. It is not known whether or not the sphere $S^3$ or the torus $S^1 \times S^1 \times S^1$ admits a metric of negative Ricci curvature. For further information see Bochner and Yano, Fischer and Wolf, Kobayashi, Milnor, and Yau.

4. *Manifolds of positive sectional curvature.* In spite of important work by Synge, Rauch, Berger, Gromoll, Cheeger, Wallach and many others, there remain a number of unanswered questions concerning compact Riemannian manifolds for which the sectional curvatures $K = \Sigma R_{ijkl} u^i v^j u^k v^l$ are all positive. Can such a manifold split topologically as a product? For example, does the product $S^m \times S^n$, with $m, n \geqslant 2$, admit a metric of strictly positive sectional curvature? (This manifold does of course admit a metric of positive Ricci curvature.) If the manifolds $M$ and $N$ have odd dimension, note that the product $M \times N$ has Euler number zero, and carries two linearly independent vector fields $u$ and $v$ which commute: $[u, v] = 0$. Thus the following two questions are related to the question just raised.

Can any compact manifold of positive sectional curvature admit two linearly independent commuting vector fields? According to Lima, the sphere $S^3$ does not admit two such vector fields, but the corresponding question for $S^7$, $S^{11}, \ldots$ remains open.

Does every even-dimensional compact manifold of positive sectional curvature have positive Euler number? In the 4-dimensional case this can be proved using the generalized Gauss-Bonnet formula. (See Chern.) However the corresponding argument in the 6-dimensional case fails, since the Gauss-Bonnet integrand can be negative at a point of positive sectional curvature. (Geroch, unpublished.)*

Similarly one can ask which manifolds admit metrics of sectional curvature $< 0$ (or $\leqslant 0$). Compare Eberlein (or Wolf).

REFERENCES

A. D. Aleksandrov, *Uniqueness theorems for surfaces in the large.* V, Vestnik Leningrad. Univ. **13** (1958), no. 19, 5–8; English transl., Amer. Math. Soc. Transl. (2) **21** (1962), 412–416. MR **21** #909; **27** #698e.

S. Aloff and N. Wallach, *An infinite family of compact simply-connected positively curved 7-manifolds,* Bull. Amer. Math. Soc. **81** (1975), 93–97.

M. Atiyah and F. Hirzebruch, *Spin-manifolds and group actions,* Essays on Topology and Related Topics (Mém. Dédiés à Georges deRham), Springer, New York, 1970, pp. 18–28. MR **43** #4064.

T. Aubin, *Métriques riemanniennes et courbure,* J. Differential Geometry **4** (1970), 383–424. MR **43** #5452.

R. L. Bishop and R. J. Crittenden, *Geometry of manifolds,* Pure and Appl. Math., vol. 15, Academic Press, New York, 1964. MR **29** #6401.

K. Yano and S. Bochner, *Curvature and Betti numbers,* Ann. of Math. Studies, no.32, Princeton Univ. Press, Princeton, N. J., 1953. MR **15**, 989.

E. Calabi, *On Kähler manifolds with vanishing canonical class,* Algebraic Geometry and Topology (A Sympos. in Honor of S. Lefschetz), Princeton Univ. Press, Princeton, N. J., 1957, pp. 78–89. MR **19**, 62.

J. Cheeger, *Some examples of manifolds of nonnegative curvature,* J. Differential Geometry **8** (1973), 623–628. MR **49** #6085.

J. Cheeger and D. Gromoll, *On the structure of complete manifolds of nonnegative curvature,* Ann. of Math. (2) **96** (1972), 413–443. MR **46** #8121.

S. Chern, *On curvature and characteristic classes of a Riemann manifold,* Abh. Math. Sem. Univ. Hamburg **20** (1955), 117–126. MR **17**, 783.

C. Delaunay, *Mémoire sur le calcul des variations,* J. École Polytech. **17** (1843), 37–120 (See p. 111).

P. Eberlein, *Some properties of the fundamental group of a Fuchsian manifold,* Invent. Math. **19** (1973), 5–13.

P. E. Ehrlich, *Local convex deformations of Ricci and sectional curvature on compact Riemannian manifolds,* Proc. Sympos. Pure Math., vol. 27, part I, Amer. Math. Soc., Providence, R. I., 1975, pp. 69–71.

H. I. Eliasson, *On variations of metrics,* Math. Scand. **29** (1971), 317–327. MR **47** #985.

A. E. Fischer and J. E. Marsden, *Manifolds of Riemannian metrics with prescribed scalar curvature,* Bull. Amer. Math. Soc. **80** (1974), 479–484. MR **49** #11561.

A. E. Fischer and J. A. Wolf, *The Calabi construction for compact Ricci flat Riemannian manifolds,* Bull. Amer. Math. Soc. **80** (1974), 92–97.

R. Geroch, *General relativity,* Proc. Sympos. Pure Math., vol. 27, part II, Amer. Math. Soc., Providence, R. I., 1975, pp. 401–441. (see Conjecture 3).

D. Gromoll, W. Klingenberg and W. Meyer, *Riemannsche Geometrie im Grossen,* Lecture Notes in Math., no. 55, Springer-Verlag, Berlin and New York, 1968. MR **37** #4751.

*[Updated reference] See R. Geroch, *Positive sectional curvature does not imply positive Gauss-Bonnet integrand,* Proc. Amer. Math. Soc. **54** (1976) 267–270.

N. Hitchin, *The space of harmonic spinors* (to appear) (see Chap. 4, Section 3).

H. Hopf, *Über Flächen mit einer Relation zwischen den Hauptkrümmungen,* Math. Nachr. **4** (1950), 232–249. MR **12**, 634.

M. Jellett, *Sur la surface dont la courbure moyenne est constante,* J. Liouville **18** (1853), 163–167.

J. L. Kazdan and F. W. Warner, *Scalar curvature and conformal deformation of Riemannian structure,* J. Differential Geometry (to appear).

J. L. Kazdan and F. W. Warner, *Prescribing curvatures,* Proc. Sympos. Pure Math., vol. 27, part II, Amer, Math. Soc., Providence, R. I., 1975, pp. 309–320.

———, *Existence and conformal deformation of metrics with prescribed Gaussian and scalar curvatures,* Ann. of Math. (to appear).

S. Kobayashi, *On compact Kaehler manifolds with positive definite Ricci tensor,* Ann. of Math. (2) **74** (1961), 570–574. MR **24** #A2922.

A. Lichnerowicz, *Spineurs harmoniques,* C. R. Acad. Sci. Paris **257** (1963), 7–9. MR **27** #6218.

E. Lima, *Commuting vector fields on* $S^3$, Ann. of Math. (2) **81** (1965), 70–81. MR **30** #1517.

J. Milnor, *A note on curvature and fundamental group,* J. Differential Geometry **2** (1968), 1–7. MR **38** #636.

T. Nagano and B. Smyth, *Minimal varieties in tori,* Proc. Sympos. Pure Math., vol. 27, part I, Amer. Math. Soc., Providence, R. I., 1975, pp, 189–190.

J. Plateau, *Statique experimentale et théorique des liquides,* Gauthier-Villars, Paris, 1873. (see Vol. 1, p. 438).

N. S. Trudinger, *Remarks concerning the conformal deformation of Riemannian structures on compact manifolds,* Ann. Scuola Norm. Sup. Pisa (3) **22** (1968), 265–274. MR **39** #2093.

N. Wallach, *Compact homogeneous Riemannian manifolds with strictly positive curvature,* Ann. of Math. (2) **96** (1972), 277–295. MR **46** #6243.

J. A. Wolf, *Growth of finitely generated solvable groups and curvature of Riemannian manifolds,* J. Differential Geometry **2** (1968), 421–446. MR **40** #1939.

H. Yamabe, *On a deformation of Riemannian structures on compact manifolds,* Osaka Math. J. **12** (1960), 21–37. MR **23** #A2847.

S. T. Yau, *Compact flat Riemannian manifolds,* J. Differential Geometry **6** (1971/72), 395–402. MR **46** #4439.

## Update* on the 1974 problem list.

**1. Mean Curvature for surfaces in 3-space.** The ***Soap Bubble Conjecture*** has been disproved by Wente, who showed that: *There exists an immersed torus in 3-space with constant mean curvature.* This has lead to quite a bit of further work on immersed surfaces of constant mean curvature; see for example Abresch, and Kapouleas.

**2. Scalar Curvature.** The ***Yamabe Conjecture*** has been proved in combined work by several authors: *Every compact Riemannian manifold is conformally equivalent to a manifold of constant scalar curvature.* As noted in the text, both Yamabe and Trudiger made substantial contributions. Aubin made further progress in 1976, and Schoen solved the remaining difficult cases in 1984.

The structure of manifolds of positive scalar curvature is now quite well understood as the result of work by Gromov and Lawson, Schoen and Yau, and by Stolz. (See also Lawson and Michelsohn, as well as Kwasik and Schultz.)

The ***Positive Mass Conjecture*** has been proved by Schoen and Yau, and also by Witten. (The published results of Schoen and Yau cover only the 3-dimensional case, although they state the general case. Witten has published details for all dimensions.) As a special case: *If an n-dimensional Riemannian manifold with scalar curvature $R \geq 0$ is isometric to the Euclidean space $\mathbf{R}^n$ throughout a neighborhood of infinity, then it is globally isometric to $\mathbf{R}^n$.*

**3. Ricci Curvature.** The ***Calabi Conjecture*** has been proved by Aubin in the special case $c_1 < 0$, and in all cases by Yau. (Both proofs depend substantially on the ideas of Calabi.) As a special case: *Every compact Kähler manifold with first Chern class zero admits a Ricci flat metric.* This yields many examples of manifolds which admit a Ricci flat metric but no flat metric. Perhaps the simplest example is a non-singular hypersurface of degree $n+1$ in $\mathbf{C}P^n$, with $n \geq 3$. (I don't know any odd-dimensional simply connected examples.)

Lohkamp has shown that: *Every manifold of dimension at least* 3 *admits a metric of negative definite Ricci curvature.* In fact such metrics are $C^0$-dense. For work on manifolds of nonnegative Ricci curvature, see Anderson, as well as Sha and Yang.

**4. Sectional Curvature.** For work on manifolds of nonnegative sectional curvature, see Gromov, as well as Grove, Petersen and Wu. However, the

*I am indebted to M. Anderson and D. Gromoll for substantial help.

Hopf conjecture on non-existence of metrics of strictly positive curvature on $S^2 \times S^2$ remains open. For manifolds of nonpositive sectional curvature, see for example Ballmann.

For surveys of recent work on curvature, see Gromoll and also Osserman.

## Bibliography


U. Abresch, *Constant mean curvature tori in terms of elliptic functions,* J. reine angew. Math. **374** (1987), 169–192.

M. Anderson, *On the topology of complete manifolds of nonnegative Ricci curvature,* Topology **29** (1990), 41–55.

———, *Short geodesics and gravitational instantons,* J. Diff. Geom. **31** (1990), 265–275.

T. Aubin, *The scalar curvature,* in "Differential Geometry and Relativity" (Cahen and Flata, eds.), Reider, 1976, pp. 5–18.

———Équations du type Monge-Ampère sur les variétés kähleriénnes compactes, Bull. Sci. Math. **102** (1978 63-95).

W. Ballmann, *Manifolds of nonpositive curvature,* Jahresb. Deut. Math. Verein. **92** (1990), 145–152.

D. Gromoll, *Spaces of nonnegative curvature,* in "Proc. Symp. Pure Math.," Part 3, 1993, pp. 337–356.

M. Gromov, *Curvature, diameter, and Betti numbers,* Comment. Math. Helv. **56** (1981), 179–185.

M. Gromov and B. Lawson, *Spin and scalar curvature in the presence of a fundamental group,* Ann. Math. **111** (1980), 209–230.

———, *The classification of simply connected manifolds of positive scalar curvature, ibid. 423–434.*

———, *Positive scalar curvature and the Dirac operator on complete riemannian manifolds,* Pub. Math. I. H. E. S. **58** (1983), 83–196.

K. Grove, P. Petersen and J. Y. Wu, *Controlled topology in geometry,* Bull. Amer. Math. Soc. **20** (1989), 181–183.

N. Kapouleas, *Compact constant mean curvature surfaces in Euclidean three-space,* J. Diff. Geom. **33** (1991), 683–715.

S. Kwasik and R. Schultz, *Positive scalar curvature and periodic fundamental groups,* Comment. Math. Helv. **65** (1990), 271–286.

B. Lawson and M.-L. Michelsohn, "Spin Geometry," Princeton Univ. Press, 1989.

J. Lohkamp, *Negatively Ricci curved manifolds,* Bull. A. M. S. **27** (1992), 288–291.

R. Osserman, *Curvature in the Eighties,* Amer. Math. Monthly **97** (1990), 731–756.

R. Schoen, *Conformal deformation of a Riemannian metric to constant scalar curvature,* J. Diff. Geom. **20** (1984), 479–495.

R. Schoen and S.-T. Yau, *On the proof of the positive mass conjecture,* Comm. Math. Phys. **64** (1979), 45–76.

———, *Existence of incompressible minimal surfaces and the topology of three dimensional manifolds of positive scalar curvature,* Ann. Math. **110** (1979), 127–142.

———, *The structure of manifolds with positive scalar curvature,* Manuscr. Math. **28** (1979), 159–183.

J.-P. Sha and D.-G. Yang, *Positive Ricci curvature on compact simply connected 4-manifolds,* in "Proc. Symp. Pure Math. (Part 3)," Amer. Math. Soc., 1993, pp. 529-538.

S. Stolz, *Simply connected manifolds of positive scalar curvature,,* Bull AMS **23** (1990), 427–432.

H. Wente, *Counterexample to a conjecture of H. Hopf,* Pac. J. Math. **121** (1986), 193–243.

E. Witten, *A new proof of the positive energy theorem,* Comm. Math. Phys. **80** (1981), 381–402.

S.-T. Yau, *On the Ricci curvature of a compact Kähler manifold and the complex Monge-Ampère equation* I, Comm. Pure Appl. Math. **31** (1978), 339–411.

# CURVATURES OF LEFT INVARIANT METRICS ON LIE GROUPS

*Reprinted from*

Advances in Mathematics

Volume 21 (1976), pp. 293–329

Lie groups with left invariant metrics form a fascinating playground for Riemannian geometry, including many exotic examples. More generally, Riemannian manifolds whose universal covering has a transitive group of isometries can be analyzed by similar methods. This field of study has been far too active[1] for me to attempt any survey of progress since 1976.

---

[1] See for example:

D. V. Alekseevskii, *Homogeneous Riemannian spaces of negative curvature*, Math. Sb. **25** (1975) 87–109.

D. V. Alekseevskii and B. N. Kim'elf'eld, *Structure of homogeneous riemannian spaces with zero Ricci curvature*, Funct. An. and Appl. **9** (1975) 97–102.

L. Berard Bergery, *Sur la courbure des métriques riemanniennes invariantes des groupes de Lie et des espaces homogènes*, Ann. Sci. Éc. Norm. Sup. Paris ($4^e$ serie) **2** (1978) 543–576.

J. E. D'Atri and I. Dotti Miatello, *Eigenvalues of the curvature operator for certain homogeneous manifolds*, Canad. J. Math. **42** (1990) 981–999.

J. E. D'Atri and W. Ziller, *Naturally reductive metrics and Einstein metrics on compact Lie groups*, Memoirs Am. Math. Soc. #215 **18** (1979).

J. E. D'Atri, *The curvature of homogeneous Siegel domains*, J. Diff. Geom. **15** (1980) 61–70.

I. Dotti Miatello, *Ricci curvature of left invariant metrics on solvable unimodular Lie groups*, Math. Z. **180** (1982) 257–263.

————, *Metrics with non positive Ricci curvature on semidirect products*, Quart. J. Math. Oxford **37** (1986) 309–314.

————, *Transitive group actions and Ricci curvature properties*, Mich. Math. J. **35** (1988) 427–434.

————, M. L. Leite and R. J. Miatello, *Negative Ricci curvature on complex simple Lie groups*, Geom. Dedicata **17** (1984) 207–218.

C. Gordon, *Naturally reductive homogeneous riemannian manifolds*, Can. J. Math. **37** (1985) 467–487.

M. Goto and K. Uesu, *Lie groups with left invariant metrics of nonnegative curvature*, Mem. Fac. Sci., Kyushu Univ. Ser. A. **35** (1981) 33–38.

M. L. Leite and I. Dotti Miatello, *Metrics of negative Ricci curvature on $SL(n, R)$, $n \geq 3$*, J. Diff. Geom. **17** (1982) 635–641.

W. Ziller, *Homogeneous Einstein metrics*, pp. 126–135 of "Global Riemannian Geometry," Willmore and Hitchin, eds., Ellis Horwood 1984.

# Curvatures of Left Invariant Metrics on Lie Groups

JOHN MILNOR

*Institute for Advanced Study, Princeton, New Jersey 08540*

This article outlines what is known to the author about the Riemannian geometry of a Lie group which has been provided with a Riemannian metric invariant under left translation.

*Contents.* 

## INTRODUCTION

When studying relationships between curvature of a complete Riemannian manifold and other topological or geometric properties, it is useful to have many examples. This paper will describe the rich collection of examples which are obtained by providing an arbitrary Lie group $G$ with a Riemannian metric invariant under left translations. (This class of examples can be enlarged substantially, with no extra work, as follows. If $\Gamma$ is any discrete subgroup of $G$, then a left invariant metric on $G$ gives rise to a metric on the quotient space $\Gamma\backslash G$, with identical curvature properties. The case where $\Gamma\backslash G$ is compact is of particular interest. Compare 4.9 and 6.2.)

The first four sections will survey the subject, giving some old results and some new results. In the 3-dimensional case the theory is essentially complete (Section 4), but in higher dimensions there remain many unsolved problems. Most proofs will be deferred until the last three sections.

The author is indebted to Nolan Wallach for extremely helpful suggestions.

## 1. Sectional Curvature

Let $G$ be an $n$-dimensional Lie group, and let $\mathfrak{g}$ be the associated Lie algebra, consisting of all smooth vector fields on $G$ which are invariant under left translations. (See for example [3, 7, 17, 20].) Choosing some basis $e_1, \ldots, e_n$ for the vector space $\mathfrak{g}$, it is easy to check that there is one and only one Riemannian metric on $G$ so that these vector fields $e_1, \ldots, e_n$ are everywhere orthonormal. More generally, given any $n \times n$ positive definite symmetric matrix $(\beta_{ij})$ of real numbers, there is one and only one Riemannian metric so that the Riemannian inner product $\langle e_i, e_j \rangle$ is everywhere equal to the constant function $\beta_{ij}$. Evidently this construction provides the most general Riemannian metric on $G$ which is left invariant (i.e., invariant under left translations of $G$). *Thus each $n$-dimensional Lie group possesses a $\frac{1}{2}n(n+1)$-dimensional family of distinct left invariant metrics.* We will see that different metrics on the same Lie group may exhibit drastically different curvature properties.

Choosing some fixed left invariant metric on $G$, note that the resulting Riemannian manifold is *homogeneous*. That is, there exists an isometry carrying any point to any other point. It follows easily that $G$ is complete. In fact, choosing $\epsilon > 0$ so that the closed $\epsilon$-ball about the identity is compact, it follows that every ball of radius $\epsilon$ is compact, hence every Cauchy sequence lies eventually within a compact set. (Compare [11, p. 176].)

The curvature of a Riemannian manifold at a point can be described most easily by the bi-quadratic *curvature function*

$$\kappa(x, y) = \langle R_{xy}(x), y \rangle.$$

(Compare Section 5.) Here $x$ and $y$ range over all tangent vectors at the given point. A given function $\kappa(x, y)$ can occur as curvature function for some Riemannian metric if and only if it is symmetric and bi-quadratic as a function of $x$ and $y$, and vanishes whenever $x = y$. (I know of no reference for this elementary fact.) The collection of all such symmetric, bi-quadratic functions with $\kappa(x, x) \equiv 0$ forms a real vector space of dimension $n^2(n^2 - 1)/12$. In other words one must prescribe $n^2(n^2 - 1)/12$ real numbers in order to describe the Riemannian curvature of an $n$-dimensional manifold at a single point.

If $u$ and $v$ are orthogonal unit vectors (or more generally if the determinant $\langle u, u \rangle \langle v, v \rangle - \langle u, v \rangle^2$ is equal to 1), then the real number $K = \kappa(u, v)$ is called the *sectional curvature* of the tangential 2-plane spanned by $u$ and $v$. Geometrically, $K$ can be described as the Gaussian curvature, at the point, of the surface swept out by all geodesics having a linear combination of $u$ and $v$ as tangent vector.

In order to study a Lie group with left invariant metric, it is best to choose an orthonormal basis $e_1, \dots, e_n$ for the left invariant vector fields. The Lie algebra structure can then be described by an $n \times n \times n$ array of *structure constants* $\alpha_{ijk}$ where

$$[e_i, e_j] = \sum_k a_{ijk} e_k$$

or equivalently

$$\alpha_{ijk} = \langle [e_i, e_j], e_k \rangle.$$

This array is skew-symmetric in the first two indices. The curvature function $\kappa$ can then be expressed as a complicated quadratic function of the $\alpha_{ijk}$. The explicit formula is usually not too useful. However, we will write down the following just to show that it can be done.

LEMMA 1.1. *With structure constants $\alpha_{ijk}$ as above, the sectional curvature $\kappa(e_1, e_2)$ is given by the formula*

$$\kappa(e_1, e_2) = \sum \big(\tfrac{1}{2}\alpha_{12k}(-\alpha_{12k} + \alpha_{2k1} + \alpha_{k12}) \\ - \tfrac{1}{4}(\alpha_{12k} - \alpha_{2k1} + \alpha_{k12})(\alpha_{12k} + \alpha_{2k1} - \alpha_{k12}) - \alpha_{k11}\alpha_{k22}\big),$$

*to be summed over $k$.*

The proof, and a more useful expression for curvature, will be found in Section 5.

At least this explicit expression shows that the curvature can be computed completely from information about the Lie algebra, together with its metric. Furthermore the curvature depends continuously on the structure constants $\alpha_{ijk}$ and vanishes whenever they vanish.

In some cases of interest, there is a great deal of cancellation in (1.1) so that we obtain a more useful formula.

Recall that the *adjoint* $L^*$ of a linear transformation $L$ between metric vector spaces is defined by the formula

$$\langle Lx, y \rangle = \langle x, L^* y \rangle.$$

The transformation $L$ is *skew-adjoint* if $L^* = -L$. For any element $x$ in a Lie algebra $\mathfrak{g}$ the linear transformation

$$y \mapsto [x, y]$$

from $\mathfrak{g}$ to itself is called $\mathrm{ad}(x)$.

Given $G$ with left invariant metric, let $u$ be a vector in the associated Lie algebra.

LEMMA 1.2. *If the linear transformation* $\mathrm{ad}(u)$ *is skew-adjoint, then*

$$\kappa(u, v) \geqslant 0$$

*for all* $v$, *where equality holds if and only if* $u$ *is orthogonal to the image* $[v, \mathfrak{g}]$.

*Proof.* We may assume without loss of generality that $u$ and $v$ are orthonormal. Choosing an orthonormal basis $e_1, \ldots, e_n$ with $e_1 = u$, $e_2 = v$, the statement that $\mathrm{ad}(e_1)$ is skew-adjoint means that the array $\alpha_{ijk}$ is skew in the last two indices for $i = 1$. Inspection then shows that the formula 1.1 reduces to

$$\kappa(e_1, e_2) = \sum_k (\alpha_{2k1})^2/4.$$

Thus $\kappa(e_1, e_2) \geqslant 0$, as asserted. ∎

The hypothesis of 1.2 depends of course on a particular choice of metric.

COROLLARY 1.3. *If* $u$ *belongs to the center of the Lie algebra* $\mathfrak{g}$, *then for any left invariant metric the inequality* $\kappa(u, v) \geqslant 0$ *is satisfied for all* $v$.

For if $u$ is central then $\mathrm{ad}(u) = 0$, and the zero transformation is certainly skew-adjoint. ∎

It may be conjectured that central elements are the only ones with this property. (Compare Section 2.5.)

Some Lie groups may possess a metric which is invariant not only under left translation but also under right translation. The basic facts about such *bi-invariant* metrics can be summarized as follows.

LEMMA. *A left invariant metric on a connected Lie group is also right invariant if and only if* $\mathrm{ad}(x)$ *is skew-adjoint for every* $x \in \mathfrak{g}$. *A connected Lie group admits such a bi-invariant metric if and only if it is isomorphic to the cartesian product of a compact group and a commutative group.*

This will be proved in 7.2 and 7.5 below.

COROLLARY 1.4. *Every compact Lie group admits a left invariant (and in fact a bi-invariant) metric so that all sectional curvatures satisfy* $K \geqslant 0$.

*Proof.* This follows from the Lemma just stated, together with 1.2. ∎

In fact we will see in 7.3 that sectional curvatures associated with a bi-invariant metric can be computed by the explicit formula

$$\kappa(u, v) = \tfrac{1}{4}\langle [u, v], [u, v]\rangle.$$

This will give an alternative proof that $\kappa(u, v) \geqslant 0$.

More generally if $G$ is the semi-direct product $AB$ of a commutative normal subgroup $A$ and a subgroup $B$ with bi-invariant metric which operates orthogonally on $A$, then $G$ possesses a left invariant metric with all sectional curvatures $K \geqslant 0$. (Compare 7.8.) It would be of interest to know whether these are the only groups which admit such a metric.

If we sharpen the inequality and require that $K > 0$, then examples become very scarce indeed.

THEOREM OF WALLACH. *The 3-sphere group $SU(2)$, consisting of $2 \times 2$ unitary matrices of determinant 1, is the only simply connected Lie group which admits a left invariant metric of strictly positive sectional curvature.*

(Compare 4.5.) Thus there are no examples at all in higher dimensions. For the proof, the reader is referred to [19].

The easiest Riemannian manifolds to understand are those which are *flat* in the sense that the sectional curvature $K$ is identically zero. A classical theorem asserts that a complete Riemannian manifold is flat if and only if its universal covering manifold is isometric to Euclidean space. In the case of a left invariant metric, the precise criterion for flatness can be stated as follows (Section 7). Recall that a Lie algebra is called *commutative* if the bracket product $[x, y]$ is identically zero. If $\mathfrak{g}$ is commutative, then it follows from 1.1 that *every* left invariant metric is flat.

THEOREM 1.5. *A Lie group with left invariant metric is flat if and only if the associated Lie algebra splits as an orthogonal direct sum* $\mathfrak{b} \oplus \mathfrak{u}$ *where* $\mathfrak{b}$ *is a commutative subalgebra,* $\mathfrak{u}$ *is a commutative ideal, and where the linear transformation* $\mathrm{ad}(b)$ *is skew-adjoint for every* $b \in \mathfrak{b}$.

Thus there exist noncommutative Lie groups with flat left invariant metric, but they are all solvable of a very sharply restricted form. The simplest example is provided by the group $E(2)$ of rigid motions of the Euclidean plane (Section 4.8).

Those Lie groups which admit left invariant metrics of strictly negative sectional curvature have been classified by Heintze. The necessary and sufficient condition is that $\mathfrak{g} = [\mathfrak{g}, \mathfrak{g}] + \mathbf{R}x$ for some $x$ such that all eigenvalues of $\mathrm{ad}(x) \mid [\mathfrak{g}, \mathfrak{g}]$ have positive real part. It follows for example that no cartesian product admits a left invariant metric of strictly negative curvature.

Those with $K \leqslant 0$ have been classified by Azencott and Wilson [1]. Since the statements are complicated we will content ourselves with the following qualitative result. Recall that a Lie group is called *unimodular* if its left invariant Haar measure is also right invariant (Section 6).

THEOREM 1.6. *If a connected Lie group G has a left invariant metric with all sectional curvatures* $K \leqslant 0$, *then it is solvable. If G is unimodular, then any such metric with* $K \leqslant 0$ *must actually be flat* ($K \equiv 0$).

In other words, every left invariant metric on a unimodular Lie group must possess some strictly positive sectional curvature unless it is completely flat as in 1.5.

*Proof.* This follows from [1, Sections 5.2, 6.2, 4.4], using 6.3 below. ∎

Here is an explicit (although rather exceptional) example of a left invariant metric with sectional curvatures $K < 0$.

SPECIAL EXAMPLE 1.7. Suppose that the Lie algebra $\mathfrak{g}$ has the property that the bracket product $[x, y]$ is always equal to a linear combination of $x$ and $y$. Assume that $\dim \mathfrak{g} \geqslant 2$, then in fact

$$[x, y] = l(x)y - l(y)x$$

where $l$ is a well defined linear mapping from $\mathfrak{g}$ to the real numbers.

Choosing any positive definite metric, the sectional curvatures are constant:

$$K = -\| l \|^2.$$

*Thus, in the noncommutative case $l \neq 0$, every possible metric on $\mathfrak{g}$ has constant negative sectional curvature.*

Here $\| l \|$ denotes the norm of the linear operator $l$. The proof will be given in Section 5.

We will see in 2.5 that these special examples can also be characterized as the only Lie algebras such that every metric has sectional curvatures of constant sign.

For other examples of Lie algebras with metrics of constant or non-constant negative curvature see 4.11.

## 2. Ricci Curvature

A somewhat cruder description of the curvature of a Riemannian manifold at a point is provided by the *Ricci quadratic form* $r(x)$. This is a real valued quadratic function of the tangent vector $x$, defined by the formula

$$r(x) = \sum_i \kappa(x, e_i) = \sum_i \langle R_{xe_i}(x), e_i \rangle$$

where $\kappa$ is the biquadratic curvature function of Section 1, and where $e_1, \ldots, e_n$ can be any orthonormal basis for the tangent space. If $u$ is a unit vector, then $r(u)$ is called the *Ricci curvature* in the direction $u$. It is equal to $n - 1$ times the average of the sectional curvatures of all tangential 2-planes containing $u$.

For computational purposes it may be more convenient to work with the self adjoint *Ricci transformation* $\hat{r}$ defined by

$$\hat{r}(x) = \sum R_{e_i x}(e_i).$$

This is related to the quadratic form $r$ by the identity

$$r(x) = \langle \hat{r}(x), x \rangle.$$

The eigenvalues of $\hat{r}$ are called the *principal Ricci curvatures*. If we choose an orthonormal basis $e_1, \ldots, e_n$ consisting of eigenvectors, note that the quadratic form is then diagonalized

$$r(\xi_1 e_1 + \cdots + \xi_n e_n) = \sum r(e_i)\, \xi_i^2.$$

In particular the numbers $r(e_i)$ can be identified with the principal Ricci curvatures, and the collection of signs $\{\operatorname{sgn} r(e_1), \ldots, \operatorname{sgn} r(e_n)\}$ can be identified with the *signature* of the quadratic form $r$.

Now let us return to the study of left invariant metrics. Here is a criterion for obtaining a direction of positive Ricci curvature.

LEMMA 2.1. *If the linear transformation* $\operatorname{ad}(u)$ *is skew-adjoint, then* $r(u) \geqslant 0$, *where equality holds if and only if* $u$ *is orthogonal to the commutator ideal* $[\mathfrak{g}, \mathfrak{g}]$.

*Proof.* This follows immediately from Lemma 1.2. ∎

As an example, if $u$ belongs to the center of $\mathfrak{g}$ then it certainly follows that $r(u) \geqslant 0$.

The criterion for everywhere positive Ricci curvature is classical and elegant.

THEOREM 2.2. *A connected Lie group admits a left invariant metric with all Ricci curvatures strictly positive if and only if it is compact with finite fundamental group.*

*Proof.* In one direction this follows from the theorem of Myers which asserts that any complete Riemannian manifold with Ricci curvatures positive and bounded away from zero must be compact with finite fundamental group. (See for example [2] or [14].) In the other direction, if $G$ is compact then we can choose a bi-invariant metric, so that each $\operatorname{ad}(x)$ is skew-adjoint. (Compare Section 1 or Section 7.) If $G$ also has finite fundamental group, so that the universal covering group $\tilde{G}$ is compact, note that $\mathfrak{g}$ must be equal to its commutator ideal $[\mathfrak{g}, \mathfrak{g}]$. For otherwise there would exist a non-trivial Lie algebra homomorphism from $\mathfrak{g}$ to the commutative Lie algebra $\mathbf{R}$. This would induce a non-trivial homomorphism from $\tilde{G}$ to the additive Lie group $\mathbf{R}$, contradicting the hypothesis that $\tilde{G}$ is compact. Now, using Lemma 2.1, it follows that all Ricci curvatures are strictly positive. ∎

*Remarks.* We will see in Section 7.7 that this result can be sharpened: If $G$ is compact with finite fundamental group, then it actually admits a bi-invariant metric with *constant* positive Ricci curvature. That is

$$r(u) = \text{constant} > 0$$

for all unit vectors $u$. Manifolds of constant Ricci curvature are often called *Einstein manifolds*. (Compare [11, p. 292].)

It would be of interest to characterize those connected Lie groups which admit left invariant metrics with all Ricci curvatures $\geqslant 0$. We will see in 6.4 that such a group $G$ must necessarily be unimodular. It will follow from 3.1 that $G$ cannot be solvable unless it is flat.

Similarly it would be of interest to know which groups admit left invariant metrics of Ricci curvature $\leqslant 0$. We will see in 4.7 that the simple group $SL(2, \mathbf{R})$ and the unimodular solvable group $E(1, 1)$ both admit non-flat left invariant metrics with Ricci curvature $\leqslant 0$. Both examples are in contrast with 1.6. It seems unlikely that any higher dimensional simple group admits such a metric.

Finally it would be interesting to know which groups admit left invariant metrics with Ricci curvature identically zero,[1] or with (constant or non-constant) strictly negative Ricci curvature.

Here is a criterion, completely analogous to 2.1, for obtaining a direction of negative Ricci curvature.

LEMMA 2.3. *If $u$ is orthogonal to the commutator ideal $[\mathfrak{g}, \mathfrak{g}]$, then $r(u) \leqslant 0$, where equality holds if and only if $\operatorname{ad}(u)$ is skew-adjoint.*

This will be proved in Section 5. Combining 2.1 and 2.3 we obtain the following sharper version of a theorem of Wolf [21].

THEOREM 2.4. *Suppose that the Lie algebra of $G$ is nilpotent but not commutative. Then for any left invariant metric there exists a direction of strictly negative Ricci curvature and a direction of strictly positive Ricci curvature.*

*Proof.* The statement that $\mathfrak{g}$ is nilpotent means that some term in the *lower central series*

$$\mathfrak{g} \supset [\mathfrak{g}, \mathfrak{g}] \supset [\mathfrak{g}, [\mathfrak{g}, \mathfrak{g}]] \supset \cdots$$

must be zero. Choosing a unit vector $u$ in the last non-zero term of this sequence of ideals, it follows that $u$ is central and contained in $[\mathfrak{g}, \mathfrak{g}]$, hence $r(u) > 0$ by 2.1.

Note that the vector space $\mathfrak{g}$ cannot be spanned by $[\mathfrak{g}, \mathfrak{g}]$ together with the center $\mathfrak{z}$. For if $\mathfrak{g} = [\mathfrak{g}, \mathfrak{g}] + \mathfrak{z}$ then $[\mathfrak{g}, \mathfrak{g}] = [\mathfrak{g}, [\mathfrak{g}, \mathfrak{g}] + \mathfrak{z}] = [\mathfrak{g}, [\mathfrak{g}, \mathfrak{g}]]$,

[1] Note added in proof. See Alekseevskii and Kimel'fel'd, *Functional Anal. Appl.* **9** (1975), 97–102.

hence the lower central series would stabilize prematurely. Therefore there exists a unit vector $v$ orthogonal to $[\mathfrak{g}, \mathfrak{g}]$ and not contained in $\mathfrak{z}$. The linear transformation $\operatorname{ad}(v)$, being nonzero and nilpotent, cannot be skew-adjoint. (For if a linear transformation $L$ is skew-adjoint with $Lx \neq 0$, then assuming inductively that $L^k x \neq 0$ it follows that

$$\langle L^{2k}x, x\rangle = \pm\langle L^k x, L^k x\rangle \neq 0$$

hence $L^{2k}x \neq 0$. Thus $L$ cannot be nilpotent.) Since $v$ is orthogonal to $[\mathfrak{g}, \mathfrak{g}]$ and $\operatorname{ad}(v)$ is not skew-adjoint, it follows that $r(v) < 0$ by Lemma 2.3. ∎

More generally we will prove the following.

THEOREM 2.5. *If the Lie algebra of $G$ contains linearly independent vectors $x$, $y$, $z$ so that*

$$[x, y] = z,$$

*then there exists a left invariant metric so that $r(x) < 0$ and $r(z) > 0$.*

Thus almost any Lie group has a left invariant metric with both positive and negative Ricci curvatures. The only exceptions are the "special examples" of Section 1.7, characterized by the property that $[x, y]$ is always a linear combination of $x$ and $y$.

*Proof.* Choose a fixed basis $b_1, \ldots, b_n$ with $b_1 = x$, $b_2 = y$, $b_3 = z$. For any real number $\epsilon > 0$, consider an auxiliary basis $e_1, \ldots, e_n$ defined by $e_1 = \epsilon b_1$, $e_2 = \epsilon b_2$, and $e_i = \epsilon^2 b_i$ for $i \geqslant 3$. Define a left invariant metric by requiring that $e_1, \ldots, e_n$ should be orthonormal. Let $\mathfrak{g}_\epsilon$ denote the Lie algebra $\mathfrak{g}$ provided with this particular metric and this particular orthonormal basis. Setting $[e_i, e_j] = \sum \alpha_{ijk} e_k$, the structure constants $\alpha_{ijk}$ are clearly functions of $\epsilon$. Now consider the limit as $\epsilon \to 0$. Inspection shows that each $\alpha_{ijk}$ tends to a well defined limit. Thus we obtain a limit Lie algebra $\mathfrak{g}_0$ with prescribed metric and prescribed orthonormal basis. Furthermore the bracket product in $\mathfrak{g}_0$ is given by

$$[e_1, e_2] = -[e_2, e_1] = e_3,$$

with $[e_i, e_j] = 0$ otherwise. Applying 2.1 and 2.3 it follows that the inequalities $r(e_1) < 0 < r(e_3)$ are satisfied in $\mathfrak{g}_0$. (For more explicit computations, see 4.6.) But these Ricci curvatures must vary continuously as we vary the structure constants, so it follows that $r(e_1) < 0 < r(e_3)$ whenever $\epsilon$ is sufficiently close to zero. ∎

## 3. Scalar Curvature

Choosing any orthonormal basis $e_1, \ldots, e_n$ for the tangent vectors at a point of a Riemannian manifold, the real number

$$\rho = r(e_1) + \cdots + r(e_n) = 2 \sum_{i<j} \kappa(e_i, e_j)$$

is called the *scalar curvature* at the point. Alternatively $\rho$ can be described as $n(n-1)$ times the average of all sectional curvatures at the point.

According to Eliasson any smooth manifold of dimension $\geqslant 3$ admits a Riemannian metric of strictly negative scalar curvature. However, metrics of positive scalar curvature do not always exist. (Compare [8, 10, 13].) For left invariant metrics, the situation can be described as follows.

THEOREM 3.1. *If the Lie group $G$ is solvable, then every left invariant metric on $G$ is either flat* (*as in* 1.5), *or else has strictly negative scalar curvature.*

This will be proved in Section 5. We will see in Section 4.7 that the corresponding statement is true also for the 3-dimensional simple group $SL(2, \mathbf{R})$. *It may be conjectured that it is true for any Lie group whose universal covering space is homeomorphic to Euclidean space.*

The following is an immediate consequence.

COROLLARY 3.2. *If $G$ is solvable and unimodular, then every left invariant metric on $G$ is either flat, or has both positive and negative sectional curvatures.*

In the nilpotent case we obtained the sharper statement that there exist positive and negative Ricci curvatures (Section 2.4). However, a solvable unimodular group does not necessarily have any directions of positive Ricci curvature (Section 4.7).

*Proof of* 3.2. If $G$ is unimodular and the metric is not flat, then there exist positive sectional curvatures by the Azencott, Wilson Theorem 1.6; while if $G$ is solvable and the metric is not flat, then there exist negative sectional curvatures by 3.1. ∎

THEOREM 3.3. *If the Lie algebra of $G$ is noncommutative, then $G$ possesses a left invariant metric of strictly negative scalar curvature.*

*Proof.* First suppose that there exist linearly independent vectors $x, y, z$ in the Lie algebra with $[x, y] = z$. As in the proof of 2.5, we can choose a basis $b_1, \ldots, b_n$ with $b_1 = x$, $b_2 = y$, $b_3 = z$, and for any $\epsilon > 0$ we can choose a metric so that the vectors

$$\epsilon b_1, \epsilon b_2, \epsilon^2 b_3, \ldots, \epsilon_2 b_n$$

are orthonormal. Denote this Lie algebra with prescribed metric and prescribed orthonormal basis by $\mathfrak{g}_\epsilon$. As $\epsilon$ tends to zero, $\mathfrak{g}_\epsilon$ tends to a well defined limit $\mathfrak{g}_0$ which is nilpotent but not commutative. Hence $\rho(\mathfrak{g}_0) < 0$ by 3.1 and 2.4, or by explicit computation as in 4.6. It follows by continuity that $\rho(\mathfrak{g}_\epsilon) < 0$ whenever $\epsilon$ is sufficiently close to zero.

On the other hand, if $x$, $y$ and $[x, y]$ are always linearly dependent, then $\mathfrak{g}$ is isomorphic to the special example of Section 1.7, hence $\mathfrak{g}$ has strictly negative curvature for any choice of metric. ∎

There remains the question as to which Lie groups admit left invariant metrics of positive scalar curvature. The following result was communicated to the author by Nolan Wallach. (Compare [12].) Let $G$ be a connected Lie group.

THEOREM 3.4 (Wallach). *If the universal covering of $G$ is not homeomorphic to Euclidean space (or equivalently if $G$ contains a compact non-commutative subgroup), then $G$ admits a left invariant metric of strictly positive scalar curvature.*

This will be proved in Section 7. As noted earlier, it is conjectured that these are the only groups which admit a left invariant metric with $\rho > 0$.

## 4. THE 3-DIMENSIONAL CASE

In order to study 3-dimensional Lie algebras we will make use of the familiar Euclidean *cross product operation.* If $u$ and $v$ are elements of a 3-dimensional vector space which is provided with a positive definite metric and with a preferred orientation, then the cross product $u \times v$ is defined. This product is bilinear and skew symmetric as a function of $u$ and $v$. The vector $u \times v$ is orthogonal to both $u$ and $v$ and has length equal to the square root of the determinant $\langle u, u\rangle\langle v, v\rangle - \langle u, v\rangle^2$. Its direction is determined by the requirement that the triple $u, v, u \times v$ is positively oriented whenever $u$ and $v$ are linearly independent.

Let $G$ be a connected 3-dimensional Lie group with left invariant metric. Choose an orientation for the Lie algebra of $G$, so that the cross product is defined.

LEMMA 4.1. *The bracket product operation in this Lie algebra* $\mathfrak{g}$ *is related to the cross product operation by the formula*

$$[u, v] = L(u \times v)$$

*where* $L$ *is a uniquely defined linear mapping from* $\mathfrak{g}$ *to itself. The Lie group* $G$ *is unimodular if and only if this linear transformation* $L$ *is self adjoint.*

All proofs will be given in Section 6.

Now let us specialize to the unimodular case. If $L$ is self adjoint, then there exists an orthonormal basis $e_1, e_2, e_3$ consisting of eigenvectors, $Le_i = \lambda_i e_i$. Replacing $e_1$ by $-e_1$ if necessary, we may assume that the basis $e_1, e_2, e_3$ is positively oriented. The bracket product operation is then given by $[e_1, e_2] = L(e_1 \times e_2) = \lambda_3 e_3$, with similar expressions for the other $[e_i, e_j]$. Thus we obtain the following normal form,

$$[e_2, e_3] = \lambda_1 e_1, \qquad [e_3, e_1] = \lambda_2 e_2, \qquad [e_1, e_2] = \lambda_3 e_3, \tag{4.2}$$

for the bracket product operation in a 3-dimensional unimodular Lie algebra with metric.

The three eigenvalues $\lambda_1, \lambda_2, \lambda_3$ are apparently well defined up to order. However, the construction was based on a choice of orientation. If we reverse the orientation of $\mathfrak{g}$, then the cross product operation will change sign, hence $L$ and its eigenvalues $\lambda_1, \lambda_2, \lambda_3$ will all change sign.

The curvature properties of the metric Lie algebra (4.2) can be described as follows. It is convenient to define numbers $\mu_1, \mu_2, \mu_3$ by the formula

$$\mu_i = \tfrac{1}{2}(\lambda_1 + \lambda_2 + \lambda_3) - \lambda_i$$

so that, for example, $\mu_1 + \mu_2 = \lambda_3$.

THEOREM 4.3. *The orthonormal basis* $e_1, e_2, e_3$, *chosen as above, diagonalizes the Ricci quadratic form, the principal Ricci curvatures being given by*

$$r(e_1) = 2\mu_2\mu_3\,, \qquad r(e_2) = 2\mu_1\mu_3\,, \qquad r(e_3) = 2\mu_1\mu_2\,.$$

In particular, it follows that scalar curvature is given by the formula

$$\rho = 2(\mu_2\mu_3 + \mu_1\mu_3 + \mu_1\mu_2).$$

Using this description of Ricci curvature, the sectional curvatures can easily be computed. In fact, at a point of any 3-dimensional manifold the explicit formula

$$\kappa(u, v) = \| u \times v \|^2 \rho/2 - r(u \times v)$$

is not difficult to verify. (There can be no such formula in dimensions $n > 3$ since the $\frac{1}{2}n(n+1)$ parameters needed to describe Ricci curvature can not suffice to determine the $\frac{1}{12}n^2(n^2-1)$ parameters needed to describe sectional curvature.)

COROLLARY 4.4. *In the 3-dimensional unimodular case, the determinant $r(e_1)\,r(e_2)\,r(e_3)$ of the Ricci quadratic form is always nonnegative. If this determinant is zero, then at least two of the principal Ricci curvatures must be zero.*

*Proof.* This follows immediately from 4.3. ∎

If this determinant $r(e_1)\,r(e_2)\,r(e_3)$ is non-zero, then it is easy to sove for $\mu_1$, $\mu_2$, $\mu_3$, and hence for the structure constants $\lambda_1$, $\lambda_2$, $\lambda_3$, as functions of the principal Ricci curvatures, well defined up to simultaneous change of sign.

Now suppose that we alter the metric, keeping the bracket product operation fixed. If we choose a new metric so that the basis

$$\eta\zeta e_1\,, \ \xi\zeta e_2\,, \ \xi\eta e_3$$

is orthonormal, then the new structure constants in formula (4.2) will clearly be

$$\xi^2\lambda_1\,, \ \eta^2\lambda_2\,, \ \zeta^2\lambda_3\,.$$

*Thus we can multiply $\lambda_1$, $\lambda_2$, $\lambda_3$ by arbitrary positive numbers without changing the underlying Lie algebra.* There are now just six distinct cases, which we tabulate as follows. By changing signs if necessary, we may assume that at most one of the structure constants $\lambda_1$, $\lambda_2$, $\lambda_3$ is negative.

| Signs of $\lambda_1$, $\lambda_2$, $\lambda_3$ | Associated Lie group | Description |
|---|---|---|
| $+, +, +$ | $SU(2)$ or $SO(3)$ | compact, simple |
| $+, +, -$ | $SL(2, \mathbf{R})$ or $O(1, 2)$ | noncompact, simple |
| $+, +, 0$ | $E(2)$ | solvable |
| $+, -, 0$ | $E(1, 1)$ | solvable |
| $+, 0, 0$ | Heisenberg group | nilpotent |
| $0, 0, 0$ | $\mathbf{R} \oplus \mathbf{R} \oplus \mathbf{R}$ | commutative |

It is not difficult to show that these six possibilities do really give rise to nonisomorphic Lie algebras. They can be distinguished for example by computing the signature of the Killing form

$$\beta(x, y) = \operatorname{trace}(\operatorname{ad}(x) \operatorname{ad}(y))$$

in each case. Here is a glossary.

$SU(2)$: group of $2 \times 2$ unitary matrices of determinant 1; homeomorphic to the unit 3-sphere.

$SO(3)$: rotation group of 3-space, isomorphic to $SU(2)/\{\pm I\}$.

$SL(2, \mathbf{R})$: group of $2 \times 2$ real matrices of determinant 1.

$O(1, 2)$: Lorentz group consisting of linear transformations preserving the quadratic form $t^2 - x^2 - y^2$. Its identity component is isomorphic to $SL(2, \mathbf{R})/\{\pm I\}$, or to the group of rigid motions of hyperbolic 2-space.

$E(2)$: group of rigid motions of Euclidean 2-space.

$E(1, 1)$: group of rigid motions of Minkowski 2-space. This group is a semi-direct product of subgroups isomorphic to $\mathbf{R} \oplus \mathbf{R}$ and to $\mathbf{R}$, where each $\tau \in \mathbf{R}$ acts on $\mathbf{R} \oplus \mathbf{R}$ by the matrix $\left[\begin{smallmatrix} e^{\tau} & 0 \\ 0 & e^{-\tau} \end{smallmatrix}\right]$.

Finally, the Heisenberg group can be described as the group of all $3 \times 3$ real matrices of the form $\left[\begin{smallmatrix} 1 & * & * \\ 0 & 1 & * \\ 0 & 0 & 1 \end{smallmatrix}\right]$

Let us study the extent to which curvature can be altered by a change of metric. The greatest freedom to alter curvature properties occurs in the compact case. (Compare [8, Section 3.3].)

COROLLARY 4.5. *Depending on the choice of left invariant metric, the Ricci quadratic form for the 3-sphere group $SU(2)$ can have signature either $(+, +, +)$ or $(+, 0, 0)$ or $(+, -, -)$; and the scalar curvature can be either positive, negative, or zero.*

*Proof.* This follows easily from 4.3. (Of course many parts of this corollary follow from theorems stated earlier, such as 2.2 and 3.3.) ∎

By way of contrast, in the commutative case where all left invariant metrics are flat, and also in the nilpotent case, curvature properties are essentially independent of the metric.

COROLLARY 4.6. *For any left invariant metric on the Heisenberg group, the Ricci quadratic form has signature $(+, -, -)$ and the scalar curvature $\rho$ is strictly negative. Furthermore the principal Ricci curvatures satisfy*

$$| r(e_1)| = | r(e_2)| = | r(e_3)| = | \rho |.$$

*Proof.* Taking $\lambda_2 = \lambda_3 = 0$ formula 4.3 shows that

$$r(e_1) = -r(e_2) = -r(e_3) = -\rho$$

is equal to $\lambda_1^2/2$. ∎

The simple group $SL(2, \mathbf{R})$ and the solvable group $E(1, 1)$ are difficult to distinguish by curvature properties.

COROLLARY 4.7. *Let $G$ be either $SL(2, \mathbf{R})$ or $E(1, 1)$. Then depending on the choice of left invariant metric the signature of the Ricci form can be either $(+, -, -)$ or $(0, 0, -)$. However, the scalar curvature $\rho$ must always be strictly negative.*

*Proof.* If $\lambda_1 = 0$ while $\lambda_2$ and $\lambda_3$ have opposite sign, then the computation

$$\rho = -\tfrac{1}{2}(\lambda_2 - \lambda_3)^2$$

shows that $\rho < 0$. If the $\lambda_i$ are all nonzero with say $\lambda_1 < 0 < \lambda_2, \lambda_3$, then the computation $\partial\rho/\partial\lambda_1 = -\lambda_1 + \lambda_2 + \lambda_3$ shows that $\rho$ is strictly monotone as a function of $\lambda_1$ (keeping $\lambda_2, \lambda_3$ fixed) for $\lambda_1 \leqslant 0$. Therefore

$$\rho(\lambda_1, \lambda_2, \lambda_3) < \rho(0, \lambda_2, \lambda_3) = -\tfrac{1}{2}(\lambda_2 - \lambda_3)^2 \leqslant 0.$$

Further details will be left to the reader. ∎

COROLLARY 4.8. *The Euclidean group $E(2)$ is non-commutative, but admits a flat left invariant metric. Every nonflat left invariant metric has Ricci form of signature $(+, -, -)$, with scalar curvature $\rho < 0$.*

The proof is easily supplied. ▮

*Remark* 4.9. It is interesting that each of these 3-dimensional unimodular groups $G$ possesses a discrete subgroup $\Gamma$ so that the quotient $\Gamma\backslash G$ is compact. (Compare 6.2.) For a precise description of the possible compact quotient manifolds, the reader is referred to [16]. In the four solvable cases it can be shown that $\Gamma\backslash G$ is always a torus bundle over a circle. In the $SL(2, \mathbf{R})$ case, $\Gamma\backslash G$ always has a finite covering space which is a non-trivial circle bundle over a surface of genus $\geqslant 2$. The topology of $\Gamma\backslash G$ determines the associated Lie algebra uniquely except in one exceptional case: Both the Euclidean group $E(2)$ and the commutative group $\mathbf{R} \oplus \mathbf{R} \oplus \mathbf{R}$ admit a discrete subgroup so that the quotient is topologically a torus $S^1 \times S^1 \times S^1$.

It is interesting that many of these quotient manifolds $\Gamma\backslash G$ occur also in the study of algebraic singularities. Compare [4, 15, 16, 18].

Now let us turn to the nonunimodular case. The possible Lie algebras can be described as follows.

LEMMA 4.10. *If the connected 3-dimensional Lie group $G$ is not unimodular, then its Lie algebra has a basis $e_1$, $e_2$, $e_3$ so that*

$$[e_1, e_2] = \alpha e_2 + \beta e_3$$

$$[e_1, e_3] = \gamma e_2 + \delta e_3$$

*with $[e_2, e_3] = 0$, and so that the matrix*

$$A = \begin{pmatrix} \alpha & \beta \\ \gamma & \delta \end{pmatrix}$$

*has trace $\alpha + \delta = 2$. If we exclude the exceptional case where $A$ is the identity matrix (compare Section 1.7), then the determinant $D = \alpha\delta - \beta\gamma$ provides a complete isomorphism invariant for this Lie algebra.*

Curvature properties can be described as follows. Consider a Lie group as in 4.10 with $A \neq I$.

THEOREM 4.11. *If the determinant $D$ is negative then every left invariant metric has Ricci quadratic form of signature $(+, -, -)$. But* *if $D \geqslant 0$ the signature* $(0, -, -)$ *is also possible, and if $D > 0$ the signature* $(-, -, -)$ *is also possible. In fact, for $D > 0$ there exists a left invariant metric of strictly negative sectional curvature and for $D > 1$ there exists a left invariant metric of constant negative curvature. In all cases the scalar curvature is strictly negative.*

Again proofs are deferred until Section 6.

In all of the non-unimodular cases, at least two of the principal Ricci curvatures are negative. *Comparing* 4.4, *we see that no* 3-*dimensional Lie group at all admits a left invariant metric with Ricci form of signature* $(+, +, -)$ *or* $(+, \pm, 0)$. I do not know whether there exist any analogous restrictions in higher dimensions.

## 5. Computations

This section contains proofs of several of the results stated earlier. Before making any actual computations it is necessary to define some basic concepts. For details the reader is referred to textbooks such as [2, 7, 11, 14].

First we consider the *Riemannian connection* $\nabla$ associated with a Riemannian metric. This connection assigns to each pair of smooth vector fields $x$ and $y$ a smooth vector field $\nabla_x y$ called the *covariant derivative* of $y$ in the direction $x$. For our purpose it suffices to know that $\nabla$ is always uniquely defined, that $\nabla_x y$ is bilinear as a function of $x$ and $y$, that it satisfies the "symmetry" condition

$$\nabla_x y - \nabla_y x = [x, y], \tag{5.1}$$

and that the identity

$$\langle \nabla_x y, z\rangle + \langle y, \nabla_x z\rangle = 0 \tag{5.2}$$

is satisfied whenever $y$ and $z$ are vector fields such that the Riemannian inner product $\langle y, z\rangle$ is a constant function. In particular, if $y$ and $z$ are left invariant vector fields on a Lie group with left invariant metric, this identity is certainly satisfied. If $x$ is also left invariant, then $\nabla_x y$ is left invariant. *Thus, for each $x$ in the Lie algebra, $\nabla_x$ is a skew-adjoint linear transformation from the Lie algebra to itself.*

If $x$, $y$, $z$ are all left invariant vector fields, then combining 5.1 and 5.2 with the various identities obtained by permuting the variables, we can solve to obtain the following formula:

$$\langle \nabla_x y, z\rangle = \tfrac{1}{2}(\langle [x, y], z\rangle - \langle [y, z], x\rangle + \langle [z, x], y\rangle). \tag{5.3}$$

In particular, choosing an orthonormal basis $e_1, \ldots, e_n$ and setting $\alpha_{ijk} = \langle [e_i, e_j], e_k \rangle$ it follows that

$$\langle \nabla_{e_i} e_j, e_k \rangle = \tfrac{1}{2}(\alpha_{ijk} - \alpha_{jki} + \alpha_{kij})$$

or in other words

$$\nabla_{e_i} e_j = \sum_k \tfrac{1}{2}(\alpha_{ijk} - \alpha_{jki} + \alpha_{kij})\, e_k . \tag{5.4}$$

The *Riemann curvature tensor* $R$ associates to each pair of smooth vector fields $x$ and $y$ the linear transformation

$$R_{xy} = \nabla_{[x,y]} - \nabla_x \nabla_y + \nabla_y \nabla_x$$

from smooth vector fields to smooth vector fields. Evidently $R_{xy} = -R_{yx}$. The transformation $R_{xy}$ is always skew-adjoint. (If we consider only left invariant vector fields, this follows from the fact that each $\nabla_x$ is skew-adjoint.) If $u$ and $v$ are orthonormal recall that the number

$$K = \kappa(u, v) = \langle R_{uv}(u), v \rangle$$

is called the *sectional curvature* associated with $u$ and $v$. The sectional curvatures depend symmetrically on $u$ and $v$, and determine the Riemann tensor $R$ uniquely. (See, for example, [11, Chapter 5].)

*Caution*: In the notation of Kobayashi, Nomizu, and of Helgason, our $R$ would be denoted by $-R$, so that the sectional curvature $K$ would equal $\langle -R_{uv}(u), v \rangle$.

We can now give an explicit formula for sectional curvature in terms of structure constants, as described in Section 1:

*Proof of Lemma* 1.1. If $e_1, \ldots, e_n$ is any orthonormal basis for the left invariant vector fields on $G$, then inserting formula 5.4 into the definition

$$\kappa(e_1, e_2) = \langle \nabla_{[e_1, e_2]} e_1 - \nabla_{e_1} \nabla_{e_2} e_1 + \nabla_{e_2} \nabla_{e_1} e_1, e_2 \rangle$$

we easily obtain the required explicit formula. ▮

To illustrate these concepts consider the following important special case. *Suppose that the Lie algebra* $\mathfrak{g}$ *contains an ideal* $\mathfrak{u}$ *of codimension* 1. Choosing a unit vector $b$ orthogonal to $\mathfrak{u}$, let

$$L : \mathfrak{u} \to \mathfrak{u}$$

denote the linear transformation ad($b$) restricted to $\mathfrak{u}$, so that $L(u) = [b, u]$. Let $L^*$ denote the adjoint transformation, and let $S = \frac{1}{2}(L + L^*)$ denote the self adjoint part of $L$.

We may also think of the ideal $\mathfrak{u}$ as a Lie algebra in its own right with the induced metric. Let $\bar{\nabla}$ denote the Riemannian connection for this metric Lie algebra $\mathfrak{u}$. The symbol $\nabla$ will be reserved for the Riemannian connection of the original Lie algebra $\mathfrak{g}$.

LEMMA 5.5. *With these notations, the covariant derivative operator* $\nabla_b$ *satisfies*

$$\nabla_b b = 0 \qquad \textit{and} \qquad \nabla_b u = \tfrac{1}{2}(L - L^*)u$$

*for each* $u \in \mathfrak{u}$. *Similarly the operator* $\nabla_u$ *satisfies*

$$\nabla_u b = -Su \qquad \textit{and} \qquad \nabla_u v = \bar{\nabla}_u v + \langle Su, v\rangle b$$

*for each* $u$ *and* $v$ *in* $\mathfrak{u}$.

*Proof.* To compute the component of $\nabla_u v$ in the $b$ direction we use formula 5.3:

$$\begin{aligned}\langle \nabla_u v, b\rangle &= \tfrac{1}{2}(\langle [u, v], b\rangle - \langle [v, b], u\rangle + \langle [b, u], v\rangle)\\ &= \tfrac{1}{2}(0 + \langle Lv, u\rangle + \langle Lu, v\rangle)\\ &= \langle Su, v\rangle.\end{aligned}$$

To compute the component orthogonal to $b$ we choose an arbitrary $w$ in $\mathfrak{u}$ and use 5.3 to compute $\langle \nabla_u v, w\rangle$. Since the computation takes place completely within $\mathfrak{u}$, this inner product is equal to $\langle \bar{\nabla}_u v, w\rangle$. Hence $\nabla_u v$ is equal to $\bar{\nabla}_u v$ plus the normal component just computed. The other arguments are completely analogous. ∎

Here is a first application. (More important applications will follow.)

*Verification of Example* 1.7. For any elements $x$ and $y$ in the Lie algebra $\mathfrak{g}$ we suppose that $[x, y]$ is a linear combination of $x$ and $y$. Fixing $x$, note that ad($x$) induces a linear mapping from the quotient vector space $\mathfrak{g}/\mathbf{R}x$ to itself, with the property that every vector maps into some multiple of itself. It follows easily that this multiple must be a constant, depending only on $x$. Calling it $l(x)$, we have

$$[x, y] \equiv l(x)y \mod \mathbf{R}x.$$

The precise function $l(x)$ can be computed by noting that

$$\text{trace ad}(x) = (n-1)\, l(x).$$

Thus $l(x)$ depends linearly on $x$. Interchanging the roles of $x$ and $y$, we can also compute $[x, y]$ modulo $\mathbf{R}y$. If $x$ and $y$ are linearly independent, these two computations combine to yield the precise formula

$$[x, y] = l(x)y - l(y)x.$$

Evidently this formula is also true, for trivial reasons, when $x$ and $y$ are linearly dependent.

Since the commutative case $l = 0$ is uninteresting, let us suppose that $l \neq 0$. Let $\mathfrak{u}$ be the kernel of the linear transformation $l$. Clearly $\mathfrak{u}$ is a commutative ideal. Choose a unit vector $b$ orthogonal to $\mathfrak{u}$, and let $\lambda = l(b)$. (Evidently $\lambda$ can be identified with the norm $\| l \|$ of the linear transformation $l$.) With notation as in 5.5, the linear transformation $L(u) = [b, u]$ is given by $L(u) = \lambda u$. Applying 5.5, it follows easily that $\nabla_b$ is the zero transformation and that

$$\nabla_u z = \lambda(b\langle u, z\rangle - u\langle b, z\rangle)$$

for any $u$ in $\mathfrak{u}$ and any $z$ in $\mathfrak{g}$. *We now assert that $R_{xy}$ is given by the formula*

$$R_{xy}(z) = \lambda^2(x\langle y, z\rangle - y\langle x, z\rangle)$$

*for any $x$, $y$, $z$ in the Lie algebra.* For example, if $x = b$ and $y = u \in \mathfrak{u}$ then $R_{bu} = \nabla_{[b,u]} - \nabla_b\nabla_u + \nabla_u\nabla_b = \lambda\nabla_u - 0 + 0$ equals the required expression; while if both $x$ and $y$ belong to $\mathfrak{u}$ then only a slightly longer computation is needed. The remaining cases follow by bilinearity.

Substituting this formula into the definition $\kappa(x, y) = \langle R_{xy}(x), y\rangle$ we see that

$$\kappa(x, y) = \lambda^2(\langle x, y\rangle^2 - \langle x, x\rangle\langle y, y\rangle).$$

Hence $\kappa(x, y)$ takes the constant value $-\lambda^2$ whenever $x$ and $y$ are orthonormal. In other words the metric has constant sectional curvature $K \equiv -\lambda^2 < 0$. ▮

Here is a further application of 5.5.

*Proof of Lemma* 2.3. The hypothesis of this lemma is that a unit vector (which we now call $b$) is orthogonal to the commutator ideal

$[\mathfrak{g}, \mathfrak{g}]$. It follows that the orthogonal complement of $b$ contains $[\mathfrak{g}, \mathfrak{g}]$ and hence is an ideal, so that 5.5 applies. We must compute the Ricci curvature $r(b)$. By definition this equals $\kappa(b, u_1) + \cdots + \kappa(b, u_{n-1})$ where $u_1, \ldots, u_{n-1}$ is any orthonormal basis for $\mathfrak{u}$. It is easiest to work with an orthonormal basis consisting of eigenvectors

$$Su_i = \lambda_i u_i$$

of the self adjoint operator $S$. For any unit vector in $\mathfrak{u}$ the sectional curvature can be computed as

$$\begin{aligned}\kappa(b, u) &= \langle R_{bu}(b), u\rangle \\ &= \langle \nabla_{[b,u]}b, u\rangle - \langle \nabla_b\nabla_u b, u\rangle + \langle \nabla_u\nabla_b b, u\rangle \\ &= \langle -SLu, u\rangle + \langle \tfrac{1}{2}(L - L^*)\, Su, u\rangle + 0.\end{aligned}$$

Taking $u$ to be an eigenvector as above, and noting that

$$\langle Lu_i\,, u_i\rangle = \langle u_i\,, L^*u_i\rangle = \lambda_i\,,$$

this reduces to $\kappa(b, u_i) = -\lambda_i{}^2$. Hence

$$r(b) = -\lambda_1{}^2 - \cdots - \lambda_{n-1}^2 = -\text{trace}(S^2).$$

Thus $r(b) \leqslant 0$, with equality if and only if $S = 0$ so that $L$ is skew-adjoint. ▮

*Caution*: It is *not* asserted that $\kappa(b, u) \leqslant 0$ for all $u$. In fact, for some particular choice of $u$ (not an eigenvector) it may well happen that $\kappa(b, u) > 0$ by 1.2. This happens, for example, in the case of the Heisenberg group.

Next let us compute the scalar curvature $\rho = \rho(\mathfrak{g})$. Thinking of $\mathfrak{u}$ as a Lie algebra in its own right with the induced metric, let $\rho(\mathfrak{u})$ denote its scalar curvature. With notations as in 5.5 we will prove the following.

LEMMA 5.6. *The scalar curvature* $\rho(\mathfrak{g})$ *associated with the metric Lie algebra* $\mathfrak{g}$ *is equal to*

$$\rho(\mathfrak{u}) - \text{trace}(S^2) - (\text{trace } S)^2.$$

*Proof*. Given orthonormal vectors $u$, $v$ in $\mathfrak{u}$, let us compare the sectional curvature

$$\kappa(u, v) = \langle \nabla_{[u,v]}u, v\rangle - \langle \nabla_u\nabla_v u, v\rangle + \langle \nabla_v\nabla_u u, v\rangle$$

as computed in $\mathfrak{g}$ with the sectional curvature

$$\bar{\kappa}(u, v) = \langle \bar{\nabla}_{[u,v]}u, v\rangle - \langle \bar{\nabla}_u \bar{\nabla}_v u, v\rangle + \langle \bar{\nabla}_v \bar{\nabla}_u u, v\rangle$$

as computed in the Lie algebra $\mathfrak{u}$. Using 5.5, inspection shows that

$$\kappa(u, v) = \bar{\kappa}(u, v) + \langle Su, v\rangle^2 - \langle Su, u\rangle\langle Sv, v\rangle.$$

Choosing an orthonormal basis consisting of eigenvectors, $Su_i = \lambda_i u_i$, it follows that

$$\kappa(u_i, u_j) = \bar{\kappa}(u_i, u_j) - \lambda_i\lambda_j$$

for $i \neq j$. Combining this with the formula

$$\kappa(b, u_i) = -\lambda_i^2,$$

we see that the Ricci curvature in the direction $u_i$ is given by

$$r(u_i) = \bar{r}(u_i) - \lambda_i \operatorname{trace}(S).$$

Summing over $i$ and adding $r(b) = -\operatorname{trace}(S^2)$, this gives the required formula

$$\rho(\mathfrak{g}) = \rho(\mathfrak{u}) - \operatorname{trace}(S)^2 - (\operatorname{trace} S^2). \quad \blacksquare$$

*Caution*: This computation does not provide a complete description of Ricci curvature, since the basis $b, u_1, \ldots, u_{n-1}$ may not diagonalize the Ricci quadratic form.

*Proof of Theorem* 3.1. If $\mathfrak{g}$ is solvable of dimension $n$, we will prove by induction on $n$ that $\rho(\mathfrak{g}) \leqslant 0$. Certainly every solvable Lie algebra contains an ideal $\mathfrak{u}$ of codimension 1. Furthermore $\mathfrak{u}$ itself is solvable, so we may assume inductively that $\rho(\mathfrak{u}) \leqslant 0$. Therefore

$$\rho(\mathfrak{g}) \leqslant -\operatorname{trace}(S^2) - (\operatorname{trace} S)^2 \leqslant 0,$$

where equality holds only if both $S = 0$ and $\rho(\mathfrak{u}) = 0$.

If both of these conditions are satisfied, we must prove that $\mathfrak{g}$ is flat. Since $S = 0$, the formulas 5.5 imply that

$$\nabla_u v = \bar{\nabla}_u v \in \mathfrak{u}$$

for any $u$ and $v$ in the ideal $\mathfrak{u}$. (In other words the ideal $\mathfrak{u}$ is "totally geodesic" in $\mathfrak{g}$.) It follows immediately that

$$R_{uv}(w) = \bar{R}_{uv}(w)$$

for any $u$, $v$, $w$ in the ideal. But we have assumed that the scalar curvature $\rho(\mathfrak{u})$ is zero, so it follows by induction that $\mathfrak{u}$ is flat. Thus $\bar{R} = 0$, hence $R_{uv}(w) = 0$.

Again applying 5.5 with $S = 0$, we have $\nabla_x b = 0$ for any $x$ in the Lie algebra, hence $R_{xy}(b) = 0$ for any $x$ and $y$. Using the symmetry property

$$\langle R_{xy}(b), z\rangle = \langle R_{bz}(x)\, y\rangle$$

of the Riemann tensor, it follows that $R_{bz} = 0$ for any $z$. Combining these statements with the fact that $R_{xy}(z)$ is trilinear as a function of $x$, $y$, $z$, it then follows easily that $R$ is identically zero. ∎

*Remark.* Here is a sketch of an alternature argument. Let $H$ be the subgroup of the universal covering $\tilde{G}$ which has Lie algebra $\mathfrak{u}$. Using the techniques of Section 7.2, one can prove that $S = 0$ if and only if the smooth mapping

$$(h, \tau) \mapsto h \exp(\tau b)$$

from $H \times \mathbf{R}$ to $\tilde{G}$ is a Riemannian isometry. Thus if $H$ is flat and $S = 0$, it follows directly that $\tilde{G}$ is flat.

## 6. Unimodular and Non-unimodular Lie Groups

Recall that a Lie group $G$ is called *unimodular* if its left invariant Haar measure is also right invariant. Here is a simple and classical criterion. Recall that each group element $g$ determines an automorphism

$$h \mapsto ghg^{-1}$$

of the group $G$. The induced automorphism of the Lie algebra is called $\mathrm{Ad}(g)$.

Lemma 6.1. *The group $G$ is unimodular if and only if the linear transformation* $\mathrm{Ad}(g)$ *has determinant* $\pm 1$ *for every $g$ in $G$.*

This is proved, for example, in [7, p. 366]. (A completely analogous argument is given in 7.1 below.) ∎

As an example, if $G$ is compact or connected semisimple then the homomorphism

$$g \mapsto |\det \mathrm{Ad}(g)|$$

from $G$ to the positive real numbers must certainly be trivial, hence $G$ is unimodular. Here is another interesting criterion.

LEMMA 6.2. *If $G$ admits a discrete subgroup $\Gamma$ with compact quotient, then $G$ is unimodular.*

*Proof.* It is not difficult to choose a compact fundamental domain $D$ for the left action of $\Gamma$ on $G$, that is a compact set $D \subset G$ so that the various left translates $\gamma D$ cover $G$ and so that the intersections $\gamma D \cap \gamma' D$ have measure zero for $\gamma \neq \gamma'$. Choosing a left invariant Haar measure $\omega$, note that the measure $\omega(D) \neq 0$ is independent of the choice of $D$. For if $E$ is another fundamental domain then

$$\omega(E) = \sum \omega(\gamma D \cap E) = \sum \omega(D \cap \gamma^{-1} E) = \omega(D),$$

where both summations extend over all elements $\gamma$ of $\Gamma$.

For any group element $g$, note that the right translate $Dg$ is also a fundamental domain for the left action of $\Gamma$ on $G$. Hence $\omega(D) = \omega(Dg)$, and it follows that the left invariant measure $\omega$ is also right invariant. ∎

*Remark.* Even if $\Gamma \backslash G$ is not compact, the left Haar measure on $G$ induces a measure on $\Gamma \backslash G$. *Whenever $\Gamma \backslash G$ has finite measure, a similar argument shows that $G$ is unimodular.*

In terms of the Lie algebra we have the following criterion.

LEMMA 6.3. *A connected Lie group is unimodular if and only if the linear transformation* $\mathrm{ad}(x)$ *has trace zero for every $x$ in the associated Lie algebra.*

As an example, if $\mathfrak{g}$ is nilpotent, then every $\mathrm{ad}(x)$ is nilpotent, and hence has trace zero.

*Proof.* We will use two different "exponential mappings." If $l$ is a linear transformation from a finite dimensional vector space to itself, let

$$e^l = \sum l^n/n!.$$

Using Jordan canonical form one sees that

$$\det(e^l) = e^{\mathrm{trace}\, l}.$$

On the other hand, for any Lie group $G$, there is a smooth mapping

$$\exp\colon \mathfrak{g} \to G$$

characterized by the property that each $x$ in $\mathfrak{g}$ gives rise to a homomorphism

$$\tau \mapsto \exp(\tau x)$$

from the Lie group of real numbers to $G$, the associated Lie algebra homomorphism being multiplication by $x$. These two exponential mappings are related by the identity

$$\mathrm{Ad}(\exp(x)) = e^{\mathrm{ad}(x)}.$$

See, for example, [7, p. 118].

Now if $|\det \mathrm{Ad}(g)|$ is identically equal to 1, it follows that

$$\det \mathrm{Ad}(\exp(x)) = \det e^{\mathrm{ad}(x)} = e^{\mathrm{trace}\, \mathrm{ad}(x)}$$

is identically 1, hence $\mathrm{trace}\, \mathrm{ad}(x) \equiv 0$. Conversely, if $\mathrm{trace}\, \mathrm{ad}(x) \equiv 0$ this argument shows that $\det \mathrm{Ad}(g) = 1$ for all $g$ in the image of the exponential mapping. Using the inverse function theorem, this includes all $g$ in some neighborhood of the identity. But $G$, being connected, is generated by any neighborhood of the identity; hence $\det \mathrm{Ad}(g)$ is identically equal to 1. ▮

A Lie algebra which satisfies this condition $\mathrm{trace}\, \mathrm{ad}(x) \equiv 0$ will be called a *unimodular Lie algebra*.

Now let $\mathfrak{g}$ be a completely arbitrary Lie algebra. Using the Jacobi identity

$$\mathrm{ad}[x, y] = \mathrm{ad}(x)\, \mathrm{ad}(y) - \mathrm{ad}(y)\, \mathrm{ad}(x)$$

we see that $\mathrm{ad}[x, y]$ has trace zero. *Therefore the linear mapping*

$$x \mapsto \mathrm{trace}\, \mathrm{ad}(x)$$

*from* $\mathfrak{g}$ *to the commutative Lie algebra* $\mathbf{R}$ *is actually a homomorphism of Lie algebras.* In particular its kernel

$$\mathfrak{u} = \{x \in \mathfrak{g} \mid \mathrm{trace}\, \mathrm{ad}(x) = 0\}$$

is an ideal, containing the commutator ideal $[\mathfrak{g}, \mathfrak{g}]$. We will call $\mathfrak{u}$ the *unimodular kernel* of $\mathfrak{g}$. It is easy to check that $\mathfrak{u}$ itself is a unimodular Lie algebra. Here is an application:

LEMMA 6.4. *If the connected Lie group G has a left invariant metric with all Ricci curvatures* $\geq 0$, *then G is unimodular.*

*Proof.* Suppose on the contrary that $G$ were not unimodular. Choosing a unit vector $b$ orthogonal to the unimodular kernel, we would have trace $\mathrm{ad}(b) \neq 0$. Hence $\mathrm{ad}(b)$ could not be skew-adjoint, and it would follow by 2.3 that $r(b) < 0$. This contradiction completes the proof. ∎

The rest of this section will be concerned with proofs of statements from Section 4 concerning 3-dimensional Lie algebras.

*Proof of Lemma* 4.1. Let $\mathfrak{g}$ be a 3-dimensional Lie algebra with positive definite metric and with preferred orientation. Choosing an oriented orthonormal basis $e_1, e_2, e_3$, define the linear transformation $L: \mathfrak{g} \to \mathfrak{g}$ by $L(e_1) = [e_2, e_3]$, $L(e_2) = [e_3, e_1]$, $L(e_3) = [e_1, e_2]$. Then the identity $L(e_i \times e_j) = [e_i, e_j]$ is true for all basis elements, hence $L(x \times y) = [x, y]$ for all $x$ and $y$. Setting

$$L(e_i) = \sum \alpha_{ij} e_j ,$$

inspection shows that

$$\begin{aligned} \text{trace ad}(e_1) &= -\alpha_{23} + \alpha_{32} \\ \text{trace ad}(e_2) &= -\alpha_{31} + \alpha_{13} \\ \text{trace ad}(e_3) &= -\alpha_{12} + \alpha_{21} . \end{aligned}$$

Thus $\mathfrak{g}$ is unimodular if and only if the matrix $(\alpha_{ij})$ is symmetric, or in other words if and only if the linear transformation $L$ is self adjoint. ∎

*Proof of Theorem* 4.3. We must compute curvature under the hypothesis that $\mathfrak{g}$ has an orthonormal basis with

$$[e_2, e_3] = \lambda_1 e_1 , \qquad [e_3, e_1] = \lambda_2 e_2 , \qquad [e_1, e_2] = \lambda_3 e_3 .$$

Using 5.3 or 5.4 we easily obtain the formula $\nabla_{e_i} v = \mu_i e_i \times v$ for any vector $v$, or briefly

$$\nabla_{e_i} = \mu_i e_i \times ,$$

where

$$\mu_1 = \tfrac{1}{2}(-\lambda_1 + \lambda_2 + \lambda_3), \quad \mu_2 = \tfrac{1}{2}(\lambda_1 - \lambda_2 + \lambda_3), \quad \mu_3 = \tfrac{1}{2}(\lambda_1 + \lambda_2 - \lambda_3).$$

For example $\nabla_{e_1} = \mu_1 e_1 \times$ is the skew-adjoint transformation defined by

$$e_1 \mapsto 0, \qquad e_2 \mapsto \mu_1 e_3\,, \qquad e_3 \mapsto -\mu_1 e_2\,.$$

Using the Jacobi identity

$$e_1 \times (e_2 \times v) - e_2 \times (e_1 \times v) = (e_1 \times e_2) \times v$$

it follows that

$$R_{e_1 e_2} = \nabla_{\lambda_3 e_3} - \nabla_{e_1}\nabla_{e_2} + \nabla_{e_2}\nabla_{e_1}$$

is equal to the linear transformation $(\lambda_3\mu_3 - \mu_1\mu_2)e_3 \times$. Hence

$$R_{e_1 e_2}(e_1) = (\lambda_3\mu_3 - \mu_1\mu_2)\, e_2\,,$$

with similar formulas for the other $R_{e_i e_j}(e_i)$. Recalling the definition

$$\hat{r}(x) = \sum R_{e_i x}(e_i)$$

from Section 2, it follows that

$$\hat{r}(e_2) = (\lambda_3\mu_3 - \mu_1\mu_2)\, e_2 + (\lambda_1\mu_1 - \mu_2\mu_3)\, e_2 = 2\mu_1\mu_3 e_2\,,$$

with similar formulas for the other $\hat{r}(e_i)$. Thus $e_1$, $e_2$, $e_3$ are eigenvectors of the Ricci transformation $\hat{r}$, with corresponding eigenvalues equal to $2\mu_2\mu_3$, $2\mu_1\mu_3$, and $2\mu_1\mu_2$, respectively. ∎

*Proof of Lemma* 4.10. We now consider a 3-dimensional Lie algebra $\mathfrak{g}$ which is *not* unimodular. Its unimodular kernel $\mathfrak{u}$, being 2-dimensional and unimodular, must clearly be commutative. Choose $e_1$ in $\mathfrak{g}$ so that trace $\mathrm{ad}(e_1) = 2$. Since $\mathfrak{u}$ is commutative, the linear transformation

$$L(u) = [e_1\,, u]$$

from $\mathfrak{u}$ to itself, with trace 2, is independent of the particular choice of $e_1$.

If $L$ maps each vector to a multiple of itself, then we are in the special case of Section 1.7 (and in fact $L$ must be the identity map). *Otherwise, the determinant $D$ of $L$ provides a complete isomorphism invariant.* For choosing $e_2$ so that the vectors $e_2$ and $L(e_2) = e_3$ are linearly independent, the conditions trace$(L) = 2$, det$(L) = D$ imply that

$$L(e_2) = e_3$$
$$L(e_3) = -De_2 + 2e_3 .$$

Thus the bracket product operation (with respect to a carefully chosen basis) is uniquely determined. ▮

Suppose that we are given a positive definite metric on the non-unimodular Lie algebra $\mathfrak{g}$. Choose an orthonormal basis $e_1$, $e_2$, $e_3$ so that $e_1$ is orthogonal to $\mathfrak{u}$, and so that the two image vectors $[e_1, e_2]$ and $[e_1, e_3]$ are mutually orthogonal. The bracket product can then be expressed as

$$[e_1, e_2] = \alpha e_2 + \beta e_3$$
$$[e_1, e_3] = \gamma e_2 + \delta e_3$$

and $[e_2, e_3] = 0$; with $\alpha + \delta \neq 0$ and $\alpha\gamma + \beta\delta = 0$.

*Remarks.* If we further normalize by requiring that $\alpha \geqslant \delta$, $\beta \geqslant \gamma$, and $\alpha + \delta > 0$, then these structure constants $\alpha$, $\beta$, $\gamma$, $\delta$ are uniquely determined. Note that the determinant invariant of 4.10 is now given by $D = 4(\alpha\delta - \beta\gamma)/(\alpha + \delta)^2$.

LEMMA 6.5. *This basis also diagonalizes the Ricci quadratic form, the principal Ricci curvatures being*

$$r(e_1) = -\alpha^2 - \delta^2 - \tfrac{1}{2}(\beta + \gamma)^2$$
$$r(e_2) = -\alpha(\alpha + \delta) + \tfrac{1}{2}(\gamma^2 - \beta^2)$$
$$r(e_3) = -\delta(\alpha + \delta) + \tfrac{1}{2}(\beta^2 - \gamma^2).$$

The proof, based on Section 5.5, is tedious but straightforward, and will be left to the reader. ▮

The *proof of Theorem* 4.11 will be given in outline only. To simplify the formulas in 6.5, let us make a scale change if necessary so that $\alpha + \delta = 2$. Then setting

$$\alpha = 1 + \xi, \qquad \beta = (1 + \xi)\eta,$$
$$\gamma = -(1 - \xi)\eta, \qquad \delta = 1 - \xi,$$

we may assume that $\xi \geqslant 0$, $\eta \geqslant 0$. The special case $\xi = \eta = 0$ (corresponding to Example 1.7) must be excluded. The expressions for principal curvatures now take the form

$$r(e_1) = -2(1 + \xi^2(1 + \eta^2)) \leqslant -2$$
$$r(e_2) = -2(1 + \xi(1 + \eta^2)) \leqslant -2$$
$$r(e_3) = -2(1 - \xi(1 + \eta^2)),$$

and the determinant becomes

$$D = (1 - \xi^2)(1 + \eta^2).$$

As examples, taking $\xi = 0$ we obtain metrics of constant negative curvature for every $D > 1$, and taking $\eta = 0$ we obtain metrics of negative sectional curvature whenever $0 < D < 1$. Further details will be left to the reader. ▌

## 7. Bi-invariant Metrics

Recall that a Riemannian metric on $G$ is called *bi-invariant* if it is invariant under both left and right translation. We will first outline the classical theory.

LEMMA 7.1. *A left invariant metric on G is also right invariant if and only if, for each group element g, the linear transformation*

$$\mathrm{Ad}(g)\colon \mathfrak{g} \to \mathfrak{g}$$

*is an isometry with respect to the induced metric on the Lie algebra* $\mathfrak{g}$.

(Compare Section 6.1.) In other words the homomorphism $g \mapsto \mathrm{Ad}(g)$ must map $G$ into the orthogonal group $O(n)$ consisting of all linear isometries of $\mathfrak{g}$.

*Proof.* Let $l_g\colon G \to G$ denote left translation by $g$ and let $r_g$ denote right translation. Thus $\mathrm{Ad}(g)$ is induced by the smooth mapping $l_g r_g^{-1}$ from $G$ to itself. Since the metric $\mu$ is left invariant, we have $l_g{}^*\mu = \mu$. If $\mu$ is also right invariant, $r_g{}^*\mu = \mu$, then evidently

$$(l_g r_g^{-1})^*\,\mu = \mu$$

so that $\mathrm{Ad}(g)$ is an isometry, and conversely. ▌

LEMMA 7.2. *In the case of a connected group $G$, a left invariant metric is actually bi-invariant if and only if the linear transformation* $\mathrm{ad}(x)$ *is skew-adjoint for every $x$ in the Lie algebra of $G$.*

*Proof.* If $g$ is sufficiently close to the identity then $g = \exp(x)$ for some uniquely defined $x$ close to zero. As in Section 6.3 we use the identity

$$\mathrm{Ad}(g) = \mathrm{Ad}(\exp(x)) = e^{\mathrm{ad}(x)}.$$

Recall that $\mathrm{Ad}(g)$ is orthogonal if and only if

$$\mathrm{Ad}(g)^{-1} = \mathrm{Ad}(g)^*.$$

Since the left side equals $e^{-\mathrm{ad}(x)}$ while the right side equals $e^{\mathrm{ad}(x)^*}$, this condition is satisfied if and only if

$$-\mathrm{ad}(x) = \mathrm{ad}(x)^*$$

so that $\mathrm{ad}(x)$ is skew-adjoint. Since a connected Lie group is generated by any neighborhood of the identity, and since products of orthogonal transformations are orthogonal, the conclusion follows. ∎

DEFINITION. It will be convenient to say that a metric on $\mathfrak{g}$ is *bi-invariant* if every $\mathrm{ad}(x)$ is skew-adjoint. Note that a bi-invariant metric on $\mathfrak{g}$ induces a bi-invariant metric on any subalgebra of $\mathfrak{g}$.

Using such a bi-invariant metric, the last two terms in formula 5.3 cancel so that we obtain simply

$$\nabla_x = \tfrac{1}{2}\mathrm{ad}(x).$$

The curvature transformation $R_{xy}$ then is equal to $\frac{1}{2}\mathrm{ad}([x, y]) - \frac{1}{4}\mathrm{ad}(x)\,\mathrm{ad}(y) + \frac{1}{4}\mathrm{ad}(y)\,\mathrm{ad}(x)$. Using the Jacobi identity $\mathrm{ad}[x, y] = \mathrm{ad}(x)\,\mathrm{ad}(y) - \mathrm{ad}(y)\,\mathrm{ad}(x)$ this reduces to

$$R_{xy} = \tfrac{1}{4}\mathrm{ad}[x, y].$$

Hence the biquadratic curvature function $\kappa(x, y) = \langle R_{xy}(x), y\rangle$ equals $\frac{1}{4}\langle[[x, y], x], y\rangle$. Using skew-adjointness once more, this can be written as

$$\kappa(x, y) = \tfrac{1}{4}\langle[x, y], [x, y]\rangle. \tag{7.3}$$

*Thus $\kappa(x, y) \geqslant 0$, with equality if and only if* $[x, y] = 0$. It follows that all Ricci curvatures satisfy $r(u) \geqslant 0$, with equality if and only if $u$ belongs to the center of $\mathfrak{g}$.

We can give a completely explicit formula for the Ricci quadratic form as follows. Recall that the *Killing form* $\beta$ is defined by $\beta(x, y) = \text{trace}(\text{ad}(x)\,\text{ad}(y))$. Since $r(x)$ can be defined as the trace of the linear transformation

$$y \mapsto R_{xy}(x) = \tfrac{1}{4}[[x, y], x] = -\tfrac{1}{4}\text{ad}(x)^2\, y,$$

it follows that $r(x) = -\frac{1}{4}\beta(x, x)$. *Thus the quadratic form $r(x)$ is independent of the particular choice of bi-invariant metric.*

One important property of bi-invariant metrics is the following. Recall that a Lie algebra is *simple* if it contains no ideals other than 0 and itself.

LEMMA 7.4. *If the metric on $\mathfrak{g}$ is bi-invariant, then the orthogonal complement of any ideal is itself an ideal. Hence $\mathfrak{g}$ can be expressed as an orthogonal direct sum of simple ideals.*

*Proof.* If $y$ is orthogonal to the ideal $\mathfrak{a}$, then we must prove that $[x, y]$ is orthogonal to $\mathfrak{a}$. But

$$\langle [x, y], a \rangle = -\langle y, [x, a] \rangle = 0$$

for any $a$ in $\mathfrak{a}$. Thus $\mathfrak{g}$ splits as a direct sum $\mathfrak{a} \oplus \mathfrak{a}^{\perp}$ of ideals. The conclusion now follows by an easy induction. ▮

If $\mathfrak{g}$ equals the orthogonal direct sum $\mathfrak{a}_1 \oplus \cdots \oplus \mathfrak{a}_k$ of simple ideals, then the simply connected Lie group $\tilde{G}$ can be expressed correspondingly as the cartesian product $A_1 \times \cdots \times A_k$ of normal subgroups. For each simply connected factor $A_i$ there are two possibilities:

*Case* 1. *If $\mathfrak{a}_i$ is commutative, hence* 1*-dimensional, then $A_i \cong \mathbf{R}$.*

*Case* 2. If $\mathfrak{a}_i$ is non-commutative, then the center of $\mathfrak{a}_i$ must be trivial, hence $A_i$ has strictly positive Ricci curvature. Applying Myers' theorem as in Section 2.2, it follows that $A_i$ is compact.

LEMMA 7.5. *The connected Lie group $G$ admits a bi-invariant metric if and only if it is isomorphic to the cartesian product of a compact group and an additive vector group.*

*Proof.* If $G$ admits a bi-invariant metric, then the argument above shows that the universal covering $\tilde{G}$ splits as the cartesian product of a compact group $H$ and a vector group $\mathbf{R}^m$. The group $G$ can be identified with the quotient $\tilde{G}/\Pi$ where $\Pi$ is a discrete normal subgroup of $\tilde{G}$. Projecting $\Pi$ into $\mathbf{R}^m$, let $V$ be the vector space spanned by its image and let $V^\perp$ be the orthogonal complement. Then $G$ is the cartesian product of the compact group $(H \times V)/\Pi$ and the vector group $V^\perp$.

The converse is straightforward. Any commutative group certainly admits a bi-invariant metric, and any compact group can be given a bi-invariant metric by starting with an arbitrary metric $\mu$ on the Lie algebra $\mathfrak{g}$ and then averaging $\mathrm{Ad}(g)^*\mu$ as $g$ varies over $G$. See, for example, [3, p. 176]. ▮

In the case of a simple group this metric is essentially unique.

LEMMA 7.6. *If the Lie algebra* $\mathfrak{g}$ *of a compact Lie group is simple, then the bi-invariant metric is unique up to multiplication by a positive constant. Such a metric necessarily has constant Ricci curvature.*

*Proof.* Let $\langle x, y\rangle$ be one bi-invariant metric on $\mathfrak{g}$. Then any other metric on $\mathfrak{g}$ can be expressed as $\langle Sx, y\rangle$ where $S$ is some self adjoint operator. If this new metric is also bi-invariant, then expressing the fact that $\mathrm{ad}(u)$ is skew-adjoint with respect to both metrics we see that $\mathrm{ad}(u)$ commutes with $S$, and hence maps each eigenspace of $S$ into itself. This implies that each eigenspace is an ideal. Since $\mathfrak{g}$ is simple, it follows that $S$ has only one eigenspace, say $Sx = \lambda x$ for all $x$. Thus the bi-invariant metric is essentially unique.

Choosing such a bi-invariant metric, consider the associated Ricci quadratic form $r(x)$. Setting $r(x) = \langle \hat{r}(x), x\rangle$ where $\hat{r}$ is self adjoint, the inner product $\langle \hat{r}(x), y\rangle$ can be considered as a Riemannian metric on $G$, since $r$ is positive definite. Evidently this metric is invariant under both left and right translations. So the argument above proves that $\hat{r}(x) = \lambda x$ and hence

$$r(u) = \langle \hat{r}(u), u\rangle = \lambda$$

for every unit vector $u$, where $\lambda > 0$ is constant. ▮

If we make a scale change, that is, multiply the metric $\langle x, y\rangle$ by a positive constant, it is easy to check that the connection $\nabla$, the Riemann tensor $R$, and the Ricci form $r(x)$ remain unchanged. Therefore, we may choose the metric so that

$$\langle x, x\rangle \equiv r(x),$$

or so that Ricci curvature is identically $+1$.

COROLLARY 7.7. *Any Lie group whose universal covering is compact admits a bi-invariant metric of constant Ricci curvature* $+1$.

*Proof.* As noted above, $\tilde{G}$ splits as a cartesian product $A_1 \times \cdots \times A_k$ of simple groups. Each simple group has a unique bi-invariant metric with Ricci curvature $+1$, and the product metric then also has Ricci curvature identically equal to 1. ∎

In fact it is easy to check that there is precisely one such bi-invariant metric, namely $\langle x, y \rangle = -\frac{1}{4}\beta(x, y)$ where $\beta(x, y) = \text{trace}(\text{ad}(x)\,\text{ad}(y))$ is the Killing form.

Next let us characterize flat left invariant metrics.

*Proof of Theorem* 1.5. Let $G$ be a simply connected Lie group which admits a flat left invariant metric. If we ignore the group structure and think of $G$ only as a complete Riemannian manifold it follows that *$G$ is isometric to Euclidean space.* As an immediate consequence, note that *every compact subgroup of $G$ is trivial.* (Compare Iwasawa's theorem quoted below.) For any compact subgroup, acting by left translation, would yield a compact group of isometries of Euclidean space. Using an averaging process, any such group of isometries would have a fixed point; but nontrivial left translations cannot have fixed points.

For any $\mathfrak{g}$ with metric, the correspondence $x \mapsto \nabla_x$ defines a linear mapping from $\mathfrak{g}$ to the Lie algebra $\mathfrak{o}(n)$ consisting of all skew-adjoint mappings from $\mathfrak{g}$ to itself. If the curvature tensor is identically zero

$$\nabla_{[x,y]} = \nabla_x \nabla_y - \nabla_y \nabla_x ,$$

then this correspondence is a homomorphism of Lie algebras. Hence its kernel $\mathfrak{u}$ is an ideal. Using the identity

$$[u, v] = \nabla_u v - \nabla_v u$$

it follows that $\mathfrak{u}$ is commutative.

Let $\mathfrak{b}$ be the orthogonal complement of $\mathfrak{u}$. For each $b \in \mathfrak{b}$ the identity

$$[b, u] = \nabla_b u - \nabla_u b = \nabla_b u$$

shows that the skew-adjoint transformation $\nabla_b$ maps the ideal $\mathfrak{u}$ into itself. Hence it maps the orthogonal complement $\mathfrak{b}$ into itself. *It follows that $\mathfrak{b}$ is a sub Lie algebra of $\mathfrak{g}$.*

Clearly $\mathfrak{b}$ maps isomorphically to a sub Lie algebra of $\mathfrak{o}(n)$. Since $\mathfrak{o}(n)$ is the Lie algebra of a compact group $O(n)$, it possesses a bi-invariant metric. Therefore $\mathfrak{b}$ possesses a bi-invariant metric, that is, one satisfying the conditions of 7.2. (Caution: This bi-invariant metric on $\mathfrak{b}$ may have nothing to do with the originally given metric.) Therefore, by 7.4, $\mathfrak{b}$ splits as a direct sum $\mathfrak{b}_1 \oplus \cdots \oplus \mathfrak{b}_k$ of simple ideals. If one of these simple ideals $\mathfrak{b}_i$ were non-commutative, then the corresponding simple Lie group $B_i$ would be compact. Hence the inclusion $\mathfrak{b}_i \subset \mathfrak{b} \subset \mathfrak{g}$ would induce a non-trivial homomorphism $B_i \to G$. Hence $G$ would contain a non-trivial compact subgroup; which is impossible. We conclude that each $\mathfrak{b}_i$ is commutative. *Therefore the Lie algebra* $\mathfrak{b}$ *is commutative.*

For each $b$ in $\mathfrak{b}$, note that $\mathrm{ad}(b)$ is skew-adjoint. For $\mathrm{ad}(b)$ restricted to $\mathfrak{b}$ is trivial, while $\mathrm{ad}(b)$ restricted to $\mathfrak{u}$ coincides with the skew-adjoint transformation $\nabla_b$.

Thus $\mathfrak{g}$ splits as an orthogonal direct sum $\mathfrak{u} \oplus \mathfrak{b}$ where $\mathfrak{u}$ is a commutative ideal, $\mathfrak{b}$ is a commutative sub algebra, and each $\mathrm{ad}(b)$ is skew-adjoint. Conversely, if these conditions are satisfied, then using 5.3 we see that

$$\nabla_u = 0, \qquad \nabla_b = \mathrm{ad}(b),$$

and it follows easily that the curvature tensor is identically zero. ∎

As a final application, we will construct metrics of positive scalar curvature. First recall the following basic result.

THEOREM OF IWASAWA. *If $G$ is a connected Lie group, then:*

(a) *Every compact subgroup is contained in a maximal compact subgroup $H$, which is necessarily a connected Lie group.*

(b) *This maximal compact subgroup is unique up to conjugation.*

(c) *As a topological space, $G$ is homeomorphic with the product of $H$ and some Euclidean space $\mathbf{R}^m$.*

This statement was proved in [9], although substantial parts of it had been obtained earlier by Cartan and by Malcev. Here is an immediate consequence.

COROLLARY. *The universal covering of $G$ is homeomorphic to Euclidean space if and only if every compact subgroup of $G$ is commutative.*

The proof is easily supplied. ∎

*Proof of Theorem* 3.4. Let $G$ be a connected Lie group, and suppose that $G$ contains a compact non-commutative subgroup $H$. We must construct a metric of strictly positive scalar curvature. By Iwasawa's theorem, we may assume that $H$ is connected. Since $H$ is compact we can construct a positive definite metric on $\mathfrak{g}$ which is invariant under the linear transformation

$$\mathrm{Ad}(h): \mathfrak{g} \to \mathfrak{g}$$

for every $h$ in $H$. Using this metric, let $e_1, \ldots, e_m$ be an orthonormal basis for the Lie algebra of $H$, and extend to an orthonormal basis $e_1, \ldots, e_n$ for $\mathfrak{g}$. Inspecting the proof of 7.2, we see that the linear transformations

$$\mathrm{ad}(e_1), \ldots, \mathrm{ad}(e_m)$$

must be skew-adjoint. (However, the remaining transformations $\mathrm{ad}(e_{m+1}), \ldots, \mathrm{ad}(e_n)$ will not usually be skew-adjoint.)

Fixing any $\epsilon > 0$, consider a new basis $e_1', \ldots, e_n'$ defined by

$$e_1' = e_1, \ldots, e_m' = e_m, \qquad e_{m+1}' = \epsilon e_{m+1}, \ldots, e_n' = \epsilon e_n .$$

Choose a new metric so that this basis $e_1', \ldots, e_n'$ is orthonormal. The symbol $\mathfrak{g}_\epsilon$ will denote the Lie algebra $\mathfrak{g}$ provided with this new metric, and provided with this specified orthonormal basis. As in the proof of 2.5, we note that the structure constants $\langle [e_i', e_j'], e_k' \rangle$ associated with $\mathfrak{g}_\epsilon$ tend to well defined limits as $\epsilon \to 0$. Hence there is a well defined limit Lie algebra $\mathfrak{g}_0$, provided with a specified metric and orthonormal basis. Evidently $\mathfrak{g}_0$ splits as an orthogonal direct sum $\mathfrak{h} \oplus \mathfrak{u}$ where $\mathfrak{h}$ is the subalgebra spanned by $e_1', \ldots, e_m'$ and $\mathfrak{u}$ is the commutative ideal spanned by $e_{m+1}', \ldots, e_n'$. Note also that $\mathrm{ad}(b)$ is skew-adjoint for each $b \in \mathfrak{h}$. Applying 5.3 we see that $\nabla_u = 0$ for every $u$ in $\mathfrak{u}$, hence $R_{xu} = 0$ and $\kappa(x, u) = 0$ for all $x$. In particular the Ricci curvature $r(u)$ is zero for $u \in \mathfrak{u}$. On the other hand for $b$ in $\mathfrak{h}$ we have $r(b) \geqslant 0$ by 2.1, where equality does not always hold since $\mathfrak{h}$ is noncommutative. *Therefore the scalar curvature* $\rho = r(e_1') + \cdots + r(e_n')$ *of the limit algebra* $\mathfrak{g}_0$ *is strictly positive.* It follows by continuity that $\rho(\mathfrak{g}_\epsilon) > 0$ whenever $\epsilon$ is sufficiently small. ∎

*Remark* 7.8. This limit algebra $\mathfrak{g}_0$ provides an interesting example of a metric Lie algebra with all sectional curvatures $K \geqslant 0$.

## REFERENCES


1. R. AZENCOTT AND E. WILSON, Homogeneous manifolds with negative curvature, Part I, *Trans. Amer. Math. Soc.* **215** (1976), 323–362.
2. R. L. BISHOP AND R. J. CRITTENDEN, "Geometry of Manifolds," Academic Press, New York, 1964.
3. C. CHEVALLEY, "Theory of Lie Groups," Princeton University Press, Princeton, N.J., 1946.
4. I. V. DOLGAČEV, Conic quotient singularities of complex surfaces, *Funct. Anal. Appl.* **8**, 2 (1974), 160–161; See also **9**, 2 (1975), 149–151.
5. H. I. ELIASSON, On variations of metrics, *Math. Scand.* **29** (1971), 317–327.
6. E. HEINTZE, On homogeneous manifolds of negative curvature, *Math. Ann.* **211** (1974), 23–34.
7. S. HELGASON, "Differential Geometry and Symmetric Spaces," Academic Press, New York, 1962.
8. N. HITCHIN, Harmonic spinors, *Advances in Math.* **14** (1974), 1–55.
9. K. IWASAWA, On some types of topological groups, *Ann. Math.* **50** (1949), 507–558.
10. J. L. KAZDAN AND F. W. WARNER, Scalar curvature and conformal deformation of Riemannian structure, *J. Diff. Geom.* **10** (1975), 113–134.
11. S. KOBAYASHI AND K. NOMIZU, "Foundations of Differential Geometry, Vol. I," Interscience, New York, 1963.
12. H. B. LAWSON AND S. T. YAU, Scalar curvature, non-abelian group actions, and the degree of symmetry of exotic spheres, *Comment. Math. Helv.* **49** (1974), 232–244.
13. A. LICHNEROWICZ, Spineurs harmoniques, *C. R. Acad. Sci. Paris* **257** (1963), 7–9.
14. J. MILNOR, Morse theory, *Ann. Math. Stud.* **51**, Princeton Univ. Press, Princeton, N.J., 1963.
15. J. MILNOR, On the 3-dimensional Brieskorn manifolds $M(p, q, r)$, *in* "Knots, Groups, and 3-Manifolds," pp. 175–225, Ann. Math. Studies, Vol. 84, Princeton Univ. Press, Princeton, N.J., 1975.
16. F. RAYMOND AND A. VASQUEZ, Closed 3-manifolds whose universal covering is a Lie group, (to appear).
17. V. S. VARADARAJAN, Lie Groups, Lie Algebras, and their Representations, Prentice-Hall, New York, 1974.
18. P. WAGREICH, Singularities of complex surfaces with solvable local fundamental group, *Topology* **11** (1972), 51–72.
19. N. WALLACH, Compact homogeneous Riemannian manifolds with strictly positive curvature, *Ann. Math.* **96** (1972), 277–295.
20. F. WARNER, "Foundations of Differentiable Manifolds and Lie Groups," Scott-Foresman, New York, 1971.
21. J. WOLF, Curvature in nilpotent Lie groups, *Proc. Amer. Math. Soc.* **15** (1964), 271–274.
22. J. WOLF, Growth of finitely generated solvable groups and curvature of Riemannian manifolds, *J. Diff. Geom.* **2** (1968), 421–446. (See Sections 5, 6.)

# ON DECIDING WHETHER A SURFACE IS PARABOLIC OR HYPERBOLIC

*Reprinted from*

The American Mathematical Monthly
Volume 84 (1977), pp. 43–46

This note is a very modest contribution to the problem of relating the Riemannian geometry of a surface to its associated conformal geometry. The original paper has been reset in TEX with minor corrections.

# ON DECIDING WHETHER A SURFACE IS PARABOLIC OR HYPERBOLIC

JOHN MILNOR

Let $M$ be a simply connected open surface, differentiable of class say $C^\infty$, with a $C^\infty$ Riemannian metric. It follows from complex function theory that $M$ is conformally diffeomorphic either to the complex plane, or to the open unit disk in the complex plane. (See for example [Springer, p. 225] together with the local argument as given in [Chern] or [Chern, Hartman, Wintner] or [Courant-Hilbert, p. 350] or [Milnor, pp. 1110, 1111].) In the first case one says that $M$ is **parabolic**, in the second case **hyperbolic**.

This note will attack the problem of deciding whether $M$ is parabolic or hyperbolic in the very special circumstance that $M$ is complete and rotationally symmetric about a point $p$, so that the Gaussian curvature $K$ can be expressed as a function of the distance $r$ from $p$. Let $2\pi g(r)$ denote the circumference of the circle of radius $r$ centered at $p$. Introducing geodesic polar coordinates $r, \theta$ the Riemannian metric takes the form $dr^2 + g(r)^2\, d\theta^2$. Computation shows that the Gauss curvature is given by

$$K = -(d^2 g/dr^2)/g$$

or briefly, $K = -\ddot{g}/g$. (See for example, [**18**, pp. 136–138] or [**16**, pp. 3B, 32–37].) Thus, the function $g$ satisfies the linear differential equation $\ddot{g} + Kg = 0$ with initial conditions $g(0) = 0$, $\dot{g}(0) = 1$.

THEOREM. *If $K \geq -1/(r^2 \log r)$ for large $r$, then the surface is conformally parabolic. But if $K \leq -(1+\varepsilon)/(r^2 \log r)$ for large $r$, and if the function $g(r)$ is unbounded, then the surface is hyperbolic.*

Here $\varepsilon > 0$ can be any positive constant.

Note that the hypothesis that $g(r)$ is unbounded is essential. In fact, if $g(r)$ is bounded, then the surface is necessarily parabolic. Compare the lemma below. An example is provided by the function $g(r) = re^{-r^2}$, with $K(r) = 6 - 4r^2$. This yields a parabolic surface since $g$ is bounded, even though $K$ tends to $-\infty$ as $r \to \infty$. (Such an example, with $g$ bounded, cannot have $K \leq 0$ for all values of $r$. For if $K \leq 0$ or equivalently $\ddot{g} \geq 0$ for all $r$, then it follows by integration that $\dot{g} \geq 1$ and $g(r) \geq r$ so that $g$ must be unbounded.)

This theorem surprised the author: It seemed strange that a parabolic surface could be converted into a hyperbolic surface simply by replacing its curvature function $K(r) = -1/(r^2 \log r)$ for $r \geq c$ by $(1+\varepsilon)K(r)$ for $r \geq c$.

The proof will be based on the following. (Compare [**1**].) Choose $a > 0$.

LEMMA. *The surface $M$ is hyperbolic if and only if the integral $\int_a^\infty dr/g(r)$ is finite.*

*Proof.* Introduce a new coordinate $\rho$ in place of $r$ by setting $\rho = \int_a^r ds/g(s)$. Then the metric takes the form

$$dr^2 + g^2\, d\theta^2 = g^2(d\rho^2 + d\theta^2).$$

(Coordinates $\rho, \theta$ for which the metric takes the latter form, conformally equivalent to the flat metric $d\rho^2 + d\theta^2$, are known classically as *isothermal coordinates*.) Now map $M$ conformally into the complex numbers by sending the point with coordinates $\rho, \theta$ to the complex number $\exp(\rho + i\theta) = e^\rho \cos\theta + ie^\rho \sin\theta$. This conformal mapping is at first not defined at the base point $p$. However, it is bounded, smooth, and conformal throughout a neighborhood of $p$, except at the point $p$ itself. Hence the apparent singularity at $p$ is *removable*, and we obtain a smooth conformal mapping which is defined and one-to-one throughout $M$.

[Expressing the construction in a different way, we map the point of $M$ with geodesic polar coordinates $(r, \theta)$ to the complex number $Re^{i\theta}$ with polar coordinates $(R, \theta)$, where $R = \exp\left(\int_a^r ds/g(s)\right) = \exp\left(\int_1^r ds/g(s) + \text{constant}\right)$. This last formula for $R$ is the most general one which yields a radial stretching factor $dR/dr$ precisely equal to the circumferential stretching factor $2\pi R/2\pi g(r)$, so that the mapping is conformal.]

Evidently the image of this mapping is either the disk of radius $\exp \int_a^\infty dr/g(r)$ or the entire complex plane according as this integral is finite or infinite. ■

REMARK. If the total absolute curvature $\iint |K|\, dA$ of a complete simply connected surface is finite, then [Blanc and Fiala] have shown that the surface is parabolic. Compare **[8]**, **[6]**. (In the rotationally symmetric case we can recover this result from the Lemma as follows. Since the Riemannian area element $dA$ is equal to $g\, dr d\theta$, it follows that

$$(2\pi)^{-1} \iint |K|\, dA = \int_0^\infty |K| g\, dr = \int_0^\infty |\ddot{g}|\, dr$$

is equal to the total variation of $\dot{g}$. If this total variation is finite, then $\dot{g}$ is bounded, hence $g(r) \le Cr$, and $\int_1^\infty dr/g = \infty$, so that the surface must be parabolic.)

Here is a basic example which illustrates the Lemma. Consider the function $g_0(r) = r \log r$ for $r \ge 2$ (choosing say $g_0(r) = r$ for $r \le 1$ and interpolating smoothly in the interval $1 \le r \le 2$). The corresponding surface is parabolic since $\int_2^\infty dr/(r \log r) = \left[\log\log r\right]_2^\infty = \infty$. The associated Gauss curvature $K_0$ is given by

$$K_0 = -\ddot{g}_0/g_0 = -1(r^2 \log r)$$

for $r \ge 2$. (In this example, the total absolute curvature $\iint |K_0|\, dA_0$ is infinite, so that the Blanc-Fiala criterion does not apply.)

*Proof of Theorem.* Let $g(r) > 0$ be any smooth function whose associated curvature $K = -\ddot{g}/g$ satisfies the inequality $K(r) \ge K_0(r)$ for $r \ge a$; where $K_0 = -1/(r^2 \log r)$ as above. (Note that $K$ may take both positive and negative values.) Multiplying $g$ by a small positive constant if necessary, we may assume that the initial inequalities

(1) $$g(a) < g_0(a),$$

(2) $$\dot{g}(a) < \dot{g}_0(a)$$

are satisfied. *It follows that* $g(r) < g_0(r)$ *for all* $r \geq a$. For otherwise, if $b$ were the smallest number greater than $a$ with $g(b) = g_0(b)$, then integrating the inequality

$$\ddot{g} = -Kg \leq -K_0 g_0 = \ddot{g}_0$$

from $a$ to $r$ and adding (2) we would obtain $\dot{g}(r) < \dot{g}_0(r)$ for $a \leq r \leq b$. Integrating this new inequality from $a$ to $b$ and adding (1) we would obtain $g(b) < g_0(b)$, which contradicts the choice of $b$. Thus $g(r) < g_0(r)$ for all $r \geq a$. Hence the integral $\int_a^\infty dr/g(r)$ is also divergent, and the surface is parabolic.

For the proof in the other direction, we use a comparison function of the form

$$g_\varepsilon(r) = r(\log r)^{1+\varepsilon} \quad \text{for} \quad r \geq 2.$$

This yields a hyperbolic surface since the integral

$$\int_2^\infty dr/r(\log r)^{1+\varepsilon} = \left[-1/\varepsilon(\log r)^\varepsilon\right]_2^\infty$$

is finite. Computation shows that the corresponding Gauss curvature is given by

$$K_\varepsilon = -(1+\varepsilon)(1+\varepsilon/\log r)/(r^2 \log r) \geq -(1+2\varepsilon)/(r^2 \log r)$$

for large $r$.

Now consider a smooth function $g(r) > 0$ whose associated Gauss curvature $K = -\ddot{g}/g$ satisfies

$$K(r) \leq -(1+2\varepsilon)/(r^2 \log r) \leq K_\varepsilon(r)$$

for $r \geq a$. If $\dot{g}(a) > 0$ then, after multiplying the function $g(r)$ by a large constant if necessary, a completely analogous argument shows that $g(r) > g_\varepsilon(r)$ for all $r \geq a$. Hence the integral $\int_a^\infty dr/g(r)$ is also convergent, and the surface is hyperbolic.

This argument will work whenever $\dot{g}(r) > 0$ for *some* sufficiently large $r$. In particular, it will certainly work whenever the function $g(r)$ is unbounded. On the other hand, if $\dot{g}(r) \leq 0$ for all large $r$, then the function $g(r)$ is bounded, and hence by the Lemma the surface is parabolic. ∎

In conclusion, here are two further problems:

(1) *Does the Theorem proved above remain valid in the more general case of a complete Riemannian surface which is not rotationally symmetric?*

The statement would still make sense as long as the base point $p$ has no conjugate points, so that there exists a global geodesic polar coordinate system with metric of the form $dr^2 + g(r,\theta)^2\, d\theta^2$. Robert Osserman points out to the author that if $K(r,\theta) \geq -1/(r^2 \log r)$ for large $r$ then, using a theorem of Ahlfors [**1**] in place of the Lemma, one can indeed prove that $M$ is parabolic. Greene and Wu [**7**] prove an analogous result in higher dimensions, assuming the sharper hypothesis that all sectional curvatures satisfy

$0 \geq K \geq -C/r^{2+\varepsilon}$. For discussion of a related problem in the hyperbolic case, see [**12**], §6.4.

(2) Consider an embedded surface in Euclidean 3-space of the form $z = f(x, y)$. *How can one decide effectively whether this surface is parabolic or hyperbolic*? This problem has been studied by Osserman [**12, 13**], H. Huber [**9**], Jenkins [**10**], and by Rüedy [**14**].

## References


**1.** L. Ahlfors, Sur le type d'une surface de Riemann, C. R. Acad. Sci., Paris, 201 (1935) 30–32. (Compare [Sario and Nakai, p. 329].)

**2.** C. Blanc and F. Fiala, Le type d'une surface et sa courbure totale, Comment. Math. Helv., 14 (1941–42) 230–233.

**3.** S. S. Chern, An elementary proof of the existence of isothermal parameters on a surface, Proc. Amer. Math. Soc., 6 (1955) 771–782.

**4.** S. S. Chern, P. Hartman, and A. Wintner, On isothermic coordinates, Comment. Math. Helv., 28 (1954) 301–309.

**5.** R. Courant and D. Hilbert, Methods of Mathematical Physics, vol. 2, Interscience, Wiley, New York, 1962.

**6.** R. Finn, On a class of conformal metrics, with application to differential geometry in the large, Comment. Math. Helv., 40 (1965–66) 1–30.

**7.** R. E. Greene and H. Wu, Curvature and complex analysis, II, Bull. Amer. Math. Soc., 78 (1972) 866–870.

**8.** A. Huber, On subharmonic functions and differential geometry in the large, Comment. Math. Helv., 32 (1957) 13–72.

**9.** H. Huber, Riemannsche Flächen von hyperbolishem Typus im euklidischen Raum, Math. Ann., 139 (1959) 140–146.

**10.** J. A. Jenkins, On hyperbolic surfaces in three-dimensional euclidean space, Michigan Math. J., 8 (1961) 1–5.

**11.** J. Milnor, A problem in cartography, this MONTHLY, 76 (1969) 1101–1112.*

**12.** R. Osserman, Riemann surfaces of class A, Trans. Amer. Math. Soc., 82 (1956) 217–245.

**13.** ——, Hyperbolic surfaces of the form $z = f(x, y)$, Math. Ann., 144 (1961) 77–79.

**14.** R. Rüedy, Embeddings of open Riemann surfaces, Comment. Math. Helv., 46 (1971) 214–225.

**15.** L. Sario and M. Nakai, Classification Theory of Riemann Surfaces, Grundlehren 164, Springer-Verlag, New York, 1970.

**16.** M. Spivak, A Comprehensive Introduction to Differential Geometry II, Publish or Perish, Inc., Boston, 1970.

**17.** G. Springer, Introduction to Riemann Surfaces, Addison-Wesley, Reading, Mass., 1957.

**18.** D. Struik, Lectures on Classical Differential Geometry, Addison-Wesley, Reading, Mass., 1950.


INSTITUTE FOR ADVANCED STUDY, PRINCETON, NJ 87540.

*This volume, pp. 161–172.

# ON FUNDAMENTAL GROUPS OF COMPLETE AFFINELY FLAT MANIFOLDS

*Reprinted from*

Advances in Mathematics
Volume 25 (1977), pp. 178–187

Let $\{\pi_1(\textit{complete affinely-flat})\}$ be the collection of all fundamental groups of manifolds with a complete affinely flat structure, and let $\{\textit{torsionfree virtually-polycyclic}\}$ be the collection of torsionfree groups which contain a polycyclic subgroup of finite index. This paper first proves that

$$\Big\{\textit{torsionfree virtually-polycyclic}\Big\} \subset \Big\{\pi_1(\textit{complete affinely-flat})\Big\}.$$

Unfortunately, it then conjectures that equality holds. This conjecture has been disproved by Margulis.[1] (See also Drumm and Goldman.[2]) In fact a 3-dimensional handlebody of genus 2 admits not only a complete affinely flat structure, but even a Lorentz metric which is complete and locally flat.

Now let $\{\pi_1(\textit{compact complete affinely-flat})\}$ be the smaller collection consisting of fundamental groups of compact manifolds which admit a complete affinely flat structure. The paper also discuss the possibility that

$$\Big\{\textit{torsionfree virtually-polycyclic}\Big\} = \Big\{\pi_1(\textit{compact complete affinely-flat})\Big\}. \qquad (?)$$

This seems to be false, but the situation is still far from clear. Boyom[3] has claimed that every nilpotent Lie group admits a left invariant complete affinely flat structure; and Nisse[4] has claimed the sharper statement that

---

[1] G. Margulis, *Complete affine locally flat manifolds with a free fundamental group*, Dokl. Akad. Nauk SSSR **272** (1983) 785–788.

[2] T. Drumm and W. Goldman, *Complete flat Lorentz 3-manifolds with free fundamental group*, Intern. J. Math. **1** (1990) 149–161.

[3] N. Boyom, *The lifting problem for affine structures in nilpotent Lie groups*, Trans. Amer. Math. Soc. **313** (1989) 347–379. (But see the erratum in J. Diff. Geom. **38** (1993) 463.)

[4] M. Nisse, *Structure affine des infranilvariétés et des infrasolvariétés*, Comptes Ren. Acad. Sci. Paris **310** (1990) 667–670.

every closed aspherical manifold of dimension $\neq 3$ with virtually polycyclic fundamental group admits a complete affinely flat structure. On the other hand a manuscript by Benoist[5] contradicts both statements. For every $n > 10$ he constructs a nilpotent Lie algebra of dimension $n$ which admits no faithful linear representation on an $(n+1)$-dimensional vector space. In particular Benoist asserts that there exist compact nilmanifolds which admit no complete flat affine structure. Assuming that Benoist is correct, there remains the possibility that

$$\{\,\textit{torsionfree virtually-polycyclic}\,\} \supsetneqq \{\,\pi_1(\textit{compact complete affinely-flat})\,\}. \qquad (?)$$

I am not aware of any progress on this question.

[5] Yves Benoist, *Une nilvariété non affine*, C. R. Acad. Sci. Paris (Ser. 1, Math.) **315** (1992) 983–986.

# On Fundamental Groups of Complete Affinely Flat Manifolds


JOHN MILNOR

*Institute for Advanced Study, Princeton, New Jersey 08540*


This note tries to decide which groups can occur as fundamental groups of complete affinely flat manifolds.

THEOREM 1.2. *If a group is torsion-free and contains a polycylic subgroup of finite index, then it is isomorphic to the fundamental group $\pi_1(M)$ for some complete affinely flat manifold $M$.*

For definitions and proof, see Section 1. The remaining sections center around two unresolved questions: *Are these the only fundamental groups which can occur* (Section 3)? *Can the manifold $M$ always be chosen to be compact* (Section 4)?

## 1. CONSTRUCTING AFFINE ACTIONS

First some definitions. A group $\Gamma$ is *virtually polycyclic* if it has a subgroup $\Gamma_0$ of finite index which is *polycyclic*, that is, admits a finite composition series $\Gamma_0 \supset \Gamma_1 \supset \cdots \supset \Gamma_k = \{1\}$ so that each quotient $\Gamma_i/\Gamma_{i+1}$ is cyclic. The number of $\Gamma_i/\Gamma_{i+1}$ which are free cyclic is an invariant called the *rank* of $\Gamma$.

More generally, given any property $P$ (such as being commutative, solvable, or torsion-free), a group is *virtually P* if it admits a subgroup of finite index with property $P$.

An action $(\gamma, x) \mapsto \gamma x$ of a group $\Gamma$ on a locally compact space $X$ is *properly discontinuous* if for each compact $K \subset X$ the set of $\gamma$ with $\gamma K \cap K \neq \varnothing$ is finite. This action is *effective* if the equation $\gamma x = x$ for all $x$ implies that $\gamma = 1$, and *free* if $\gamma x = x$ for any $x$ implies that $\gamma = 1$.

THEOREM 1.1. *Any virtually polycyclic group admits an effective and properly discontinuous action by affine transformations $x \mapsto Ax + b$ of some Euclidean space $\mathbf{R}^n$.*

*Proof.* Let $\Gamma_0 \subset \Gamma$ be a polycyclic subgroup of finite index. Auslander [2] and Swan [26] have shown that $\Gamma_0$ can be embedded into the general linear group $\mathrm{GL}(m, \mathbf{Z})$ for some $m$. *Hence $\Gamma_0$ can be embedded as a discrete subgroup of the complex general linear group* $\mathrm{GL}(m, \mathbf{C})$.

Consider the Zariski closure of $\Gamma_0$, that is, the smallest complex algebraic variety in GL($m$, **C**) which contains $\Gamma_0$. This closure is both a smooth algebraic variety, and a solvable complex Lie group. (Compare [7].) Its identity component $S$ is a subgroup of finite index, hence the intersection $\Delta = S \cap \Gamma_0$ is a subgroup of finite index in $\Gamma_0$.

By Lie's theorem, $S$ is conjugate to a subgroup of the standard Borel group $B$ consisting of upper triangular matrices in GL($m$, **C**). (Compare [17, p. 50] or [15, p. 134].) Hence, after applying an inner automorphism, we may assume that $\Delta \subset B$.

Recall that $B$ splits as a semidirect product $UD$, where $U$ is the group of uni-triangular matrices and $D$ is the commutative group consisting of diagonal matrices. Note that $U$, being a complex $m(m-1)/2$-dimensional plane in the vector space of $m \times m$ matrices, has the structure of a complex affine space. Since $U$ is a normal subgroup of $B$, there is a short exact sequence

$$1 \longrightarrow U \longrightarrow B \xrightarrow{\pi} D \longrightarrow 1.$$

Using this projection homomorphism $\pi$, each group element $b$ of $B$ operates on $U$ by the affine transformation

$$u \mapsto bu\pi(b)^{-1}.$$

(Or, equivalently, we could say that $B$ operates on the coset space $B/D \cong U$ by left translation.) Therefore, the discrete subgroup $\Delta \subset B$ also operates by affine transformations of $U$. This action need not be either effective or properly discontinuous. However, the subgroup $\Delta \cap U$, which operates on $U$ just by left translation, certainly acts effectively and properly discontinuously.

Next consider the quotient group $\Delta/(\Delta \cap U)$ which is finitely generated (since $\Delta$ is) and commutative (since it embeds in $D$). This quotient splits as a direct sum $\Lambda_1 \oplus \cdots \oplus \Lambda_p$ of cyclic groups. Each cyclic group operates effectively and properly discontinuously either by translations or by rotations of the complex numbers **C**. Hence $\Delta/(\Delta \cap U)$ acts effectively and properly discontinuously by affine transformations of the product $\mathbf{C} \times \cdots \times \mathbf{C} = \mathbf{C}^p$.

The diagonal action of $\Delta$ on the product affine space $U \times \mathbf{C}^p$ now has all of the required properties. It is clearly an effective action by affine transformations, and it is not difficult to show that it is properly discontinuous.

Finally we must pass from the subgroup $\Delta$ to the larger group $\Gamma$. Let $q$ be the index of $\Delta$ in $\Gamma$. Given a representation of $\Delta$ by affine transformations of $\mathbf{R}^k$, we can form the induced representation of $\Gamma$ by affine transformations of the space $\mathrm{Hom}_\Delta(\Gamma, \mathbf{R}^k)$. By definition, this is the $kq$-dimensional affine space consisting of all functions $f: \Gamma \to \mathbf{R}^k$ which are $\Delta$-equivariant:

$$f(\delta\gamma) = \delta f(\gamma) \qquad \text{for} \quad \delta \in \Delta, \gamma \in \Gamma.$$

If $\gamma_1, \ldots, \gamma_q$ are coset representatives, so that $\Gamma = \Delta\gamma_1 \cup \cdots \cup \Delta\gamma_q$, note that $f$

is uniquely determined by the values $f(\gamma_1), \ldots, f(\gamma_q) \in \mathbf{R}^k$, which can be prescribed arbitrarily.

Each group element $\varphi$ in $\Gamma$ acts on this affine space $\mathrm{Hom}_\Delta(\Gamma, \mathbf{R}^k)$ by the affine transformation $f \mapsto {}^\varphi f$, where ${}^\varphi f(\gamma) = f(\gamma\varphi)$. If the action of $\Delta$ on $\mathbf{R}^k$ is effective and properly discontinuous, then it is easy to check that this induced action of $\Gamma$ on $\mathrm{Hom}_\Delta(\Gamma, \mathbf{R}^k)$ is also. ∎

It is now easy to prove Theorem 1.2, as stated in the Introduction.

First the necessary definitions. An *affinely flat* structure on an $n$-dimensional manifold $M$ can be specified by a collection of coordinate homeomorphisms

$$f_\alpha : U_\alpha \to V_\alpha \subset \mathbf{R}^n,$$

where the $U_\alpha$ are open sets covering $M$ and the $V_\alpha$ are open subsets of $\mathbf{R}^n$. Whenever $U_\alpha \cap U_\beta \neq \varnothing$, it is required that the change of coordinate homeomorphism

$$f_\beta f_\alpha^{-1} : f_\alpha(U_\alpha \cap U_\beta) \to f_\beta(U_\alpha \cap U_\beta)$$

should extend to an affine transformation $x \mapsto Ax + b$ from $\mathbf{R}^n$ to itself.

One can define more restrictive kinds of flatness by requiring that the matrix $A$ should belong to some proper subgroup of $\mathrm{GL}(n, \mathbf{R})$. For example, if $A$ is required to be orthogonal, then we obtain the concept of a *Riemannian-flat manifold*, locally isometric to Euclidean space. Similarly, if $A$ must belong to the Lorentz group $O(n-1, 1)$, we obtain the concept of a *Lorentz-flat* manifold (cf. [5]), and if $A$ is required to have determinant $\pm 1$ we obtain the concept of an *area-preserving affinely flat* manifold.

By a *geodesic* in an affinely flat manifold $M$ is meant a mapping $\mathbf{R} \to M$ compatible with the affine structure. The manifold $M$ is said to be *complete* if every geodesic segment $[0, 1] \to M$ can be extended to a full geodesic, or equivalently if the universal covering manifold $\tilde{M}$ is affinely diffeomorphic to $\mathbf{R}^n$.

In order to actually construct a complete affinely flat manifold with given fundamental group $\Gamma$, it is usually easiest to start with an action of $\Gamma$ by affine transformations of $\mathbf{R}^n$. If this action is free and properly discontinuous, then the quotient $M = \Gamma \backslash \mathbf{R}^n$ is the required manifold. For details, the reader is referred to [29, pp. 39, 45].

*Proof of* 1.2. If $\Gamma$ is virtually polycyclic, then it admits a properly discontinuous, affine action on some $\mathbf{R}^n$ by Theorem 1.1. Note that the subgroup fixing any point of $\mathbf{R}^n$ must be finite. Now suppose that the group $\Gamma$ is also *torsion-free*, i.e., suppose that every finite subgroup is trivial. Then the action of $\Gamma$ on $\mathbf{R}^n$ must be free, and the conclusion follows. ∎

If we sharpen the requirements, and ask that $\Gamma$ act by Euclidean isometries on $\mathbf{R}^n$, then the situation was thoroughly analyzed by Bieberbach more than 60 years ago. Here is an outline. (Compare [20].)

THEOREM 1.3. *The group $\Gamma$ can occur as the fundamental group of a complete Riemannian-flat manifold $M$ if and only if it is torsion-free, finitely generated, and contains a commutative subgroup of finite index. The manifold $M$ can always be chosen to be compact.*

In fact, if $\Gamma$ is finitely generated and virtually commutative, then it contains a free Abelian subgroup, $\Delta \cong \mathbf{Z}^k$, of finite index $q$. Letting $\Delta$ operate on $\mathbf{R}^k$ by translation, it follows that $\Gamma$ operates on the Euclidean space $\mathrm{Hom}_\Delta(\Gamma, \mathbf{R}^k) \cong \mathbf{R}^{kq}$ by Euclidean isometries. This action is effective and properly discontinuous. If $\Gamma$ is also torsion-free, then the action is free, and $M = \Gamma\backslash\mathbf{R}^{kq}$ is the required complete Riemannian-flat manifold.

Given any group $\Gamma$ which acts properly discontinuously by Euclidean isometries of $\mathbf{R}^n$, Bieberbach showed that there always exists a Euclidean subspace $\mathbf{R}^k$ on which $\Gamma$ acts with compact fundamental domain. If $\Gamma$ is torsion-free, then it follows that the quotient $\Gamma\backslash\mathbf{R}^k$ is a compact flat manifold, embedded as a deformation retract in the complete manifold $\Gamma\backslash\mathbf{R}^n$. (In the language of Cheeger and Gromoll, this compact totally geodesic submanifold is called the "soul" of $\Gamma\backslash\mathbf{R}^n$.)

Finally, Bieberbach showed that $\Gamma$ contains, as subgroup of finite index, a free Abelian group of rank $k$ which acts on $\mathbf{R}^k$ by translation. For details, the reader is referred to [6; 29, pp. 99–106]. ∎

## 2. Lemmas Concerning Lie Groups and Discrete Subgroups

This section proves three lemmas, for use later. The first is an immediate consequence of a theorem of Tits [27].

LEMMA 2.1. *Let $\Gamma$ be a discrete subgroup of a Lie group $G$ which has finitely many components. Then either $\Gamma$ is virtually polycyclic or $\Gamma$ contains a subgroup $\mathbf{Z} * \mathbf{Z}$ which is free on two generators.*

These two possibilities are mutually exclusive.

It is natural to ask which Lie groups have the property that *every* discrete subgroup is virtually polycyclic. Recall that a topological group $G$ is said to be *amenable* if the space of bounded continuous functions $f\colon G \to \mathbf{R}$ admits a *left invariant mean*, that is, a linear real-valued function $f \mapsto m(f)$ which is invariant under left translation and satisfies

$$\inf f \leqslant m(f) \leqslant \sup f.$$

Using Lemma 2.1, the following statement is more or less well known. (Cf. [12; 22].)

LEMMA 2.2. *If $G$ is a Lie group with finitely many components, then the following four conditions are equivalent*:

(a) *$G$ is amenable*,

(b) *every discrete subgroup of $G$ is virtually polycyclic*,

(c) *every semisimple connected subgroup of $G$ is compact*,

(d) *$G$ contains a solvable normal subgroup with compact quotient.*

*Proof of Lemma* 2.1. If $\Gamma$ is any group which admits a faithful finite dimensional representation, so that

$$\Gamma \subset \mathrm{GL}(n, \mathbf{R})$$

for some $n$, then Tits' theorem asserts that either $\Gamma$ is virtually solvable or $\Gamma$ contains a free noncyclic subgroup.

The same dichotomy holds for any $\Gamma$ which can be embedded in a Lie group $G$ with finitely many components. For, according to Ado's theorem [17, p. 202], the Lie algebra of $G$ admits a faithful linear representation. Hence some group $G'$ which is locally isomorphic to $G$ embeds in $\mathrm{GL}(n, \mathbf{R})$ for some $n$. Thus there exist homomorphisms

$$G \leftarrow \tilde{G}_0 \rightarrow G'$$

with discrete central kernel, where $\tilde{G}_0$ is the universal covering of the identity component of $G$. For any $\Gamma \subset G$, we can construct an associated subgroup $\Gamma' \subset G'$ by first intersecting with the identity component $G_0$, then passing to the full inverse image in $\tilde{G}_0$, and finally projecting into $G'$. It is easy to check that $\Gamma$ is virtually solvable [or contains a free noncyclic subgroup] if and only if $\Gamma'$ is virtually solvable [or contains a free noncyclic subgroup].

To conclude the proof, we make use of Mostow's theorem which asserts that any discrete solvable subgroup of a connected Lie group must actually be polycyclic. Hence, for discrete groups, virtually solvable implies virtually polycyclic. ∎

*Proof of Lemma* 2.2. The proof that (d) $\Rightarrow$ (a) $\Rightarrow$ (b) is based on theorems from Greenleaf [13]. Clearly, every compact group is amenable. Since every commutative group is amenable, and every extension of an amenable group by an amenable group is amenable [13, pp. 5, 30], it follows that (d) $\Rightarrow$ (a).

Since discrete subgroups of amenable groups are amenable, but free noncyclic groups are not amenable [13, pp. 30, 6], it follows using Lemma 2.1 that (a) $\Rightarrow$ (b).

Proof that (b) ⇒ (c). Let $H$ be any semisimple subgroup of $G$. We make use of the Iwasawa decomposition

$$\mathfrak{h} = \mathfrak{k} \oplus \mathfrak{a} \oplus \mathfrak{n},$$

where $\mathfrak{h}$ is the Lie algebra of $H$. (Compare [14, p. 222].) Here $\mathfrak{k}$ is a subalgebra with the property that the associated Lie group $K \subset H$ is compact, at least when $H$ has finite center. The subalgebra $\mathfrak{a}$ is commutative, and the subalgebra $\mathfrak{n}$ is nilpotent, with the property that $\mathrm{ad}(x)\colon \mathfrak{h} \to \mathfrak{h}$ is nilpotent for each $x \in \mathfrak{n}$. Closely related is the *Cartan decomposition* $\mathfrak{h} = \mathfrak{k} \oplus \mathfrak{p}$, where $\mathfrak{p}$ is a vector space containing $\mathfrak{a}$ and satisfying the condition $[\mathfrak{k}, \mathfrak{p}] \subset \mathfrak{p}$ (see [14, p. 157]).

We first show that the nilpotent summand $\mathfrak{n}$ must be trivial; for otherwise the Lie algebra $\mathfrak{h}$ would contain an element $x \neq 0$ with $\mathrm{ad}(x)$ nilpotent. By the Jacobson–Morozov theorem, $x$ would be contained in a subalgebra $\mathfrak{s}$ isomorphic to $\mathfrak{sl}(2, \mathbf{R})$ (see [17, pp. 100, 203, 79, 14]). Let $S \subset H$ be the corresponding Lie group, and let $\bar{S}$ be the closure of $S$ in $G$. Then evidently the adjoint action of $S$ on $\mathfrak{s}$ extends to a continuous action of $\bar{S}$ on $\mathfrak{s}$. In this way we obtain a continuous homomorphism from $\bar{S}$ onto the group $PSL(2, \mathbf{R})$ of automorphisms of $\mathfrak{s}$. Now choosing a discrete subgroup isomorphic to $\mathbf{Z} * \mathbf{Z}$ in $PSL(2, \mathbf{R})$, and lifting both generators to $\bar{S} \subset G$, we obtain a free noncyclic discrete subgroup of $G$, contradicting (b).

Hence we may assume that $\mathfrak{n} = 0$. It follows that the two vectors spaces $\mathfrak{a} \subset \mathfrak{p}$ have the same dimension, and hence are equal. Therefore $\mathfrak{a} = \mathfrak{p}$ is a commutative ideal in the semisimple Lie algebra $\mathfrak{h}$, hence $\mathfrak{a} = \mathfrak{p} = 0$. This means that $\mathfrak{h} = \mathfrak{k}$, so that the Lie group $H$, modulo its center, must be compact. By a theorem of Weyl [14, p. 123], this implies that $H$ itself is compact, proving (c).

Proof that (c) ⇒ (d). Let $N \subset G$ be the largest connected solvable normal subgroup. (See, for example, [8, p. 241].) By Levi's theorem, there exists a complementary semisimple subgroup $H \subset G$ (see [17, p. 91]). Then $H$ maps onto the identity component of the quotient $G/N$, so if $H$ is compact (or is contained in a compact subgroup of $G$) then it follows that $G/N$ is compact, proving (d), and completing the proof of 2.2. ∎

We need one further lemma, concerning a Lie group which acts affinely on Euclidean space.

Lemma 2.3. *If the Lie group $G$ is compact, or connected and semisimple, then any smooth representation of $G$ by affine transformations of $\mathbf{R}^n$ admits a fixed point.*

*Proof.* This is an immediate consequence of the complete reducibility theorem for linear representations of such groups (see, for example, [8, p. 246]). We identify the affine space $\mathbf{R}^n$ with the hyperplane $\mathbf{R}^n \times 1$ in $\mathbf{R}^{n+1}$. Clearly any representation of $G$ by affine transformations of $\mathbf{R}^n \times 1$ extends uniquely to a representation by linear transformations of $\mathbf{R}^{n+1}$. Since the linear subspace $\mathbf{R}^n \times 0$ is invariant, there must exist a complementary invariant linear subspace $L$. The intersection $L \cap (\mathbf{R}^n \times 1)$ is now the required fixed point. ∎

## 3. Are There Other Properly Discontinuous Affine Groups?

Let $M$ be a complete affinely flat manifold. Then one can conjecture that

(1) *the fundamental group* $\pi_1(M)$ *must be virtually polycyclic.*

If this is true, then it follows in particular that:

(2) *the group* $\pi_1(M)$ *must be finitely generated, and*

(3) *the Euler characteristic* $\chi(M) = \sum (-1)^i \operatorname{rank} H_i(M)$ *must be defined and equal to zero, except for the special case* $M \cong \mathbf{R}^n$, *with* $\chi(\mathbf{R}^n) = 1$.

The proof that (1) implies (2) and (3) is not difficult. (Compare [29, p. 106].)

Conversely, if (1) is false, then it follows from 2.1 that (2) and (3) are false also. For the group $\pi_1(M)$ certainly embeds discretely in the Lie group consisting of all affine transformations of $\tilde{M} \cong \mathbf{R}^n$. Hence, if $\pi_1(M)$ is not virtually polycyclic, then it must contain a subgroup $\Phi \cong \mathbf{Z} * \mathbf{Z}$ which is free on two generators. The quotient $M' = \Phi \backslash \mathbf{R}^n$ would then be a complete affinely flat manifold having the homotopy type of a figure 8. Its Euler characteristic $\chi(M')$ would be $-1$. The maximal Abelian covering manifold of $M'$ would have a fundamental group $[\Phi, \Phi] \cong \mathbf{Z} * \mathbf{Z} * \cdots$ which is not even finitely generated. ∎

I do not know whether such a manifold exists even in dimension 3. One could try to construct a Lorentz-flat example by starting with a discrete subgroup $\mathbf{Z} * \mathbf{Z} \subset O(2, 1)$, then adding translation components to obtain a group of isometries of Lorentz 3-space; but it seems difficult to decide whether the resulting group action is properly discontinuous.

If we consider affinely flat manifolds which are not complete, then the case $\pi_1(M) \supset \mathbf{Z} * \mathbf{Z}$ definitely can occur. As an example, the compact manifold

$$M = S^1 \times (\text{surface of genus 2})$$

admits an affinely flat structure having the open cone $x > (y^2 + z^2)^{1/2}$ in $\mathbf{R}^3$ as universal covering space. The proof (which I learned from M. Hirsch) is not difficult.

It would be interesting to know whether the Euler characteristic of such a compact, but not complete, flat manifold must be zero (Cf. [30]). If $M$ is compact *and* complete, then Kostant and Sullivan have shown that $\chi(M) = 0$. On the other hand, using a weaker concept of flatness, Smillie has constructed compact examples with $\chi(M) \neq 0$.

It would also be interesting to know whether a compact manifold with an area-preserving flat-affine structure is necessarily complete.

As a partial justification for conjecture (1) above, we can consider the corresponding situation for a connected Lie group $G$ which acts smoothly by affine transformations of $\mathbf{R}^n$. Such a group is said to act *properly* if for each compact set $K \subset \mathbf{R}^n$ the set of group elements $g$ with $gK \cap K \neq \varnothing$ is compact.

THEOREM 3.1. *If the connected Lie group $G$ acts properly by affine transformations of $\mathbf{R}^n$, then $G$ is amenable, hence every discrete subgroup is virtually polycyclic.*

*Proof.* Let $H$ be any semisimple subgroup of $G$. Then $H$ has a fixed point $p$ in $\mathbf{R}^n$ by Lemma 2.3. Since the action is proper, the subgroup consisting of all elements in $G$ which fix $p$ is compact. Thus $H$ is contained in a compact group, and it follows [14, p. 128] that $H$ itself is compact. Using Lemma 2.2, the conclusion follows. ∎

If we require that $G$ acts freely on $\mathbf{R}^n$, then the situation can be described even more precisely.

THEOREM 3.2. *A connected Lie group acts freely by affine transformations of some $\mathbf{R}^n$ if and only if it is simply connected and solvable.*

For, if $G$ admits such an action, then every semisimple subgroup is trivial by Lemma 2.3, hence $G$ is solvable. Furthermore, its maximal compact subgroup is also trivial by Lemma 2.3, so $G$ is simply connected.

Conversely, if $G$ is solvable, then using Ado's theorem we can embed a locally isomorphic group into $\mathrm{GL}(m, \mathbf{C})$, and hence into its Borel subgroup. The argument is then analogous to the proof of Theorem 1.1. ∎

## 4. The Compact Case

Let $\Gamma$ be torsion-free and virtually polycyclic of rank $k$. Then we have seen that $\Gamma$ is isomorphic to the fundamental group of a *complete* affinely flat manifold. Furthermore, Johnson has shown that $\Gamma$ is isomorphic to the fundamental group of a *compact* $k$-dimensional manifold, with universal covering $\tilde{M}$ diffeomorphic to $\mathbf{R}^k$. (See [4, 18]. This is proved by embedding $\Gamma$ as a discrete cocompact subgroup of a Lie group $G$ having finitely many components, and then letting $\Gamma$ operate on the coset space $G/K \cong \mathbf{R}^k$, where $K$ is a maximal compact subgroup.)

Now let us ask whether we can combine these two results, constructing a manifold $M$ which is both compact and complete affinely flat with $\pi_1(M) \cong \Gamma$. Many examples are known (Cf. Theorem 1.3, as well as [1, 5, 23]), but the general case seems quite difficult.

As a first step, we can ask a corresponding question for a $k$-dimensional Lie group.

*Does every solvable Lie group G admit a complete affinely flat structure which is invariant under left translation, or equivalently, does the universal covering group* $\tilde{G}$ *operate simply transitively by affine transformations of* $\mathbf{R}^k$?

If $G$ admits such a structure, then evidently, for any discrete subgroup $\Gamma$, the coset space $\Gamma\backslash G$ is a complete affinely flat manifold. In many cases, $\Gamma$ can be chosen so that $\Gamma\backslash G$ is also compact.

All that is known about this question is a list of a few special cases:

If the Lie algebra of $G$ admits a nonsingular derivation, then an affirmative answer has been given by Scheuneman [23]. Such a Lie algebra $\mathfrak{g}$ is necessarily nilpotent. (Compare 10, 11, 16.) As an example, if $\mathfrak{g}$ is graded by positive integers, with $[\mathfrak{g}_i, \mathfrak{g}_j] \subset \mathfrak{g}_{i+j}$, then a nonsingular derivation can be constructed by setting $Dx = jx$ for $x \in \mathfrak{g}_j$.

Similarly, if $G$ is 3-step nilpotent, then an affirmative answer has been given by Scheuneman [24].

Another interesting class of nilpotent examples has been studied by Auslander [3]. Let $A$ be an associative algebra, finite-dimensional over $\mathbf{R}$, which is nilpotent. Introducing the new associative product operation

$$a, b \mapsto a + b + ab,$$

we make the underlying flat manifold $A$ into a Lie group, denoted by $A^*$. Note that the given complete affinely flat structure on $A^*$ is invariant under both left and right translation. (In fact, this is the most general example of a simply connected Lie group with such a left and right invariant complete flat structure.)

Here is an explicit case. Suppose that $A$ is commutative, with basis $a, a^2, \ldots, a^k$ where $a^{k+1} = 0$. Then the Lie group $A^*$ is also commutative. Choosing any lattice $\Gamma \cong \mathbf{Z}^k$ in $A^*$, we can form the quotient $k$-torus $\Gamma\backslash A^*$, which is provided with an exotic complete affinely flat structure. (Compare [28].)

## Acknowledgments

I want to thank the University of Warwick for its hospitality during the course of this work, and M. Hirsch, J. Smillie, and W. Thurston for stimulating discussions.

## References

1. L. Auslander, Examples of locally affine spaces, *Ann. of Math.* **64** (1956), 255–259.
2. L. Auslander, On a problem of Phillip Hall, *Ann. of Math.* **86** (1967), 112–116.
3. L. Auslander, Simply transitive groups of affine motions, *Amer. J. Math.*, to appear.*
4. L. Auslander and F. E. A. Johnson, On a conjecture of C. T. C. Wall, *J. London Math. Soc.* **14** (1976), 331–332.

*[Updated reference] Volume **99** (1977) 809–826.

5. L. Auslander and L. Markus, Flat Lorentz 3-manifolds, *Mem. Amer. Math. Soc.* **30** (1959).
6. L. Bieberbach, Über die Bewegungsgruppen der Euklidischen Raume, *Math. Ann.* **70** (1911), 297–336. (Also **72** (1912), 400–412.)
7. A. Borel, "Linear Algebraic Groups," Benjamin, New York, 1969.
8. N. Bourbaki, *Elem. Math.* **37**; "Groupes et algèbres de Lie, II, III," Hermann, Paris, 1972.
9. J. Cheeger and D. Gromoll, On the structure of complete manifolds of nonnegative curvature, *Ann. of Math.* **96** (1972), 413–443.
10. J. Dixmier and W. G. Lister, Derivations of nilpotent Lie algebras, *Proc. Amer. Math. Soc.* **8** (1957), 155–158.
11. J. Dyer, A nilpotent Lie algebra with nilpotent automorphism group, *Bull. Amer. Math. Soc.* **76** (1970), 52–56.
12. H. Furstenberg, A Poisson formula for semi-simple Lie groups, *Ann. of Math.* **77** (1963), 335–386.
13. F. P. Greenleaf, Invariant Means on Topological Groups, *Van Nostrand Math. Studies* **16**, 1969.
14. S. Helgason, "Differential Geometry and Symmetric Spaces," Academic Press, New York, 1962.
15. J. E. Humphreys, "Linear Algebraic Groups," Springer Pub., New York, 1975.
16. N. Jacobson, A note on automorphisms and derivations of Lie algebras, *Proc. Amer. Math. Soc.* **6** (1955), 281–283.
17. N. Jacobson, "Lie Algebras," Interscience, New York, 1962.
18. F. E. A. Johnson, "On a Conjecture of C. T. C. Wall," mimeographed, 1974.
19. B. Kostant and D. Sullivan, The Euler characteristic of an affine space form is zero, *Bull. Amer. Math. Soc.* **81** (1975), 937–938.
20. J. Milnor, Hilbert's problem 18: on crystalographic groups, fundamental domains, and on sphere packing, *in* "Proceedings Symposia Pure Math.," Vol. 28, pp. 491–506, Amer. Math. Soc., Providence, R.I., 1976.
21. G. D. Mostow, On the fundamental group of homogeneous spaces, *Ann. of Math.* **66** (1957), 249–255.
22. N. W. Rickert, Amenable groups and groups with the fixed point property, *Trans. Amer. Math. Soc.* **127** (1967), 221–232.
23. J. Scheuneman, Examples of compact locally affine spaces, *Bull. Amer. Math. Soc.* **77** (1971), 589–592.
24. J. Scheuneman, Affine structures on three-step nilpotent Lie algebras, *Proc. Amer. Math. Soc.* **46** (1974), 451–454.
25. J. Smillie, "Flat Manifolds with Non-Zero Euler Characteristic," preprint, University of Chicago, 1976.*
26. R. Swan, Representations of polycyclic groups, *Proc. Amer. Math. Soc.* **18** (1967), 573–574.
27. J. Tits, Free subgroups in linear groups, *J. Algebra* **20** (1972), 250–270.
28. A. T. Vasquez, to appear.
29. J. Wolf, "Spaces of Constant Curvature," McGraw-Hill, New York, 1967, and Publish or Perish, Boston, Mass., 1974.
30. M. Hirsch and W. Thurston, Foliated bundles, invariant measures and flat manifolds, *Ann. of Math.* **101** (1975), 369–390.

*[Updated reference] Comment. Math. Helv. **52** (1977) 453–455.

# Algebraic Geometry and Topology

# ON THE BETTI NUMBERS OF REAL VARIETIES

*Reprinted from*

Proceedings of the
American Mathematical Society
Volume 15 (1964), pp. 275–280

This note gives upper bounds for the topological complexity of a real algebraic variety in terms of the degrees of the defining equations. Similar results were obtained by Thom[1] at about the same time. More recently, these results have been substantially sharpened by Khovansky,[2] and also by Risler[3] et al.,[4] with a view towards applications in theoretical computer science.

---

[1] R. Thom, *Sur l'homologie des variétés réelles*, pp. 255–265 of "Differential and Combinatorial Topology, A Symposium in Honor of Marston Morse", Cairns ed., Princeton U. Press 1965.

[2] Compare J.-J. Risler, *Complexité et géométrie réelle (d'apres A. Khovansky)*, Sem. Bourbaki 1984–85 n$^o$ 637, Astérisque **133–34** (1986) 89–100.

[3] J.-J. Risler, *Some aspects of complexity in real algebraic geometry*, J. Symb. Comp. **5** (1988) 109–119.

[4] R. Benedetti, F. Loeser and J.-J. Risler, *Two bounds for the number of connected components of a real algebraic set*, pp. 22–35 of "Real Analytic and Algebraic Geometry, Trento 1988", Lecture Notes Math **1420**, Springer 1990; and also *Bounding the number of connected components of a real algebraic set*, Discr. & Comp. Geom. **6** (1991) 191–209.

# ON THE BETTI NUMBERS OF REAL VARIETIES

J. MILNOR

The object of this note will be to give an upper bound for the sum of the Betti numbers of a real affine algebraic variety. (*Added in proof*. Similar results have been obtained by R. Thom [10].)

Let $V$ be a variety in the real Cartesian space $\boldsymbol{R}^m$, defined by polynomial equations

$$f_1(x_1, \cdots, x_m) = 0, \cdots, f_p(x_1, \cdots, x_m) = 0.$$

The $q$th *Betti number* of $V$ will mean the rank of the Čech cohomology group $H^q(V)$, using coefficients in some fixed field $F$.

THEOREM 2. *If each polynomial* $f_i$ *has degree* $\leqq k$, *then the sum of the Betti numbers of* $V$ *is* $\leqq k(2k-1)^{m-1}$.

Analogous statements for complex and/or projective varieties will be given at the end.

I wish to thank W. May for suggesting this problem to me.

REMARK A. This is certainly not a best possible estimate. (Compare Remark B.) In the examples which I know, the sum of the Betti numbers of $V$ is always $\leqq k^m$. Consider for example the $m$ polynomials

$$f_i(x_1, \cdots, x_m) = (x_i - 1)(x_i - 2) \cdots (x_i - k),$$

where $i = 1, 2, \cdots, m$. These define a zero-dimensional variety, consisting of precisely $k^m$ points.

The proofs of Theorem 2 and of Theorem 1 (which will be stated later) depend on the following.

LEMMA 1. *Let* $V_0 \subset \boldsymbol{R}^m$ *be a zero-dimensional variety defined by polynomial equations* $f_1 = 0, \cdots, f_m = 0$. *Suppose that the gradient vectors* $df_1, \cdots, df_m$ *are linearly independent at each point of* $V_0$. *Then the number of points in* $V_0$ *is at most equal to the product* $(\deg f_1)(\deg f_2) \cdots (\deg f_m)$.

PROOF. Approximate $f_1, \cdots, f_m$ by real polynomials $F_1, \cdots, F_m$ of the same degrees whose coefficients are algebraically independent. Now consider the variety $V_C$ in the complex Cartesian space $\boldsymbol{C}^m$ defined by the equations $F_1 = 0, \cdots, F_m = 0$. It follows from van der Waerden [9, §41] that the number of points in $V_C$ is equal to $(\deg f_1)(\deg f_2) \cdots (\deg f_m)$. Since each point of $V_0$ lies close to some real point of $V_C$; this proves Lemma 1.

Received by the editors January 17, 1963.

Next let $W$ be a compact, nonsingular hypersurface in $\boldsymbol{R}^m$, defined by the equation $f=0$, where $f$ is a polynomial of degree $2k$.

THEOREM 1. *The sum of the Betti numbers of $W$ is $\leqq 2k(2k-1)^{m-1}$.*

Similar estimates have been given by Olneĭnik [5].

The proof follows. Consider the function $\boldsymbol{n}: W \to S^{m-1}$ which assigns to each point of $W$ its unit normal vector $\operatorname{grad} f / |\operatorname{grad} f|$. According to a theorem of Sard, the set of critical values of $\boldsymbol{n}$ has measure zero in $S^{m-1}$. (See [7] or [6, p. 10].) Hence, after rotating the coordinate system if necessary, we may assume that the two points $(0, \cdots, 0, \pm 1)$ of $S^{m-1}$ are not critical values of $\boldsymbol{n}$. In terms of local coordinates $u_1, \cdots, u_{m-1}$ on $W$, this means that the matrix $(\partial \boldsymbol{n}_i / \partial u_j)_{i,j<m}$ is nonsingular whenever $\boldsymbol{n} = (0, \cdots, 0, \pm 1)$.

*Now note that the "height function" $(x_1, \cdots, x_m) \to x_m$ on $W$ has no degenerate critical points.* In fact near each critical point of this function we can choose local coordinates $u_1, \cdots, u_{m-1}$ so that

$$x_1 = u_1, \cdots, u_{m-1} = u_{m-1}, \qquad x_m = h(u_1, \cdots, u_{m-1}),$$

where $h$ is the height function. Then

$$\boldsymbol{n}(u_1, \cdots, u_{m-1}) = \pm (\partial h/\partial u_1, \cdots, \partial h/\partial u_{m-1}, -1)/\sqrt{(\textstyle\sum (\partial h/\partial u_j)^2 + 1)};$$

hence

$$\partial \boldsymbol{n}_i / \partial u_j = \pm \partial^2 h / \partial u_i \partial u_j$$

at each critical point. Therefore the matrix $(\partial^2 h / \partial u_i \partial u_j)$ is nonsingular; as was asserted.

Now applying Morse theory to the height function $h: W \to \boldsymbol{R}$, it follows that *the sum of the Betti numbers of $W$ is less than or equal to the number of critical points of $h$.* (See for example Seifert and Threlfall [8, §§5 and 8].)

The critical points of $h$ can be characterized as the solutions of the $m$ polynomial equations

$$\partial f/\partial x_1 = 0, \cdots, \partial f/\partial x_{m-1} = 0, \qquad f = 0.$$

Note that at each critical point of $h$ the $m$ gradient vectors

$$d(\partial f/\partial x_1), \cdots, d(\partial f/\partial x_{m-1}), df$$

are linearly independent.

PROOF. If we differentiate the identity

$$f(u_1, \cdots, u_{m-1}, h(u_1, \cdots, u_{m-1})) = 0$$

twice, and then evaluate at a critical point, we obtain

$$\partial^2 f/\partial x_i \partial x_j + (\partial f/\partial x_m)(\partial^2 h/\partial u_i \partial u_j) = 0;$$

which shows that the matrix $(\partial^2 f/\partial x_i \partial x_j)_{i,j<m}$ is nonsingular. From this the linear independence of the $m$ gradient vectors follows easily.

Now Lemma 1 asserts that the number of critical points of $h$ is less than or equal to the product

$$(\deg \partial f/\partial x_1) \cdots (\deg \partial f/\partial x_{m-1})(\deg f) = (2k-1)^{m-1}\, 2k.$$

Therefore the sum of the Betti numbers of $W$ is less than or equal to this number. This proves Theorem 1.

REMARK B. Using a more complicated argument, which applies Morse theory directly to the function $f$, one can prove the following slightly sharper inequality:

$$\text{rank } H^*W \leqq (2k-1)^m + 1.$$

Using this, Theorem 2 can be sharpened to the form

$$\text{rank } H^*V \leqq \tfrac{1}{2}((2k-1)^m + 1).$$

PROOF OF THEOREM 2. Let $rD^m$ denote the disk of radius $r$ in $\boldsymbol{R}^m$. For each $r$ we will show that $H^*(V \cap rD^m)$ has rank $\leqq k(2k-1)^{m-1}$.

Given numbers $\epsilon$, $\delta > 0$ let $K(\epsilon, \delta)$ denote the set of points in $\boldsymbol{R}^m$ for which

$$f_1^2 + \cdots + f_p^2 + \epsilon^2(x_1^2 + \cdots + x_m^2) \leqq \delta^2.$$

This is a compact set, since it is contained in the disk of radius $\delta/\epsilon$. If $r \leqq \delta/\epsilon$ note that $V \cap rD^m \subset K(\epsilon, \delta)$. The set of points in $\boldsymbol{R}^m$ for which the equality

$$f_1^2 + \cdots + f_p^2 + \epsilon^2 | x |^2 = \delta^2$$

holds will be denoted by $\partial K(\epsilon, \delta)$, and called the boundary of $K$. This is an algebraic hypersurface of degree $2k$, where $k = \text{Max}\{\deg f_1, \cdots, \deg f_p\} \geqq 1$. This hypersurface is nonsingular if and only if $\delta^2$ is a regular value of the function $f_1^2 + \cdots + f_p^2 + \epsilon^2 |x|^2$. Hence, keeping $\epsilon$ fixed, it follows from Sard's theorem that this hypersurface is nonsingular for almost all values of $\delta$. We will assume henceforth that $\epsilon$ and $\delta$ are chosen so that this is the case.

Applying Theorem 1 to $\partial K$ we see that $H^*(\partial K)$ has rank $\leqq 2k(2k-1)^{m-1}$. But the Alexander duality theorem implies that

$$\text{rank } H^*(K) = \tfrac{1}{2} \text{ rank } H^*(\partial K).$$

(See for example [3, p. 279].) Hence

$$\text{rank } H^*K \leqq k(2k-1)^{m-1}.$$

Now choose sequences $\{\epsilon_i\}$ and $\{\delta_i\}$ so that $\{\epsilon_i\}$ decreases monotonely with limit zero; so that $\{\delta_i/\epsilon_i\}$ decreases monotonely with limit $r$; and so that each $\partial K(\epsilon_i, \delta_i)$ is nonsingular. Then

$$K(\epsilon_1, \delta_1) \supset K(\epsilon_2, \delta_2) \supset K(\epsilon_3, \delta_3) \supset \cdots$$

with intersection $V \cap rD^m$. (These inclusion relations become clear if we put the defining inequality for $K(\epsilon, \delta)$ in the form

$$\sum f_j^2/\delta^2 + |x|^2 \epsilon^2/\delta^2 \leqq 1,$$

and note that $\delta_{i+1} \leqq \delta_i \epsilon_{i+1}/\epsilon_i \leqq \delta_i$.) Therefore the Čech cohomology group $H^*(V \cap rD^m)$ is equal to the direct limit of the groups $H^*K(\epsilon_i, \delta_i)$. (Compare Eilenberg and Steenrod [1, p. 261].) Hence

$$\text{rank } H^*(V \cap rD^m) \leqq \lim\sup(\text{rank } H^*K(\epsilon_i, \delta_i)) \leqq k(2k-1)^{m-1}.$$

There remains one small but tricky point: the passage to the limit as $r \to \infty$. One approach is to make use of the theorem that $V$ can be triangulated. (See for example Lefschetz and Whitehead [2].) It follows that the Betti numbers of $V$ based on singular homology theory are the same as those based on Čech cohomology theory. But the singular homology group $H_*(V)$ is equal to the direct limit of the singular homology groups of the compact subsets[1] of $V$. If $P \subset V \cap rD^m \subset V$ is any compact sub-polyhedron, then the restriction homomorphism $H^*V \to H^*P$ has rank $\leqq k(2k-1)^{m-1}$; hence the corresponding homology homomorphism $H_*P \to H_*V$ also has rank $\leqq k(2k-1)^{m-1}$. Therefore the limit $H_*V$ has rank $\leqq k(2k-1)^{m-1}$; which completes the proof.

An alternative approach which does not assume triangulability can be sketched as follows. By an argument similar to that in [4, p. 338] one constructs an exact sequence

$$0 \to \mathcal{L}'H^{q-1}(V \cap rD^m) \to H^qV \to \mathcal{L}H^q(V \cap rD^m) \to 0,$$

where $\mathcal{L}$ denotes the inverse limit functor, and $\mathcal{L}'$ denotes its first derived functor. Since each $H^{q-1}(V \cap rD^m)$ is a finite-dimensional

---

[1] In contrast note that the Čech homology group of a space need not be the direct limit of the Čech homology groups of its compact subsets. A counter-example is provided by the union of the curve $y = \sin(1/x)/x$ and the $y$-axis.

vector space, an argument similar to that used by Eilenberg and Steenrod in [1, p. 226] shows that the $\mathcal{L}'$ term is zero. Therefore

$$\operatorname{rank} H^*V = \operatorname{rank} (\mathcal{L}H^*(V \cap rD^m)) \leqq k(2k - 1)^{m-1},$$

which completes the alternative proof.

As corollaries one obtains similar estimates, presumably rather crude, for complex or projective varieties.

COROLLARY 1 (THE COMPLEX AFFINE CASE). *If* $V_1 \subset \mathbf{C}^m$ *is defined by polynomial equations of degree* $\leqq k$, *then* $\operatorname{rank} H^* V_1 \leqq k(2k-1)^{2m-1}$.

COROLLARY 2 (THE COMPLEX PROJECTIVE CASE). *If* $V_2 \subset \mathbf{P}_m(\mathbf{C})$ *is defined by homogeneous polynomial equations of degree* $\leqq k$, *then* $\operatorname{rank} H^* V_2 \leqq mk(2k-1)^{2m+1}$.

COROLLARY 3 (THE REAL PROJECTIVE CASE). *If* $V_3 \subset \mathbf{P}_m(\mathbf{R})$ *is defined by homogeneous polynomial equations of degree* $\leqq k$, *then* $\operatorname{rank} H^* V_3 \leqq mk(2k-1)^m$. *Here the initial factor* $m$ *in this expression is needed only in the special case when the coefficient field* $F$ *has characteristic* 2.

In order to prove Corollary 1 it is only necessary to think of $V_1$ as a real variety in $\mathbf{R}^{2m}$. To prove Corollary 2 note that the same homogeneous equations can be used to define an affine variety $V_2' \subset \mathbf{C}^{m+1}$. The intersection $V_2' \cap S^{2m+1}$ is the total space of a circle bundle over $V_2$. Using the Gysin sequence of this bundle, one arrives at the given estimate. Corollary 3 is proved in the same way. In this case $V_3' \cap S^m$ is a 2-fold covering of $V_3$; so that the prime 2 plays a special role.

*Added in proof.* A rather different problem arises when one considers a locus which is defined by polynomial inequalities. In this case any estimate must depend not only on the degrees of the inequalities but also upon their number. As an example the $p$ quadratic inequalities

$$(x-1)^2 \geqq \frac{1}{10}, \qquad (x-2)^2 \geqq \frac{1}{10}, \cdots, (x-p)^2 \geqq \frac{1}{10}$$

on the real line define a locus with $p+1$ components.

THEOREM 3. *If* $X \subset \mathbf{R}^m$ *is defined by polynomial inequalities of the form*

$$f_1 \geqq 0, \cdots, f_p \geqq 0$$

*with total degree* $d = \deg f_1 + \cdots + \deg f_p$, *then*

$$\operatorname{rank} H^*X \leqq \tfrac{1}{2}(2+d)(1+d)^{m-1}.$$

PROOF. Again we first replace $X$ by $X \cap rD^m$. This can be done by simply adjoining the additional inequality

$$f_0 = r^2 - |x|^2 \geqq 0.$$

Given numbers $\epsilon \geqq \epsilon^{p+1} \geqq \delta > 0$, let $L(\epsilon, \delta)$ denote the set of points in for which

$$f_0 + \epsilon \geqq 0, \cdots, f_p + \epsilon \geqq 0$$

and

$$(f_0 + \epsilon)(f_1 + \epsilon) \cdots (f_p + \epsilon) - \delta \geqq 0.$$

This set is compact. Its boundary is obtained by setting only the last expression equal to zero.

Given $\epsilon$ we may assume that $\delta$ is chosen so that the boundary is nonsingular. Therefore the proof (if not the exact statement) of Theorem 1 shows that

$$\operatorname{rank} H^*(\partial L(\epsilon, \delta)) \leqq (2 + d)(1 + d)^{m-1};$$

and hence that

$$\operatorname{rank} H^* L(\epsilon, \delta) \leqq \tfrac{1}{2}(2 + d)(1 + d)^{m-1}.$$

Now choose numbers $\epsilon_i$ tending monotonely to zero and choose $0 < \delta_i < (\epsilon_i - \epsilon_{i+1})^{p+1}$; so that $L(\epsilon_1, \delta_1) \supset L(\epsilon_2, \delta_2) \supset \cdots$ with intersection $X \cap rD^m$. As in the proof of Theorem 2, it follows that

$$\operatorname{rank} H^* X \leqq \limsup (\operatorname{rank} H^*(X \cap rD^m)) \leqq \tfrac{1}{2}(2 + d)(1 + d)^{m-1}.$$

## REFERENCES

1 S. Eilenberg and N. Steenrod, *Foundations of algebraic topology*, Princeton Univ. Press, Princeton, N. J., 1952.

2. S. Lefschetz and J. H. C. Whitehead, *On analytical complexes*, Trans. Amer. Math. Soc. **35** (1933), 510–517.

3. J. Milnor, *On the immersion of n-manifolds in (n+1)-space*, Comment. Math. Helv. **30** (1956), 275–284.

4. ———, *On axiomatic homology theory*, Pacific J. Math. **12** (1962), 337–341.

5. O. A. Oleĭnik, *Estimates of the Betti numbers of real algebraic hypersurfaces*, Mat. Sb. (N.S.) **28** (70) (1951), 635–640. (Russian)

6. G. de Rham, *Variétés différentiables*, Hermann, Paris, 1955.

7. A. Sard, *The measure of the critical values of differentiable maps*, Bull. Amer. Math. Soc. **48** (1942), 883–897.

8. H. Seifert and W. Threlfall, *Variationsrechnung im Grossen*, Teubner, Berlin, 1938.

9. B. van der Waerden, *Einführung in die algebraische Geometrie*, Spinger, Berlin, 1939.

10. R. Thom, *Sur l'homologie des variétés algébriques réelles*, (to appear).

PRINCETON UNIVERSITY

# ISOLATED SINGULARITIES DEFINED BY WEIGHTED HOMOGENEOUS POLYNOMIALS

**(*with Peter Orlik*)**

*Reprinted from*

Topology

Volume 9 (1970), pp. 385–393

A complex polynomial $f(z_1, \dots, z_m)$ is called ***weighted homogeneous*** if there are positive integers $k_i$ and $k$ so that

$$f(t^{k_1} z_1, \dots, t^{k_m} z_m) = t^k f(z_1, \dots, z_m).$$

An equivalent condition would be that the multiplicative group $\mathbf{C}^* = \mathbf{C} \setminus \{0\}$ acts on the hypersurface $V \subset \mathbf{C}^m$ defined by the equation $f = 0$, by an action of the form[1]

$$t \cdot (z_1, \dots, z_m) = (t^{k_1} z_1, \dots, t^{k_m} z_m).$$

Now suppose that this variety $V$ has an isolated singularity at the origin.[2] This note shows that the essential topological features of such a singularity are uniquely and effectively determined by the rational numbers $w_i = k/k_i$.

---

[1] Compare P. Orlik and P. Wagreich, *Isolated singularities of algebraic surfaces with* **C*** *action*, Annals of Math. **93** (1971) 205–228.

[2] One important example is the variety $z_1^{n_1} + \cdots + z_m^{n_m} = 0$ studied by Pham and Brieskorn, with $n_i \geq 2$. In this case we can take $k$ to be the least common multiple of the $n_i$ and $k_i = k/n_i$. Compare Milnor, *On the 3-dimensional Brieskorn manifolds* $M(p, q, r)$, pp. 175–225 of "Knots, Groups and 3-Manifolds, papers dedicated to the memory of R. H. Fox", L. Neuwirth, ed., Annals of Math. Studies **84**, Princeton U. Press 1975.

# ISOLATED SINGULARITIES DEFINED BY WEIGHTED HOMOGENEOUS POLYNOMIALS

JOHN MILNOR and PETER ORLIK†

(*Received* 20 *October* 1969)

## §1. INTRODUCTION

THE OBJECT of this paper is to give simple explicit formulas for the multiplicity $\mu$ and the characteristic polynomial $\Delta(t)$ associated with a weighted homogeneous polynomial in several complex variables, having an isolated critical point at the origin.

Let $\mathbf{C}^m$ denote complex $m$-space and $f(\mathbf{z}) = f(z_1, z_2, \ldots, z_m)$ a polynomial function in $m$ complex variables. The locus of zeros of $f(\mathbf{z})$

$$V = \{\mathbf{z} \in \mathbf{C}^m \mid f(\mathbf{z}) = 0\}$$

is called the associated *algebraic variety* (hypersurface).

The topology of $V$ at a point $\mathbf{x}$ was studied by Milnor [2]. Intersect $V$ with a small sphere $S_\varepsilon = S_\varepsilon^{2m-1}$ centered at $\mathbf{x}$ and consider the set

$$K = V \cap S_\varepsilon .$$

It is shown [2, § 4.8] that the mapping $\Phi(\mathbf{z}) = f(\mathbf{z})/|f(\mathbf{z})|$ from $S_\varepsilon - K$ to the unit circle is the projection map of a smooth fiber bundle. Each fiber $F_\theta = \Phi^{-1}(e^{i\theta}) \subset S_\varepsilon - K$ is a smooth, parallelizable $(2m-2)$-manifold.

The point $\mathbf{x}$ is *singular* if all partial derivatives $\partial f/\partial z_i$ vanish at $\mathbf{x}$, i.e. if $f$ has $\mathbf{x}$ as critical point. Otherwise $\mathbf{x}$ is a *simple* point. A singularity is called *isolated* if in some neighborhood of $\mathbf{x}$ all other points are simple. If $\mathbf{x}$ is an isolated singularity of $V$, then $K$ is a closed $(m-3)$-connected $(2m-3)$-manifold and $F_\theta$ has the homotopy type of a bouquet of $(m-1)$-spheres. Their number, $\mu$, is strictly positive [2, §6.5 and §7.1].

We obtain $S_\varepsilon - K$ from $F \times [0, 2\pi]$ by identifying $F \times 0$ and $F \times 2\pi$ by a homeomorphism $h: F \to F$ called the characteristic map. Since $F$ has the homotopy type of $S_1^{m-1} \vee \cdots \vee S_\mu^{m-1}$ the induced map

$$h_*: H_{m-1}(F; \mathbf{C}) \to H_{m-1}(F; \mathbf{C})$$

is of particular interest. Denote its characteristic polynomial by

$$\Delta(t) = \det(tI_* - h_*),$$

where $I$ is the identity map of $F$.

† Partially supported by NSF Grant GP-13285.

Brieskorn [1] and Pham [4] computed $\mu$ and $\Delta(t)$ for varieties defined by polynomials of the form

$$f(z) = z_1^{a_1} + z_2^{a_2} + \cdots + z_m^{a_m}.$$

The aim of this paper is to find $\mu$ and $\Delta(t)$ for the larger class of polynomials defined below.

Let $w_1, w_2, \ldots, w_m$ be positive rational numbers. The polynomial $f(z_1, z_2, \ldots, z_m)$ is *weighted homogeneous* of type $(w_1, w_2, \ldots, w_m)$ if it can be expressed as a linear combination of monomials $z_1^{i_1} z_2^{i_2} \cdots z_m^{i_m}$ for which

$$i_1/w_1 + i_2/w_2 + \cdots + i_m/w_m = 1.$$

*Remark.* Clearly every weighted homogeneous polynomial must vanish at the origin. If the origin is a simple point of $f^{-1}(0)$, then some $w_i$ must equal 1. Otherwise, if the origin is a critical point, it is clear that $w_1 > 1, \ldots, w_m > 1$.

THEOREM 1. *Let $f(z_1, \ldots, z_m)$ be a weighted homogeneous polynomial of type $(w_1, \ldots, w_m)$ having an isolated critical point at the origin. Then $H_{m-1}(F_\theta; \mathbf{Z})$ is free abelian of rank*

$$\mu = (w_1 - 1)(w_2 - 1) \cdots (w_m - 1).$$

Thus this product is an integer, even though the $w_i$ themselves may not be integers.

For a classification of those weighted homogeneous polynomials which have an isolated critical point at the origin, we refer to Orlik and Wagreich [3].

In order to state our result on the characteristic polynomial $\Delta(t)$ we need the following definitions.

To each monic polynomial $(t - \alpha_1)(t - \alpha_2) \cdots (t - \alpha_k)$ with $\alpha_1, \alpha_2, \ldots, \alpha_k \in \mathbf{C}^*$, the multiplicative group of nonzero complex numbers, assign the divisor

$$\operatorname{divisor}((t - \alpha_1) \cdots (t - \alpha_k)) = \langle \alpha_1 \rangle + \cdots + \langle \alpha_k \rangle$$

thought of as an element of the integral group ring $\mathbf{ZC}^*$. Introduce the special notation

$$\Lambda_n = \operatorname{divisor}(t^n - 1) = \langle 1 \rangle + \langle \xi \rangle + \cdots + \langle \xi^{n-1} \rangle$$

where $\xi = \exp(2\pi i/n)$. It will also be convenient to introduce the idempotent element

$$E_n = n^{-1} \Lambda_n$$

which belongs to the rational group ring $\mathbf{QC}^*$. The ring identity element $\Lambda_1 = E_1 = \langle 1 \rangle$ will be written briefly as 1.

Given the integers $a_1, a_2, \ldots, a_k$ denote by $[a_1, \ldots, a_k]$ their least common multiple and by $(a_1, \ldots, a_k)$ their greatest common divisor. Note the multiplication rules

$$\Lambda_a \Lambda_b = (a, b) \Lambda_{[a,b]} \quad \text{and} \quad E_a E_b = E_{[a,b]}.$$

THEOREM 4. *Let $f(z_1, \ldots, z_m)$ be a weighted homogeneous polynomial of type $(w_1, \ldots, w_m)$ having an isolated critical point at the origin. Express the weights in irreducible form as $w_i = u_i/v_i$, $i = 1, 2, \ldots, m$. Then the characteristic polynomial $\Delta(t)$ of the linear transformation*

$$h_*: H_{m-1}(F; \mathbf{C}) \to H_{m-1}(F; \mathbf{C})$$

*is determined by*

$$\text{divisor } \Delta = (w_1 E_{u_1} - 1)(w_2 E_{u_2} - 1) \cdots (w_m E_{u_m} - 1)$$
$$= (v_1^{-1} \Lambda_{u_1} - 1)(v_2{}^{-1}\Lambda_{u_2} - 1) \cdots (v_m{}^{-1}\Lambda_{u_m} - 1).$$

In Section 4 this result will be used to describe the homology of $K$.

## §2. THE RANK OF $H_{m-1}(F_\theta; Z)$

We interpret $\mu$ as in [2, §7] as the local degree at $\mathbf{x}$ of the mapping

$$f' : (\mathbf{C}^m, \mathbf{x}) \to (\mathbf{C}^m, \mathbf{0})$$

defined by

$$(z_1, \ldots, z_m) \mapsto (\partial f/\partial z_1, \ldots, \partial f/\partial z_m).$$

LEMMA 1. *If a polynomial mapping* $G: \mathbf{C}^m \to \mathbf{C}^m$ *is such that the i-th component of* $G(\mathbf{z})$ *is a homogeneous polynomial of degree* $d_i$, *and if* $G^{-1}(\mathbf{0}) = \mathbf{0}$ *then the local degree of* $G$ *at* $\mathbf{0}$ *is equal to* $d_1 d_2 \cdots d_m$.

*Proof.* The hypersurfaces $G_i{}^{-1}(\mathbf{0}) \subset \mathbf{C}^m$ extend to projective hypersurfaces $H_i \subset P^m(\mathbf{C})$ of the same degrees. The total intersection multiplicity of $H_1, \ldots, H_m$ is equal to the product $d_1 \cdots d_m$ of degrees. But the intersection $H_1 \cap \cdots \cap H_m$ clearly consists of the single point $\mathbf{0} \in \mathbf{C}^m \subset P^m(\mathbf{C})$.

THEOREM 1. *Let* $f(z_1, \ldots, z_m)$ *be a weighted homogeneous polynomial of type* $(w_1, \ldots, w_m)$ *having an isolated critical point at the origin. Then* $H_{m-1}(F_\theta; \mathbf{Z})$ *is free abelian of rank*

$$\mu = (w_1 - 1)(w_2 - 1) \cdots (w_m - 1).$$

*Proof.* Express each weight as an irreducible fraction $w_i = u_i/v_i$, $i = 1, 2, \ldots, m$. Let $d = [u_1, \ldots, u_m]$ denote the least common multiple of the numerators and define the integers $q_i = d/w_i$, $i = 1, 2, \ldots, m$. We introduce the auxiliary map

$$G(z_1, \ldots, z_m) = (z_1{}^{q_1}, \ldots, z_m{}^{q_m})$$

so that $f \circ G$ is homogeneous of degree $d$. Then the $i$-th component of $f' \circ G$ is a homogeneous polynomial of degree $d - q_i$. So the local degree of $f' \circ G$ at $\mathbf{0}$ is $(d - q_1)(d - q_2) \cdots (d - q_m)$. Dividing by the local degree $q_1 q_2 \cdots q_m$ of $G$ at $\mathbf{0}$, we see that the local degree $\mu$ of $f'$ is equal to

$$\frac{d - q_1}{q_1} \cdot \frac{d - q_2}{q_2} \cdots \frac{d - q_m}{q_m} = (w_1 - 1)(w_2 - 1) \cdots (w_m - 1)$$

as required.

## §3. THE CHARACTERISTIC POLYNOMIAL

Let

$$\Delta(t) = \det(tI_* - h_*)$$

denote the characteristic polynomial of the linear transformation

$$h_* : H_{m-1}(F; \mathbf{C}) \to H_{m-1}(F; \mathbf{C})$$

defined in Section 1. Assume that $V$ is defined by a weighted homogeneous polynomial having an isolated critical point at $\mathbf{0}$. It is proved in [2, §9.4] that the fiber $F$ of the fibration $F \to S_\varepsilon - K \to S^1$ is diffeomorphic to the non-singular hypersurface

$$F' = \{\mathbf{z} \in \mathbf{C}^m \mid f(\mathbf{z}) = 1\}.$$

As characteristic homeomorphism from $F'$ to itself one can choose the periodic unitary transformation

$$h(z_1, z_2, \ldots, z_m) = (\exp(2\pi i/w_1)z_1, \exp(2\pi i/w_2)z_2, \ldots, \exp(2\pi i/w_m)z_m).$$

The characteristic polynomial of $h$ is computed by Milnor [2, §9.5 and 9.6] as follows. Let $h^j : F' \to F'$ denote the composition of $h$ with itself $j$ times. Denote by $\chi_j$ the euler number of the fixed point manifold of $h^j$. Define a sequence of integers $s_1, s_2, s_3, \ldots$ by the formula $\chi_n = \sum_{j|n} s_j$. It is not difficult to show that the numbers $s_j$ are almost all zero. Furthermore it is shown in [2] that $s_j$ is always divisible by $j$.

THEOREM [2, §9.6]. *The euler numbers $\chi_j$ of the fixed point manifolds of the various iterates $h^j : F' \to F'$ are related to the characteristic polynomial $\Delta(t)$ of $h_*$ by the formula*

$$\Delta(t)^{(-1)^m} = (t-1)/\prod_j (t^j - 1)^{s_j/j}$$

*where* $\chi_n = \sum_{j|n} s_j$.

In other words

$$\begin{aligned}(-1)^m \text{ divisor } \Delta &= 1 - \textstyle\sum (s_j/j)\Lambda_j \\ &= 1 - \textstyle\sum s_j E_j .\end{aligned}$$

The proof in [2] is based on the Weil zeta function of the mapping $h$.

Using Theorem 1 we can now give a more explicit expression in terms of the weights $w_1, \ldots, w_m$, where $w_i = u_i/v_i$, in irreducible form. Define a sequence $\chi_1, \chi_2, \ldots$ of numbers by the formula

(1) $$1 - \chi_n = \prod \{1 - w_i \mid n \equiv 0 \bmod u_i\}.$$

(By definition $\chi_n = 0$ if there is no $w_i$ with $n \equiv 0 \bmod u_i$.) Define the numbers $s_1, s_2, \ldots$ by the formula

(2) $$\chi_n = \sum_{j|n} s_j .$$

THEOREM 2. *The numbers $s_1, s_2, \ldots$ defined above are integers, almost all zero, and*

$$\text{divisor } \Delta = (-1)^m(1 - \textstyle\sum s_j E_j).$$

*Proof.* It is sufficient to show that $\chi_n$, as defined by (1) is equal to the euler number of the variety $F(n)$ defined by the equations

$$\begin{aligned} f(z_1, z_2, \ldots, z_m) &= 1 \\ \exp(2\pi i n/w_1)z_1 &= z_1 \\ &\vdots \\ \exp(2\pi i n/w_m)z_m &= z_m . \end{aligned}$$

Renumbering the variables, if necessary, we may assume that

$$n \equiv 0 \bmod u_i \text{ for } i \leq k.$$
$$n \not\equiv 0 \bmod u_i \text{ for } i > k.$$

Then the last equations become $z_{k+1} = \cdots = z_m = 0$; hence $F(n) = \{(z_1, \ldots, z_k) \mid f(z_1, \ldots, z_k, 0, \ldots, 0) = 1\}$. Note that the polynomial $f(z_1, \ldots, z_k, 0, \ldots, 0)$ is weighted homogeneous in $k$ variables. It has no critical points other than the origin. For by hypothesis, the gradient of $f(z_1, \ldots, z_m)$ at any point $(z_1, \ldots, z_k, 0, \ldots, 0) \neq \mathbf{0}$ is non-zero; and this gradient vector, being invariant under the linear transformation

$$h^n(z_1, \ldots, z_m) = (z_1, \ldots, z_k, \exp(2\pi in/w_{k+1})z_{k+1}, \ldots, \exp(2\pi in/w_m)z_m),$$

must be tangent to the plane $z_{k+1} = \cdots = z_m = 0$. (Compare [2, p. 22].) Thus, thinking of $f(z_1, \ldots, z_k, 0, \ldots, 0)$ as a weighted homogeneous polynomial in $k$ variables, the associated fiber in $\mathbf{C}^k$ is precisely $F(n)$.

By Theorem 1 therefore

$$\mu(F(n)) = (w_1 - 1) \cdots (w_k - 1)$$

and hence

$$\chi(F(n)) = 1 + (-1)^{k-1}\mu(F(n)) = 1 - (1 - w_1) \cdots (1 - w_k).$$

Evidently, this completes the proof.

Given a completely arbitrary $m$-tuple $(w_1, \ldots, w_m)$ of positive rationals, define rational numbers $\chi_j(w_1, \ldots, w_m)$ and $s_j(w_1, \ldots, w_m)$ by the formulas (1) and (2) above. In addition, define the element

$$\delta(w_1, \ldots, w_m) \in \mathbf{QC}^*$$

by the formula

(3) $$\delta = (-1)^m(1 - \sum_{j \geq 1} s_j(w_1, \ldots, w_m)E_j).$$

Thus, if $(w_1, \ldots, w_m)$ is the $m$-tuple of weights of Theorem 2, then $\delta(w_1, \ldots, w_m) = \text{divisor } \Delta$.

THEOREM 3. *The identity*

$$\delta(w_1, \ldots, w_m)\delta(w_{m+1}, \ldots, w_p) = \delta(w_1, \ldots, w_p)$$

*is valid in* $\mathbf{QC}^*$.

*Proof.* Let $\chi_j, s_j$ be the sequences associated with $w_1, \ldots, w_m$; let $\chi_j', s_j'$ be associated with $w_{m+1}, \ldots, w_p$; and let $\chi_j'', s_j''$ be associated with the composite $p$-tuple.

The formula $(1 - \chi_j'') = (1 - \chi_j)(1 - \chi_j')$, or equivalently

$$\chi_n'' = \chi_n + \chi_n' - \chi_n \chi_n',$$

follows from definition (1). Substituting $\chi_n = \sum_{j|n} s_j$ this becomes

($4_n$) $$\sum_{j|n} s_j'' = \sum_{j|n} s_j + \sum_{j|n} s_j' - \sum_{a|n} \sum_{b|n} s_a s_b'.$$

Next the formula

$$(5_n) \qquad s_n'' = s_n + s_n' - \sum_{[a,\,b]=n} s_a s_b'$$

will be proved by induction on $n$. It is certainly true for $n = 1$. Adding the alleged equations $(5_j)$ over all divisors $j$ of $n$ we clearly get the true equation $(4_n)$. It follows by induction that $(5_n)$ is true.

Now multiply $(5_n)$ by the element $E_n$ and sum over $n$. Since $(s_a E_a)(s_b' E_b) = s_a s_b' E_{[a,b]}$, this yields

$$(6) \qquad \Sigma s_n'' E_n = \Sigma s_n E_n + \Sigma s_n' E_n - (\Sigma s_a E_a)(\Sigma s_b' E_b)$$

or in other words

$$1 - \Sigma s_n'' E_n = (1 - \Sigma s_a E_a)(1 - \Sigma s_b' E_b).$$

Multiplying the two factors on the right by $(-1)^m$ and $(-1)^{p-m}$ respectively, this proves Theorem 3.

THEOREM 4. *Let $f(z_1, \ldots, z_m)$ be a weighted homogeneous polynomial of type $(w_1, \ldots, w_m)$ having an isolated critical point at the origin. Express the weights in irreducible form as $w_i = u_i/v_i$, $i = 1, 2, \ldots, m$. Then the characteristic polynomial $\Delta(t)$ of the linear transformation*

$$h_*: H_{m-1}(F; \mathbf{C}) \to H_{m-1}(F; \mathbf{C})$$

*is determined by*

$$\begin{aligned} \text{divisor } \Delta &= (w_1 E_{u_1} - 1)(w_2 E_{u_2} - 1) \cdots (w_m E_{u_m} - 1) \\ &= (v_1^{-1}\Lambda_{u_1} - 1)(v_2^{-1}\Lambda_{u_2} - 1) \cdots (v_m^{-1}\Lambda_{u_m} - 1). \end{aligned}$$

*Proof.* More generally, for any $m$-tuple $(w_1, \ldots, w_m)$ of positive rationals, we will show that the expression

$$\delta(w_1, \ldots, w_m) = (-1)^m(1 - \Sigma s_j(w_j, \ldots, w_m)E_j)$$

is equal to $(w_1 E_{u_1} - 1) \cdots (w_m E_{u_m} - 1)$. Together with Theorem 2, this will complete the proof.

First suppose that $m = 1$. Then clearly $\chi_j(w_1)$ is equal to either $w_1$ or 0 according as $j \equiv 0 \bmod u_1$ or $j \not\equiv 0 \bmod u_1$. Hence

$$\begin{aligned} s_j(w_1) &= w_1 \quad \text{if} \quad j = u_1 . \\ s_j(w_1) &= 0 \quad\ \ \text{if} \quad j \neq u_1. \end{aligned}$$

This proves that

$$\delta(w_1) = -(1 - w_1 E_{u_1}) = w_1 E_{u_1} - 1,$$

and it follows by induction that

$$\delta(w_1, \ldots, w_m) = (w_1 E_{u_1} - 1) \cdots (w_m E_{u_m} - 1)$$

as required.

Note that Theorem 4 can be reformulated as follows.

COROLLARY. *divisor* $\Delta = \sum (-1)^{m-s} w_{i_1} \cdots w_{i_s} E_{[u_{i1}, \ldots, u_{is}]}$

$$= \sum (-1)^{m-s} \frac{w_i \cdots w_{i_s}}{[u_{i_1}, \ldots, u_{i_s}]} \Lambda_{[u_{i1}, \ldots, u_{is}]}$$

*to be summed over all the* $2^m$ *subsets* $\{i_1, \ldots, i_s\}$ *of* $\{1, \ldots, m\}$.

## §4. THE HOMOLOGY OF $K$

Consider the manifold

$$K = f^{-1}(0) \cap S_\varepsilon$$

of dimension $(2m - 3)$ associated with an isolated singularity in $\mathbf{C}^m$. The homology of $K$ is related to the characteristic polynomial $\Delta(t)$ as follows.

THEOREM [2, §8]. *The manifold K is a homology sphere (and when* dim $K \neq 3$ *actually a topological sphere) if and only if the integer* $\Delta(1)$ *is equal to* $\pm 1$.

In fact it is shown in [2] that $K$ is $(m - 3)$-connected, and that the order of $H_{m-2}(K)$ (using reduced homology if $m = 2$) is equal to $|\Delta(1)|$. Here the "order" of an infinite group is defined to be zero.

If the divisor of a monic polynomial $\Delta$ is known as a linear combination of the divisors $\Lambda_r$ (compare Theorem 4 and its Corollary), then $\Delta(1)$ can be computed easily as follows.

LEMMA 2. *Setting divisor* $\Delta = a_1\Lambda_1 + \cdots + a_s\Lambda_s$, *define two numbers* $\kappa(\Delta)$ *and* $\rho(\Delta)$ *by the formulas*

$$\kappa = a_1 + \cdots + a_s, \; \rho = 2^{a_2}3^{a_3} \cdots s^{a_s}.$$

*Then* $\kappa$ *and* $\rho$ *are non-negative integers, and*

$$\Delta(1) = \rho \quad \textit{if} \quad \kappa = 0,$$
$$\Delta(1) = 0 \quad \textit{if} \quad \kappa > 0.$$

It follows that $\Delta(1) \geq 0$.

*Proof.* Since

$$\text{divisor } \Delta = \kappa\Lambda_1 + \Sigma a_r(\Lambda_r - \Lambda_1)$$

where $\Lambda_r - \Lambda_1$ is the divisor of the polynomial

$$(t^r - 1)/(t - 1) = 1 + t + \cdots + t^{r-1}$$

it follows that

$$\Delta -(t) = (t\,1)^{\kappa}\prod(1 + t + \cdots + t^{r-1})^{a_r}.$$

Substituting $t = 1$ the conclusion follows.

Note that $\kappa(\Delta)$ is just the highest power to which $(t - 1)$ divides $\Delta(t)$. Alternatively, $\kappa(\Delta)$ can be described as the $(m-1)$-st or $(m-2)$-nd Betti number of the manifold $K$. In fact, since the linear transformation $h_*$ has finite period, it can certainly be diagonalized by a suitable choice of basis for $H_{m-1}(F; \mathbf{C})$. Proceeding as in [2, pp. 67–68], the conclusion follows.

The following may help to simplify the computation of $\kappa$ and $\rho$.

LEMMA 3. *Suppose that the divisor of the polynomial* $\Delta$ *is equal to* $\delta(w_1, \ldots, w_m)$ *where each* $w_i = u_i/v_i$ *is a rational number greater than* 1; *and where each of the numerators* $u_1, \ldots, u_k$ *is relatively prime to each of* $u_{k+1}, \ldots, u_m$. *Then there exist polynomials* $\Delta_1$ *and* $\Delta_2$ *with*

$$\text{divisor } \Delta_1 = \delta(w_1, \ldots, w_k), \text{ divisor } \Delta_2 = \delta(w_{k+1}, \ldots, w_m).$$

*Furthermore* $\kappa(\Delta)$ *and* $\rho(\Delta)$ *are determined by the integers* $\kappa_i = \kappa(\Delta_i)$ *and* $\rho_i = \rho(\Delta_i)$, $i = 1, 2$, *according to the formulas*

$$\kappa(\Delta) = \kappa_1 \kappa_2, \qquad \rho(\Delta) = \rho_1^{\kappa_2} \rho_2^{\kappa_1}.$$

*Proof.* Setting

$$\delta(w_1, \ldots, w_k) = \sum a_i \Lambda_i \quad \text{and} \quad \delta(w_{k+1}, \ldots, w_m) = \sum b_j \Lambda_j$$

with $a_i, b_j \in \mathbf{Q}$, note that each $i$ with $a_i \neq 0$ is prime to each $j$ with $b_j \neq 0$. Hence $\Lambda_i \Lambda_j = \Lambda_{ij}$, so that

$$\text{divisor } \Delta = \delta(w_1, \ldots, w_k)\, \delta(w_{k+1}, \ldots, w_m) = \sum a_i b_j \Lambda_{ij}. \tag{7}$$

Thus the products $a_i b_j$ must all be integers. Since $a_1 = (-1)^k$ and $b_1 = (-1)^{m-k}$ it follows that the $a_i$ and $b_j$ are also integers.

To show that $\Sigma a_i \Lambda_i$ is actually the divisor of a polynomial (rather than just of a rational function) it is necessary to carry out an analogous argument, using the divisors of the cyclotomic polynomials (in place of the $\Lambda_i$) as a basis. Details are left to the reader.

*Remark.* If $\Delta(t)$ is actually the characteristic polynomial associated with $f(z_1, \ldots, z_m)$ as in Theorems 2 and 4, then making use of Orlik and Wagreich [3], one can prove the following sharper statement. The polynomial $f$ can be deformed, without changing the $w_i$, into one of the form

$$f_1(z_1, \ldots, z_k) + f_2(z_{k+1}, \ldots, z_m),$$

where $f_1$ and $f_2$ again satisfy the hypotheses of Theorems 2 and 4. Hence $\Delta_1$ and $\Delta_2$ can be now interpreted concretely as the characteristic polynomials associated with $f_1$ and $f_2$.

To complete the proof of Lemma 3, it is only necessary to note that the formulas $\kappa = \kappa_1 \kappa_2$ and $\rho = \rho_1^{\kappa_2} \rho_2^{\kappa_1}$ follow easily from formula (7).

We conclude with a numerical example. If

$$f(z_1, z_2, z_3) = z_1^5 + z_1 z_2^6 + z_3^7$$

with $w_1 = 5$, $w_2 = 15/2$, $w_3 = 7$, then the rank of the homology group $H_2(F^4)$ is equal to $(w_1 - 1)(w_2 - 1)(w_3 - 1) = 156$. Lemma 3 clearly applies; in fact

$$\text{divisor } \Delta_1 = (\Lambda_5 - 1)(\tfrac{1}{2}\Lambda_{15} - 1) = 2\Lambda_{15} - \Lambda_5 + 1$$

so that $\kappa_1 = 2$, $\rho_1 = 15^2/5 = 45$; and divisor $\Delta_2 = \Lambda_7 - 1$, so that $\kappa_2 = 0$, $\rho_2 = 7$. Hence the order $\Delta(1)$ of the homology group $H_1(K^3)$ is equal to

$$45^0 \cdot 7^2 = 49.$$

This value of $\Delta(1)$ could also be determined by first carrying out the explicit computation

$$\Delta(t) = (t^{105} - 1)^2 (t^{35} - 1)^{-1} (t^{15} - 1)^{-2} (t^7 - 1)(t^5 - 1)(t - 1)^{-1}.$$

## REFERENCES

1. E. BRIESKORN: Beispiele zur Differentialtopologie von Singularitäten, *Invent. Math*, **2** (1966), 1–14.
2. J. MILNOR: Singular points of complex hypersurfaces. *Ann. Math. Stud.* **61**, (1968).
3. P. ORLIK and P. WAGREICH: *Isolated singularities of algebraic surfaces with* **C*** *action*, *Ann. Math.* (to appear). Institute for Advanced Study, Princeton, 1969.
4. F. PHAM: Formules de Picard–Lefschetz généralisées et ramification des intégrales, *Bull. Soc. Math. Fr.* **93** (1965), 333–367.

*Massachusetts Institute of Technology and*
*University of Wisconsin, Madison*

# ISOLATED CRITICAL POINTS OF COMPLEX FUNCTIONS

*Reprinted from*

Differential Geometry
Proceedings of Symposia in Pure Mathematics
Volume 27, Part I
American Mathematical Society, 1975, pp. 381–382

A very brief exposition of work by Arnol′d, Brieskorn, Thom, and Tjurina.

# ISOLATED CRITICAL POINTS OF COMPLEX FUNCTIONS

JOHN MILNOR

A complex valued complex analytic function $f(z_1, \cdots, z_n)$ is said to have a *critical point* wherever $\partial f/\partial z_1 = \cdots = \partial f/\partial z_n = 0$. Let $V$ be the vector space consisting of all germs of analytic functions $f(z_1, \cdots, z_n)$ which vanish, and have a critical point, at the origin. Let $G$ be the group consisting of all germs of complex analytic diffeomorphisms from a neighborhood of the origin to a neighborhood of the origin in $C^n$. Then $G$ operates on $V$ by composition, and an orbit $f \circ G \subset V$ is called *a right equivalence class* of germs. Such an orbit has finite codimension in $V$ if and only if the origin is isolated as a critical point of $f$.

Following V. Arnol'd an orbit $f \circ G$ is called *simple* if it possesses a neighborhood intersecting only finitely many other orbits. Arnol'd shows that every simple orbit is represented by one and only one of the following polynomials:

$$\begin{array}{lll} A_\mu & z_1^{\mu+1} + z_2^2 + \cdots + z_n^2, & \mu \geqq 1, \\ D_\mu & z_1^{\mu-1} + z_1 z_2^2 + z_3^2 + \cdots + z_n^2, & \mu \geqq 4, \\ E_6 & z_1^4 + z_2^3 + z_3^2 + \cdots + z_n^2, & \\ E_7 & z_1^3 z_2 + z_2^3 + z_3^2 + \cdots + z_n^2, & \\ E_8 & z_1^5 + z_2^3 + z_3^2 + \cdots + z_n^2. & \end{array}$$

Here the subscript $\mu$ is equal to the codimension plus one. Those orbits with codimension $\mu - 1 \leqq 4$ (or rather their real analogues) play a central role in Thom's theory of elementary catastrophes. In Thom's terminology, $A_2$, $A_3$, $A_4$, and $A_5$ are called repectively the "fold", "cusp", "swallow's tail", and "butterfly";

*AMS (MOS) subject classifications* (1970). Primary 57D70, 14B05.

while $D_4$ is called an "elliptic or hyperbolic umbilic", and $D_5$ is called a "parabolic umbilic".

In the special case $n = 3$ these polynomials (or the related algebraic singularities) have been listed earlier by Hirzebruch, Artin, Tjurina and others. Tjurina's characterization is particularly elegant. Let $f(z_1,\cdots, z_n)$ have an isolated critical point at the origin; and suppose that $n$ is odd, say $n = 2m + 1$. Let $X_f$ be the compact $4m$-dimensional manifold-with-boundary obtained by intersecting a small disk $|z_1|^2 + \cdots + |z_n|^2 \leqq \varepsilon$ with the nonsingular hypersurface $f^{-1}(c)$, where $|c|$ is very small but not zero. Then the middle homology group $H_{2m}(X_f; \boldsymbol{Z})$ is free abelian of rank say $\mu$, and is spanned by elements $\alpha$ whose self-intersection number $\alpha\cdot\alpha$ is either $+2$ (for $m$ even) or $-2$ (for $m$ odd). Tjurina shows that this intersection pairing is positive definite (for $m$ even) or negative definite (for $m$ odd) if and only if $f$ is right equivalent to one of the polynomials listed above. The corresponding homology lattice $H_{2m}(X_f; Z)$ is isomorphic to the lattice generated by the corresponding classical root system $A_\mu$, $D_\mu$, $E_6$, $E_7$, or $E_8$.

In the case $n = 3$ these same simple orbits have been characterized by Brieskorn as the only ones for which the 3-dimensional manifold $\partial X_f$ has finite fundamental group.

I want to thank Arnol'd for communicating his research to me before publication.

## References

V. I. Arnol'd, *Singularities of smooth mappings*, Uspehi Mat. Nauk **23** (1968), no. 1 (139), 3–44 = Russian Math. Surveys **23** (1968), no. 1, 1–43. MR **37** # 2243.

———, *Normal forms for functions near degenerate critical points, the Weyl groups of $A_k$, $D_k$, $E_k$ and Lagrangian singularities*, Funkcional. Anal. i Priložen. **6** (1972), no. 4, 3–25 = Functional Anal. Appl. **6** (1972), 254–272.

———, *Lectures on bifurcations in versal families*, Uspehi Mat. Nauk **27** (1972), no. 5 (167), 119–183 = Russian Math. Surveys **27** (1972), no. 5, 54–123.

M. Artin, *On isolated rational singularities of surfaces*, Amer. J. Math. **88** (1966), 129–136. MR **33** # 7340.

N. Bourbaki, *Éléments de mathématique*. Fasc. XXXIV. *Groupes et algèbres de Lie*. Chaps. 4–6, Actualités Sci. Indust., no. 1337, Hermann, Paris, 1968. MR **39** # 1590.

E. Brieskorn, *Rationale Singularitäten komplexer Flächen*, Invent. Math. **4** (1967/68), 336–358. MR **36** # 5136.

F. Hirzebruch, *The topology of normal singularities of an algebraic surface (d'après Mumford)*, Séminaire Bourbaki: 1962/63, Exposé 250, Benjamin, New York, 1966. MR **33** # 5420j.

J. Milnor, *Singular points of complex hypersurfaces*, Ann. of Math. Studies, no. 61, Princeton Univ. Press, Princeton, N. J., 1968. MR **39** # 969.

R. Thom, *Topological models in biology*, Topology **8** (1969), 313–335. MR **39** # 6629.

G. N. Tjurina, *Topological properties of isolated singularities of complex spaces of codimension one*, Izv. Akad. Nauk SSSR Ser. Mat. **32** (1968), 605–620 = Math. USSR Izv. **2** (1968), 557–571. MR **37** # 3053.

P. Wagreich, *Singularities of complex surfaces with solvable local fundamental group*, Topology **11** (1972), 51–72.

INSTITUTE FOR ADVANCED STUDY

# Euclidean and Non-Euclidean Geometry

# A PROBLEM IN CARTOGRAPHY

*Reprinted from*

The American Mathematical Monthly

Volume 76 (1969), pp. 1101–1112

Let $f : X \to E$ be an embedding of one metric space in another satisfying Lipschitz inequalities

$$0 < c_1 \leq \frac{\mathrm{dist}_E(f(x), f(x'))}{\mathrm{dist}_X(x, x')} \leq c_2 < \infty$$

for $x \neq x'$. If we choose the largest possible value for $c_1$ and the smallest possible value for $c_2$, then the number

$$\delta = \log(c_2/c_1) \geq 0$$

is called the ***distortion***[1] of $f$. In the special case where $X$ is a region bounded by a circle in the 2-sphere and $E$ is the Euclidean plane, this note shows that there is an essentially unique map which minimizes distortion. It is known to cartographers as the ***azimuthal equidistant projection*** and to differential geometers as the inverse of the ***exponential map*** from the tangent plane at the center point of $X$. It has the property that the "minimum infinitesimal scale" at a point $x_0$

$$\mathrm{Scale}_{\min}(x_0) = \liminf_{x \to x_0} \frac{\mathrm{dist}_E(f(x), f(x_0))}{\mathrm{dist}_X(x, x_0)}$$

is independent of $x_0$, while the "maximum infinitesimal scale"

$$\mathrm{Scale}_{\max}(x_0) = \limsup_{x \to x_0} \frac{\mathrm{dist}_E(f(x), f(x_0))}{\mathrm{dist}_X(x, x_0)}$$

[1] Such a map with finite distortion is called a ***quasi-isometry*** or a ***bi-Lipschitz homeomorphism***. Compare D. Sullivan, *Hyperbolic geometry and homeomorphisms*, pp. 543–555 of "Geometric Topology", Georgia Topology Conference 1977, Acad. Press 1979.

is maximized precisely along the boundary of $X$. Quite likely there is a similar result for the case where $X$ is a disk in the spherical or hyperbolic space of arbitrary dimension and $E$ is the corresponding Euclidean space. As one possible application, such minimal distortion maps might provide a useful way of visualizing polyhedra in hyperbolic 3-space.

However, the statement leaves many unanswered questions. Take for example an arbitrary geodesically convex region on the 2-sphere. Is there a unique embedding in the plane which minimizes distortion? For a distortion minimizing map, is $\text{Scale}_{\min}(x)$ constant? Is $\text{Scale}_{\max}(x)$ maximized along the boundary?

I am not aware of any progress on these questions, although the corresponding problems for conformal maps were solved by Chebyshef in the last century. It is interesting to note that Chebyshef's methods have recently been applied to practical cartography by Mitchell Feigenbaum, as consultant for Hammond.[2]

[2] *Atlas of the World,* Hammond 1993. See I. Mundell, *Maps that shape the world,* New Scientist **139** #1880 (July 3, 1993) 21–23.

# A PROBLEM IN CARTOGRAPHY

JOHN MILNOR, Massachusetts Institute of Technology

**1. Introduction.** The central problem of mathematical cartography is the problem of representing a portion of the curved surface of the earth on a flat piece of paper without introducing any more distortion than is absolutely necessary. This note will propose a quantitative definition for the term "distortion," and then study the mathematical problem of choosing a method of mapping which minimizes distortion.

To simplify the problem we first replace the rather irregular surface of the earth by a perfect sphere.

DEFINITIONS. Let $S$ be the sphere of radius $r$ consisting of all points $x$ in the 3-dimensional euclidean space with distance $r$ from the origin, and let $U$ be any nondegenerate subset of $S$. (By "nondegenerate" we mean that $U$ must contain at least two distinct points.)

*A map projection $f$ on the domain $U$ will mean a function which assigns to each point $x$ of $U$ some point $f(x)$ of the euclidean plane $E$.*

Let $d_S(x, y)$ denote the geodesic distance between two points $x$ and $y$ of the sphere $S$. By definition, this is equal to the length of the shorter great circle arc joining $x$ to $y$. The euclidean distance between two points $a$ and $b$ of the plane $E$ will be denoted analogously by $d_E(a, b)$.

*The scale of a map projection $f$ with respect to a pair of distinct points $x$ and $y$ in the domain $U$ is defined to be the ratio*

$$d_E(f(x), f(y))/d_S(x, y).$$

Ideally we would like this scale to be the same for all pairs of points $x$ and $y$ in $U$, but this is not usually possible. So we must introduce the *minimum scale* $\sigma_1$, defined to be the infimum of the ratio $d_E(f(x), f(y))/d_S(x, y)$ as $x$ and $y$ vary over all pairs of distinct points in $U$, and the *maximum scale* $\sigma_2$, defined to be the supremum of the ratio $d_E(f(x), f(y))/d_S(x, y)$. In other words $\sigma_1$ and $\sigma_2$ are the

Professor Milnor was graduated from Princeton in 1951 and received his Ph.D. in 1954 under Ralph Fox. After a year as Higgins Lecturer, he was appointed Assistant Professor and also was made an Alfred P. Sloan Research Fellow. He was promoted to Associate Professor in 1957, to Professor in 1960, and to Putnam Professor in 1962. He was a visiting professor at the University of California, Berkeley for 1959–60, at U.C.L.A. for 1967–68, and presently is Sloan Professor at MIT. He received honorary Sc.D. degrees from Syracuse University in 1965 and the University of Chicago in 1967. He is a member of the American Academy of Arts and Sciences and the Philosophical Society.

In 1962 Professor Milnor received the Fields Medal, the highest honor for a mathematician, at the International Congress of Mathematicians. In 1963 he was elected to the National Academy of Sciences, one of the youngest ever thus honored. In 1966 he received the President's National Medal of Science, with the citation: "For clever and ingenious approaches in topology which have solved long outstanding problems and opened new exciting areas in this active branch of mathematics." *Editor.*

"best" possible constants such that the inequality

$$\sigma_1 d_S(x, y) \leqq d_E(f(x), f(y)) \leqq \sigma_2 d_S(x, y)$$

holds for all points $x$ and $y$ in $U$.

To measure the extent to which scale fails to be constant we propose the following:

DEFINITION. *The distortion of the map projection $f$ is the natural logarithm*

$$\delta = \log(\sigma_2/\sigma_1)$$

*of the ratio of maximum scale to minimum scale.*

Thus $0 \leqq \delta \leqq \infty$, where $\delta$ is finite if and only if both $\sigma_1$ and $\sigma_2$ are positive and finite numbers. If $\delta$ is finite, notice that the function $f$ is continuous and one-to-one.

We would like to find a map projection $f$ with no distortion at all ($\delta = 0$). Since this is not possible except in a few special and uninteresting cases (e.g., the case of a domain $U$ consisting of only three points), the best we can actually do is to try to find a map projection for which $\delta$ is as small as possible.

DEFINITION. *A minimum distortion map projection $f_0$ on $U$ will mean a map projection whose distortion $\delta_0$ is less than or equal to the distortion of every other map projection on $U$.*

PRELIMINARY THEOREM. *For every nondegenerate set of points $U$ on the sphere there exists a minimum distortion map projection $f_0$ with domain $U$.*

The proof of this theorem, which is quite elementary, will be deferred until Appendix A.

Unfortunately the proof will fail to suggest answers to a number of relevant questions: Is this minimum distortion map $f_0$ unique in some sense? Is $f_0$ differentiable (assuming that $U$ is a nice enough set so that differentiability makes sense)? How can one actually construct $f_0$, or even a reasonable approximation to $f_0$? How can one estimate the minimum possible distortion $\delta_0$ associated with a given set $U$?

This note will succeed in answering these questions only in one very special case, namely, the case of the region bounded by a circle on $S$.

Given a fixed point $x_0$ of $S$, let $D_\alpha$ denote the closed *disk* of geodesic radius $r\alpha$, consisting of all points $x$ in $S$ for which $d_S(x, x_0) \leqq r\alpha$. Here $\alpha$ can be any number in the interval $0 < \alpha < \pi$.

MAIN THEOREM. *There is one and, up to similarity transformations of the plane, only one minimum distortion map projection $f_0$ on the domain $D_\alpha$. This map projection is infinitely differentiable, and has distortion $\delta_0$ equal to* $\log(\alpha/\sin\alpha)$.

This minimum distortion projection $f_0$, known to cartographers as the "azimuthal equidistant projection," can be characterized by the fact that it preserves both distances and directions from the central point $x_0$. The explicit formula $\delta_0 = \log(\alpha/\sin\alpha)$ shows that the distortion $\delta_0$ is small for small values of $\alpha$, being asymptotically equal to

$$\alpha^2/6 \sim \tfrac{2}{3} \text{ area } D_\alpha/\text{area } S$$

as $\alpha \to 0$. However $\delta_0$ tends to infinity as $\alpha \to \pi$.

This theorem will be proved in Section 2. The problem of estimating the $\delta_0$ associated with a more general domain $U$ is discussed in Section 3. There are two appendices, one proving that minimum distortion map projections exist, and a second discussing a corresponding problem for conformal map projections, following Chebyshef.

**2. The azimuthal equidistant projection.** Again let $D_\alpha$ denote a spherical disk of geodesic radius $r\alpha$ centered at $x_0$.

LEMMA 1. *The distortion $\delta$ for any map projection $f$ with domain $D_\alpha$ satisfies $\delta \geqq \log(\alpha/\sin\alpha)$.*

*Proof.* We may assume that $f$ has finite distortion. Hence the "Lipschitz inequality"

$$d_E(f(x), f(y)) \leqq \sigma_2 d_S(x, y) \tag{1}$$

is satisfied, where $\sigma_2$ is a finite constant, and it follows that $f$ is continuous. Furthermore $f$ is one-to-one.

Let $C_\alpha$ denote the boundary of the disk $D_\alpha$. Clearly the image $f(C_\alpha)$ is a simple closed curve in the plane. We shall first prove:

ASSERTION A. *Every half-line emanating from the point $f(x_0)$ in the plane must intersect the simple closed curve $f(C_\alpha)$ at least once.*

*Proof.* The Jordan Curve Theorem asserts that the simple closed curve $f(C_\alpha)$ cuts the plane into two components

$$E - f(C_\alpha) = E_1 \cup E_2,$$

one of these components, say $E_1$, being bounded, and the second unbounded. But the bounded component $E_1$ is just the image, under the continuous one-to-one function $f$, of the interior of the disk $D_\alpha$. This is proved, for example, in Newman [**10**, Theorem 12.2, p. 121]. In particular it follows that the point $f(x_0)$ must belong to the bounded component $E_1$. Hence every half-line emanating from $x_0$ must cross $f(C_\alpha)$, since otherwise it would lie completely within the bounded set $E_1$ which is impossible. This proves Assertion A.

Since the curve $C_\alpha$ on $S$ has finite length $2\pi r \sin\alpha$, it follows easily from the Lipschitz inequality (1) that $f(C_\alpha)$ also has finite length $L$, where

$$L \leqq 2\pi\sigma_2 r \sin\alpha. \tag{2}$$

(The *length* of a not necessarily smooth curve is defined for example in [6, p. 36].)

Now let us make use of the inequality

$$d_E(f(x), f(y)) \geqq \sigma_1 d_S(x, y). \tag{3}$$

Since every point of $C_\alpha$ has geodesic distance exactly $r\alpha$ from $x_0$ it follows that every point of $f(C_\alpha)$ has euclidean distance $\geqq \sigma_1 r\alpha$ from $f(x_0)$.

Thus $f(C_\alpha)$ is a simple closed curve of finite length $L$ which lies outside an open disk $D^*$ of radius $\sigma_1 r\alpha$ in the plane, and cuts every half-line through the center of this disk.

ASSERTION B. *This implies that* $L \geqq 2\pi\sigma_1 r\alpha$, *where equality holds if and only if* $f(C_\alpha)$ *is precisely equal to the boundary of* $D^*$.

*Proof.* Cut $f(C_\alpha)$ by a straight line through the center of $D^*$ and choose intersection points, say $a$ and $b$, which lie on opposite sides of $D^*$. Let $A$ be either one of the two arcs of $f(C_\alpha)$ from $a$ to $b$. Introducing polar coordinates $\rho$ and $\theta$ about the center of $D^*$, first assume that the arc $A$ can be described, in terms of a parameter $t$, by piecewise smooth functions

$$\rho = \rho(t), \qquad \theta = \theta(t).$$

Then

$$\text{length } A = \int (\dot\rho^2 + \rho^2\dot\theta^2)^{1/2} dt \geqq \int \rho \,|\,\dot\theta\,|\, dt,$$

where the dot denotes differentiation. Since

$$\rho \geqq \sigma_1 r\alpha \quad \text{and} \quad \int |\,\dot\theta\,|\, dt \geqq \left| \int \dot\theta dt \right| \geqq \pi,$$

this proves that length $A \geqq \pi\sigma_1 r\alpha$, and therefore $L \geqq 2\pi\sigma_1 r\alpha$, as required.

If $A$ is not piecewise smooth, then an extra step is needed. For each $\epsilon > 0$ it is possible to approximate $A$ by a polygonal path $A'_\epsilon$ from $a$ to $b$ which lies outside the disk of radius $\sigma_1 r\alpha - \epsilon$ and satisfies

$$\text{length } A \geqq \text{length } A'_\epsilon \geqq \pi(\sigma_1 r\alpha - \epsilon).$$

Letting $\epsilon \to 0$, we obtain length $A \geqq \pi\sigma_1 r\alpha$, as before.

Now suppose that the length of $A$ is precisely equal to $\pi\sigma_1 r\alpha$. Then any portion of $A$ which has distance greater than $\sigma_1 r\alpha$ from the center of $D^*$ must be a straight line segment. Otherwise, replacing some small portion of $A$ by a straight line segment we could decrease its length, which is impossible.

Any maximal line segment $A_0$ which forms a part of $A$ must lead from one of the end points $a$ or $b$ of $A$ to a point of the circle bounding $D^*$. The only other possibility would be that both end points of $A_0$ lie on the circle, which is impossible. Thus $A$ consists of a line segment (possibly degenerate) from $a$ to the circle, followed by a circle arc, followed by a line segment to $b$. Elementary geometry now shows that the minimal length $\pi\sigma_1 r\alpha$ is achieved only if $A$ is the semicircle. Hence $L$ can equal $2\pi\sigma_1 r\alpha$ only if $f(C_\alpha)$ is the full circle. This completes the proof of Assertion B.

Combining Assertion B with the inequality (2) we obtain

$$2\pi\sigma_1 r\alpha \leqq 2\pi\sigma_2 r \sin \alpha$$

or

$$\alpha/\sin \alpha \leqq \sigma_2/\sigma_1$$

and hence $\log(\alpha/\sin \alpha) \leqq \delta$, which completes the proof of Lemma 1.

LEMMA 2. *If the distortion of $f$ is precisely equal to* $\log(\alpha/\sin \alpha)$, *then $f$ is an azimuthal equidistant projection.*

By definition this means that $f$ carries each great circle passing through $x_0$ into a straight line in the plane, the angle between two great circles being equal to the angle between the corresponding straight lines, and that $f$ carries each circle $C$ centered at $x_0$ to a circle $f(C)$ centered at $f(x_0)$, the radius of $f(C)$ being proportional to the geodesic radius of $C$.

To differential geometers, this means that $f$ is the inverse of the so called exponential map. It follows that $f$ is infinitely differentiable, even at $x_0$. See for example [9, p. 147].

*Proof of Lemma* 2. If $\delta = \log(\alpha/\sin \alpha)$, then according to Assertion B the image $f(C_\alpha)$ must be precisely equal to the circle of radius

$$\sigma_1 r\alpha = \sigma_2 r \sin \alpha$$

centered at $f(x_0)$. Hence the image $f(D_\alpha)$ must be precisely the closed disk bounded by this circle. (Compare the proof of Assertion A.)

Now consider an arbitrary point $x$ of $D_\alpha$. Construct a great circle segment from $x_0$ through $x$ to a point $\bar{x}$ on the boundary $C_\alpha$ of $D_\alpha$. If $c$ denotes the geodesic distance $d_S(x_0, x)$, note that $x$ has geodesic distance precisely $r\alpha - c$ from $\bar{x}$, and geodesic distance strictly greater than $r\alpha - c$ from every other point of $C_\alpha$. Hence, using inequality (3), the image $f(x)$ must

(a) have distance at least $\sigma_1 c$ from $f(x_0)$,

(b) have distance at least $\sigma_1(r\alpha - c)$ from $f(\bar{x})$, and

(c) have distance greater than $\sigma_1(r\alpha - c)$ from every other point of $f(C_\alpha)$.

Clearly there is one and only one point in the disk $f(D_\alpha)$ which satisfies these three conditions: namely, the point which lies at distance $\sigma_1 c$ along the line segment from $f(x_0)$ to $f(\bar{x})$. Thus the map projection $f$ on $D_\alpha$ is completely determined by what is does to boundary points of $D_\alpha$.

To complete the proof of Lemma 2 we need only verify that $f$ carries the circle $C_\alpha$ to the circle $f(C_\alpha)$ by a similarity transformation which multiplies all lengths by the constant factor $\sigma_2$. Suppose that we cut $C_\alpha$ into two arcs $A$ and $A'$, so that

$$\text{length } A + \text{length } A' = \text{length } C_\alpha = 2\pi r \sin \alpha.$$

The Lipschitz inequality (1) implies that

$$\text{(4)} \qquad \text{length } f(A) \leqq \sigma_2 \text{ length } A, \qquad \text{length } f(A') \leqq \sigma_2 \text{ length } A'.$$

But

$$\text{length } f(A) + \text{length } f(A') = \text{length } f(C_\alpha)$$

is precisely equal to $\sigma_2$ times the length $2\pi r \sin \alpha$ of $C_\alpha$. So both of the inequalities (4) must actually be equalities. This proves Lemma 2.

Now we must prove the converse.

LEMMA 3. *The azimuthal equidistant projection on the disk $D_\alpha$ has distortion $\delta$ precisely equal to* $\log(\alpha/\sin\alpha)$.

*Proof.* Centering $D_\alpha$ at the north pole, we will use the longitude $0 \leqq \theta \leqq 2\pi$ and the colatitude $0 \leqq \gamma \leqq \alpha$ as coordinates. Suppose that $f$ maps the point with colatitude $\gamma$ and longitude $\theta$ to the point with cartesian coordinates $(r\gamma \cos \theta, r\gamma \sin \theta)$ in the plane. The length of any smooth curve $\gamma = \gamma(t)$, $\theta = \theta(t)$ in $D_\alpha$ is given by the integral

$$L = r \int (\dot{\gamma}^2 + \dot{\theta}^2 \sin^2 \gamma)^{1/2} dt,$$

and the length of the corresponding curve in $f(D_\alpha)$ is

$$L' = r \int (\dot{\gamma}^2 + \dot{\theta}^2 \gamma^2)^{1/2} dt.$$

But, since $\gamma/\sin\gamma$ is a monotone increasing function of $\gamma$, we have

$$\sin \gamma \leqq \gamma \leqq (\alpha/\sin \alpha) \sin \gamma,$$

from which it follows easily that

$$\text{(5)} \qquad L \leqq L' \leqq (\alpha/\sin \alpha) L.$$

Starting from this inequality (5) we will prove that

$$d_S(x, y) \leqq d_E(f(x), f(y)) \leqq (\alpha/\sin \alpha) d_S(x, y)$$

for every $x$ and $y$ in $D_\alpha$. Clearly this will imply that $\delta \leqq \log(\alpha/\sin \alpha)$ and hence, by Lemma 1, that $\delta = \log(\alpha/\sin\alpha)$.

*Proof that $d_S(x, y) \leqq d_E(f(x), f(y))$*. Join $f(x)$ to $f(y)$ within the convex set $f(D_\alpha)$ by a line segment of length $L'$ precisely equal to $d_E(f(x), f(y))$. The corresponding curve in $D_\alpha$ will have length $L \geqq d_S(x, y)$. Since $L \leqq L'$, we obtain $d_S(x, y) \leqq d_E(f(x), f(y))$, as required.

*Proof that* $d_E(f(x), f(y)) \leqq (\alpha/\sin\alpha) d_S(x, y)$. First suppose that $\alpha \leqq \pi/2$, so that the disk $D_\alpha$ is "geodesically convex." Then the proof is quite analogous. Join $x$ to $y$, *within* $D_\alpha$, by a great circle segment $A$ of length $L = d_S(x, y)$. Then $f(A)$ has length $L' \geqq d_E(f(x), f(y))$, so the inequality $L' \leqq (\alpha/\sin\alpha)L$ implies that $d_E(f(x), f(y)) \leqq (\alpha/\sin\alpha) d_S(x, y)$, as required.

If $\alpha > \pi/2$, so that the disk $D_\alpha$ is not geodesically convex, then a more complicated argument is necessary. Suppose that the shortest great circle arc from $x$ to $y$ does not lie completely within $D_\alpha$, but rather crosses out of $D_\alpha$ at a boundary point $\bar{x}$, and then crosses back in at another boundary point $\bar{y}$. We shall show that

$$d_E(f(x), f(\bar{x})) \leqq (\alpha/\sin\alpha) d_S(x, \bar{x}), \tag{6}$$

$$d_E(f(\bar{x}), f(\bar{y})) \leqq (\alpha/\sin\alpha) d_S(\bar{x}, \bar{y}), \tag{7}$$

$$d_E(f(\bar{y}), f(y)) \leqq (\alpha/\sin\alpha) d_S(\bar{y}, y). \tag{8}$$

Adding these three inequalities, we shall clearly obtain the required inequality.

But (6) and (8) can be proved by the argument above. To prove (7) we introduce an auxiliary azimuthal equidistant projection $g$ whose domain is the complementary disk $D'_{\pi-\alpha}$ centered at the south pole. Since $\pi - \alpha \leqq \pi/2$ we have

$$d_E(g(\bar{x}), g(\bar{y})) \leqq ((\pi - \alpha)/\sin(\pi - \alpha)) d_S(\bar{x}, \bar{y}).$$

Multiplying this by $\alpha/(\pi - \alpha)$ we obtain the required inequality (7). This completes the proof of Lemma 3.

Clearly Lemmas 1, 2, and 3 imply the "Main Theorem" of Section 1.

**3. Discussion.** How can one estimate the minimum possible distortion $\delta_0$ for map projections on a given set $U$? Here is a crude estimate. Define the *angular width* $w$ of a set $U$ as follows. Choose a smallest possible "lune" (figure bounded by two great semicircles) containing $U$, and let $w$ be the angle at the vertex of this lune.

ASSERTION. *Any set with angular width* $w < \pi$ *possesses a map projection with distortion* $\delta \leqq \log \sec(w/2)$.

This is proved by rotating so that the lune is centered on the equator, and then using the latitude and longitude of $x$ as the cartesian coordinates of $f(x)$. The computations are similar to those in the proof of Lemma 3.

It is conjectured that this estimate gives the right order of magnitude in the case of a small geodesically convex region, in the sense that $\delta_0$ is greater than say one sixth of $\log \sec(w/2)$. But $\log \sec(w/2)$ is not a really good estimate for $\delta_0$, except perhaps in the case of a long narrow region.

It would be more interesting to find a relation between $\delta_0$ and area.

PROBLEM. *Among all geodesically convex regions of given area, does the disk* $D_\alpha$ *require the largest distortion?*

In other words, if area$(U)=$area$(D_\alpha)$ does it follow that $U$ has a map projection with distortion $\delta \leqq \log(\alpha/\sin\alpha)$? If true this would imply the existence of map projection with smaller distortion than any which are actually known for many regions on the sphere. A test case which would be particularly interesting would be that of a small "rectangular" region on the sphere.

Slightly cruder is the following possible estimate.

PROBLEM. *Does every geodesically convex region $U$ possess a map projection with distortion less than the normalized area,*

$$\delta < \text{area } U/\text{area } S?$$

As an example, for the continental United States with about 1.5 percent of the earth's area, does there exist a map projection with scale errors of no more than 1.5 percent (or perhaps $1.5+\epsilon$ to allow for the lack of geodesic convexity)? All standard map projections for the continental United States seem to have scale errors of at least 2.2 percent.

**Appendix A. Minimum distortion projections always exist.** We shall first prove the following. Let $U$ be a subset of the sphere $S$ and let $\overline{U}$ denote the topological closure of $U$.

LEMMA 4. *Any map projection $f$ on $U$ with distortion $\delta < \infty$ extends uniquely to a map projection $\bar{f}$ on $\overline{U}$ having the same distortion $\delta$.*

*Proof.* The inequalities

$$\sigma_1 d_S(x, y) \leqq d_E(f(x), f(y)) \leqq \sigma_2 d_S(x, y)$$

show that $f$ is uniformly continuous, and hence extends uniquely to a continuous function $\bar{f}$ on $\overline{U}$. (See [**3**, p. 55].) Clearly $\bar{f}$ will also satisfy these inequalities.

Now, given some fixed set $U$, consider all possible map projections $f$ with domain $U$, and let $\delta_0$ denote the infimum of the corresponding distortions $\delta(f)$. We must construct a map projection $f_0$ whose distortion is precisely equal to $\delta_0$. We may assume that $\delta_0 < \infty$, since otherwise there is nothing to prove.

REMARK. Note that there exists a map projection with finite distortion on $U$ if and only if the closure $\overline{U}$ is not the entire sphere. For if $U$ is not everywhere dense on $S$ then $U$ is contained in some disk $D_{\pi-\epsilon}$ and hence possesses a map projection with distortion $\delta \leqq \log((\pi-\epsilon)/\sin(\pi-\epsilon)) < \infty$. But if $\overline{U}=S$ then a map projection with finite distortion on $U$ would extend to a map projection with finite distortion on $S$, which is impossible since $S \supset D_\alpha$ for all $\alpha$, or since $S$ is not homeomorphic to any subset of $E$. (See for example [**10**, p. 122].)

Choose a sequence of map projections $\{f_1, f_2, f_3, \cdots\}$ on $U$ so that the corresponding sequence $\{\delta_1, \delta_2, \delta_3, \cdots\}$ of distortions tends to the limit $\delta_0$. We may assume that each $f_i$ has been chosen so as to have maximum scale equal to 1, and so that the image $f_i(U)$ contains the origin.

Choose a countable dense subset

$$U' = \{x_1, x_2, x_3, \cdots\}$$

of $U$. Since the points $f_1(x_1), f_2(x_1), \cdots$ all have distance $\leqq \pi r$ from the origin, we can choose a convergent subsequence. That is there exists an infinite set $I_1$ of positive integers so that the sequence of points $f_i(x_1)$, where $i$ tends to infinity through the set $I_1$, converges to some limit $a_1$ in $E$. Similarly we can find an infinite set $I_2 \subset I_1$ so that the limit

$$\lim\{f_i(x_2) \mid i \to \infty, i \in I_2\}$$

exists. Call this limit $a_2$. Continuing inductively we can define a function $f$ from $U'$ to the plane by $f(x_j) = a_j = \lim\{f_i(x_j) \mid i \to \infty, i \in I_j\}$. Since the inequalities

$$e^{-\delta_i} d_S(x, y) \leqq d_E(f_i(x), f_i(y)) \leqq d_S(x, y)$$

hold for all $i$, it follows, taking the limit as $i$ tends to infinity through an appropriate $I_j$, that

$$e^{-\delta_0} d_S(x, y) \leqq d_E(f(x), f(y)) \leqq d_S(x, y)$$

for all $x$ and $y$ in $U'$. Thus $f$ is a map projection on $U'$ with distortion $\delta_0$.

Now applying Lemma 4 we obtain the required map projection on $U$ with distortion $\delta_0$.

**Appendix B. Conformal map projections.** Recall that a map projection $f$, defined on an open set $U$, is called *conformal* (cartographers prefer the term "orthomorphic") if it is differentiable and preserves angles. (That is, $f$ transforms any pair of curves in $U$, whose tangent vectors at a point of intersection span the angle $\alpha$ into a pair of curves in $E$, whose tangent vectors at the corresponding intersection point span the same angle $\alpha$.)

It follows that $f$ has a well defined *infinitesimal-scale* $\sigma(x)$ at each point $x$ of $U$. By definition $\sigma(x)$ is the limit of the ratio $d_E(f(x), f(y))/d_S(x, y)$ as $y$ tends to the limit $x$. (Compare [**1**, p. 74].)

We shall make use of the *Laplace-Beltrami operator* $\Delta$, a second order partial differential operator which assigns to each twice differentiable real valued function $g$ on a Riemannian manifold a new real valued function $\Delta g$. In euclidean space this is the familiar Laplace operator. We shall use $\Delta$ only on the sphere $S$ of radius $r$. Using latitude $\lambda$ and longitude $\theta$ as coordinates, the operator $\Delta$ on the sphere takes the form

$$r^2 \Delta g = g_{\lambda\lambda} - g_\lambda \tan \lambda + g_{\theta\theta} \sec^2 \lambda.$$

(Compare [**14**, p. 160]. The subscripts denote partial derivatives.)

Suppose now that $U$ is a simply connected open subset of the sphere $S$.

LEMMA 5. *The infinitesimal-scale function $\sigma(x)$ associated with a conformal map projection $f$ on $U$ determines $f$ up to an (orientation preserving or reversing)* *rigid motion of the plane. A given positive real valued function $\sigma$ on $U$ is the infinitesimal-scale function associated with some conformal $f$ if and only if $\sigma$ is twice differentiable and satisfies the differential equation $r^2\Delta \log \sigma = 1$.*

As an example, the function $\sigma(x) = \sec(\text{latitude } x)$ provides a solution to this equation $r^2\Delta \log \sigma = 1$, except at the north and south poles. The corresponding $f$ turns out to be the familiar Mercator projection.

(Note that our differential equation cannot have any solution which is defined and smooth throughout the entire sphere, since the condition $\Delta \log \sigma > 0$ implies easily that $\sigma$ cannot have any local maximum.)

*Proof of Lemma 5.* More generally, consider a smooth surface $M$ provided with a Riemannian metric, expressed in terms of local coordinates $u$ and $v$ as $ds^2 = Edu^2 + 2Fdudv + Gdv^2$. Let $\Delta$ denote the associated Laplace-Beltrami operator, and let $K$ denote the Gaussian curvature of $M$. Consider a second Riemannian metric of the form $\sigma^2 ds^2$ on $M$, where $\sigma$ is a positive twice differentiable function. Computation (using for example [**14**, pp. 113, 160]) shows that the Gaussian curvature $K'$ associated with this new Riemannian metric is given by the formula $K' = (K - \Delta \log \sigma)/\sigma^2$.

If $\sigma$ is the infinitesimal-scale function associated with a conformal mapping $f$ from $M$ to $M'$, then clearly $K'(x)$ is just the Gaussian curvature of $M'$ at $f(x)$. Thus if $M'$ is the euclidean plane, with $K' \equiv 0$, we see that the differential equation

$$\Delta \log \sigma = K$$

must be satisfied. In particular, taking $M$ to be the open subset $U$ of $S$, with $K \equiv 1/r^2$, we obtain the required equation

$$r^2\Delta \log \sigma = 1.$$

Conversely, given any solution $\sigma$ to the differential equation $\Delta \log \sigma = K$, the Riemannian metric $\sigma^2 ds^2$ has curvature $K'$ identically zero. Hence any sufficiently small connected open subset of $M$, with the metric $\sigma^2 ds^2$, can be mapped isometrically onto an open subset of the plane ([**14**, p. 145]). This isometry is unique up to rigid motions of the plane, since any isometry $\phi$ from one connected open subset of the plane to another extends to an isometry of the entire plane. (Assuming that $\phi$ preserves orientation, we can think of $\phi$ as a complex analytic function [**1**, p. 74] with $|d\phi/dz| \equiv 1$. Hence $d\phi/dz$ is constant and $\phi(z) = cz + c'$ with $|c| = 1$.)

Now if $M$ is simply connected then a monodromy argument shows that these local isometries can be chosen so as to fit together to yield a smooth mapping $f$ from all of $M$ to $E$.

(Compare [**8**, p. 1297]. The "Monodromy Theorem" says that if we are given connected open sets $U_\alpha$ covering a simply connected manifold $M$, and for each $U_\alpha$ a collection $F_\alpha$ of functions from $U_\alpha$ to $Y$ satisfying the following condition, then there exists a function from $M$ to $Y$ whose restriction to each $U_\alpha$ belongs to $F_\alpha$. The condition is that for each $f_\alpha$ in $F_\alpha$ and each $x$ in $U_\alpha \cap U_\beta$ there should exist one and only one $f_\beta$ in $F_\beta$ which coincides with $f_\alpha$ throughout some neighborhood of $x$. Compare [1, p. 285], [12, p. 67].)

In the large, this mapping $f$ from $M$ to $E$ may not be one-to-one, but locally it carries $M$, with the metric $\sigma^2 ds^2$, isometrically to $E$. Hence it carries $M$ with the original metric $ds^2$ conformally to $E$, the infinitesimal-scale function of $f$ being precisely equal to $\sigma$. This completes the proof of Lemma 5.

Chebyshef [2] studied conformal map projections, using the ratio $\sup \sigma(x)/\inf \sigma(x)$ of maximum infinitesimal-scale to minimum infinitesimal-scale as a measure of distortion.

REMARK. If the domain $U$ is geodesically convex, note that the maximum infinitesimal-scale $\sup \sigma(x)$ is equal to the maximum scale $\sigma_2$ of Section 1. (Compare the proof of Lemma 3.) Similarly, if $f$ is one-to-one and $f(U)$ is convex, then $\inf \sigma(x) = \sigma_1$.

CHEBYSHEF THEOREM. *If $U$ is a simply connected region bounded by a twice differentiable curve, then there exists one and, up to a similarity transformation of $E$, only one conformal map projection which minimizes this ratio* $\sup \sigma/\inf \sigma$. *This "best possible" conformal map projection is characterized by the property that its infinitesimal-scale function $\sigma(x)$ is constant along the boundary of $U$.*

This result has been available for more than a hundred years, but to my knowledge it has never been used by actual map makers.

*Proof.* Setting $g(x) = \log \sigma(x)$, first note that the differential equation $r^2 \Delta g = 1$ has a unique solution satisfying the boundary condition $g(x) = 0$ for $x \in bd(U)$. See for example [5, p. 288]. If $h$ is any other function which is twice differentiable and satisfies the equation $r^2 \Delta h = 1$ throughout the interior of $U$, then we shall show that

$$\sup h - \inf h \geqq \sup g - \inf g, \tag{9}$$

where equality holds if and only if

$$h = g + \text{constant}.$$

(Note that $\sup g - \inf g$ is just the logarithm of the ratio $\sup \sigma/\inf \sigma$ which we want to minimize.) Clearly this will complete the proof.

Since $\Delta g > 0$, an easy argument shows that the function $g$ cannot attain its maximum at any interior point of $U$. Since $g$ must achieve a maximum at some point of the compact set $\overline{U}$, it follows that the maximum must be attained on $bd(U)$. Thus $\sup g(x) = 0$.

The difference $h - g$ satisfies the homogeneous equation $\Delta(h - g) = 0$, and so cannot achieve its maximum at an interior point of $U$ unless $h - g = \text{constant}$. (See [5, p. 232].) Hence any sequence of points $x_1, x_2, \cdots$ for which

$$\lim_{i \to \infty} (h(x_i) - g(x_i)) = \sup(h - g)$$

must be a sequence tending to the boundary of $U$, unless $h-g=\text{constant}$.

Setting $c=\sup h$ (we may assume that $c$ is finite since otherwise (9) would trivially be satisfied), we have

$$g(x_i) \to 0, \qquad h(x_i) \leqq c,$$

hence

$$\sup(h - g) = \lim(h(x_i) - g(x_i)) \leqq c,$$

or in other words

$$h(x) \leqq g(x) + c$$

for all $x$. Therefore $\inf h \leqq \inf g+c$, which proves (9).

If equality holds, then at the interior point $x_0$ of $U$ where $g$ achieves its minimum we have

$$h(x_0) = g(x_0) + c.$$

Thus $h-g$ achieves its maximum $c$ at an interior point, and hence is constant. This completes the proof.

Remark. The "best possible" conformal map projection $f$, although locally well behaved, may not be one-to-one in the large. However, if $U$ is geodesically convex, then it can be shown that $f$ is one-to-one and that $f(U)$ is also convex.

### References


1. L. V. Ahlfors, Complex Analysis, 2nd ed., McGraw-Hill, New York, 1966.
2. P. L. Chebyshef (or Tchebychef), Sur la construction des cartes géographiques, Oeuvres I, Chelsea, New York, 1962, pp. 233–236 and 239–247.
3. J. Dieudonné, Foundations of Modern Analysis, Academic Press, New York, 1960.
4. L. Driencourt and J. Laborde, Traité des projections des cartes géographiques, 4 volumes, Hermann, Paris, 1932.
5. P. Garabedian, Partial Differential Equations, Wiley, New York, 1964.
6. E. Hille, Analytic Function Theory, vol. 1, Ginn, New York, 1959.
7. A. R. Hinks, Map Projections, Cambridge Univ. Press, New York, 1921.
8. T. Klotz, A complete $R_\Lambda$-harmonically immersed surface in $E^3$ on which $H\neq 0$, Proc. Amer. Math. Soc., 19 (1968) 1296–1298.
9. S. Kobayashi and K. Nomizu, Foundations of Differential Geometry, Interscience, New York, 1963.
10. M. H. A. Newman, Elements of the Topology of Plane Sets of Points, 2nd ed., Cambridge University Press, New York, 1951.
11. A. H. Robinson, Elements of Cartography, Wiley, New York, 1960.
12. N. E. Steenrod, The Topology of Fibre Bundles, Princeton University Press, Princeton, N. J., 1951.
13. J. A. Steers, An Introduction to the Study of Map Projections, University of London Press, London, 1957.
14. D. J. Struik, Lectures on Classical Differential Geometry, Addison-Wesley, Reading, Mass., 1950.

# HILBERT'S PROBLEM 18: ON CRYSTALLOGRAPHIC GROUPS, FUNDAMENTAL DOMAINS, AND ON SPHERE PACKING.

*Reprinted from*

Mathematical Developments arising from Hilbert Problems
(Conference held in De Kalb, Illinois, May 1974)
Proceedings of Symposia in Pure Mathematics
Volume XXVIII, Part 2, Felix Browder, ed.
American Mathematical Society 1976, pp. 491–506

As part of his 75th anniversary symposium on the Hilbert Problems, Felix Browder assigned me the job of discussing Problem Number 18. This turned out to involve three different questions, which were barely related to each other. The first question, on so-called "crystallographic groups" was solved very early by Bieberbach.[1] It generalizes however to questions about lattices in Lie groups which still form a very important topic of study.[2]

The second part, involving tiling of Euclidean space, is also a very lively topic today.[3] Important developments have been the theory of Penrose tilings[4] and the closely related physical problem of anomalous crystal structure,[5] as well as Thurston's theory of self-similar fractal tilings.[6] There is still a great deal of life in the subject.

---

[1] Compare L. Charlap, *Bieberbach Groups and Flat Manifolds*, Springer 1986.

[2] See for example M. S. Raghunathan, *Discrete Subgroups of Lie Groups*, Springer 1972, as well as G. A. Margulis, *Discrete Subgroups of Semisimple Lie Groups*, Ergebnisse Math. (3) **17**, Springer 1991.

[3] See B. Grünbaum and G. C. Shephard, "Tilings and Patterns", Freeman 1987.

[4] R. Penrose, *Pentaplexity*, Eureka **39** (1978) 16–22; or Math. Intell. **2** (1979) 32–37.

[5] See for example M. Jaric, "Introduction to the Mathematics of Quasicrystals", Acad. Press 1989.

[6] W. Thurston, *Groups, Tilings and Finite State Automata*, Colloquium Lectures, Amer. Math. Soc., Providence 1989; Research report GCG-1, Geometry Supercomputer Project, Univ. Minn., Minneapolis 1989.

On the third topic, that of sphere packings, there has also been much progress[7], most dramatically with Hsiang's announcement[8] of a proof of Kepler's conjecture on the densest packing in 3-space. The subject of sphere packing in higher dimensions is closely related to important problems in applied mathematics, and will surely be a serious field of study for many years.

The original text has been re-typeset via TeX. One error at the end of §1 has been corrected, and two references filled in.

---

[7] See J. H. Conway and N. J. A. Sloane, "Sphere Packings, Lattices and Groups", Grundlehren Math. **290**, Springer 1988.

[8] Wu-Yi Hsiang, *On the sphere packing problem and the proof of Kepler's conjecture,* Internat. J. Math. **4** (1993) 739–831.

# HILBERT'S PROBLEM 18: ON CRYSTALLOGRAPHIC GROUPS, FUNDAMENTAL DOMAINS, AND ON SPHERE PACKING

**Abstract**

This paper is divided into three completely independent sections, corresponding to three different questions asked by Hilbert. The first section describes Bieberbach's theorems on discrete groups of Euclidean motions. The second describes Heesch's tiling of the plane by a polygon which is not a fundamental domain. The third studies the densest packing of Euclidean space by balls of fixed diameter.

I am indebted to I. Kaplansky and R. M. Robinson for helpful information.

## §1. "Is there in $n$-dimensional euclidean space ... only a finite number of essentially different kinds of groups of motions with a [compact] fundamental region?"

Let $E(n)$ be the Euclidean group, consisting of all isometries of the $n$-dimensional Euclidean space $\mathbf{R}^n$. We look for discrete subgroups $\Gamma \subset E(n)$ such that the coset space $\Gamma \backslash E(n)$ is compact. This is equivalent to the requirement that there exist a compact fundamental domain, that is a compact region $D$ in $\mathbf{R}^n$ such that the various congruent copies $\gamma D$ with $\gamma \in \Gamma$ cover $\mathbf{R}^n$ and have only boundary points in common.

Hilbert's question was answered affirmatively by Bieberbach in 1910. Bieberbach first proved the following important statement:

(1) *Every discrete subgroup $\Gamma \subset E(n)$ with compact fundamental domain contains n linearly independent translations.*

He then showed easily that for any such $\Gamma$:

(2) *The subgroup $T$ consisting of all translations in $\Gamma$ forms a free abelian normal subgroup of finite index. That is, $\Gamma$ admits a short exact sequence of the form*

$$1 \longrightarrow Z^n \xrightarrow{\ i\ } \Gamma \longrightarrow \Phi \longrightarrow 1,$$

*with $\Phi$ finite. Furthermore, the image $T = i(\mathbf{Z}^n)$ is a maximal abelian subgroup of $\Gamma$. In other words, if we let $\Phi$ act on $\mathbf{Z}^n$ by inner automorphism, we obtain a faithful representation of $\Phi$ in* $\mathrm{GL}(n, \mathbf{Z})$.

It follows that this group $T$ of translations in $\Gamma$ can be characterized as the union of all abelian normal subgroups of $\Gamma$. For modern proofs of (1) and (2) the reader is referred to Auslander or Wolf. Finally Bieberbach showed that:

(3) *For each fixed n, there are only finitely many isomorphism classes of groups $\Gamma$ which can be obtained in this way as extensions of $\mathbf{Z}^n$ by finite subgroups of* $\mathrm{GL}(n, \mathbf{Z})$.

The embedding of $\Gamma$ in the Euclidean group $E(n)$ is not uniquely determined in general. (We see this already in the case where $\Gamma$ is free abelian.) However in 1912, inspired by a paper of Frobenius, Bieberbach showed that the embedding of $\Gamma$ in $E(n)$ is uniquely determined up to conjugation by an element of the larger group consisting of all one-to-one affine transformations of $\mathbf{R}^n$.

Many years later, Zassenhaus pointed out that a converse theorem is true also:

*Any group $\Gamma$ which can be described as an extension of $\mathbf{Z}^n$ by a finite subgroup of* $\mathrm{GL}(n, \mathbf{Z})$ *can actually be embedded as a discrete subgroup of the Euclidean group $E(n)$ so as to act on $\mathbf{R}^n$ with compact fundamental domain.*

We will call such groups $\Gamma$ ***Bieberbach*** groups.

For the proof of (3), Bieberbach first recalled Minkowski's statement that there is an upper bound, depending only on $n$, for the orders of finite subgroups of $\mathrm{GL}(n, \mathbf{Z})$. In fact, Minkowski had noted that any finite subgroup of $\mathrm{GL}(n, \mathbf{Z})$ maps injectively into the finite group $\mathrm{GL}(n, \mathbf{Z}/k)$ for $k \geq 3$. (A brief proof is given in Newman.)

Given a finite group $\Phi$, he also needed to know that there are only finitely many embeddings of $\Phi$ in $\mathrm{GL}(n, \mathbf{Z})$, up to inner automorphism. In fact, in Bieberbach's words:

*"Eine gegebene [endliche] Gruppe kann nur auf endlich viele inäquivalente Weisen durch ganzzahlige unimodulare Substitutionen von n Variablen dargestellt werden."*

Although this statement follows easily from the work of Jordan or of Minkowski, Bieberbach seems to have been the first to state it explicitly, and to realize its importance. This theorem has recently become known in the literature as the "Jordan-Zassenhaus Theorem," since Zassenhaus published a different proof in 1938.

Finally, given the action of $\Phi$ on $\mathbf{Z}^n$, one needs to know that there are only finitely many extensions $1 \to \mathbf{Z}^n \to \Gamma \to \Phi \to 1$. Nowadays we would say these extensions are classified by elements of the finite cohomology group $H^2(\Phi; \mathbf{Z}^n)$.

Here is a table, listing the numbers of distinct Bieberbach groups $\Gamma$ in dimensions up to 4.

| dimension $n$ | number of distinct groups |
|---|---|
| 1 | 2 |
| 2 | 17 |
| 3 | 219 |
| 4 | 4783 |

Thus for $n = 1$ there are just two distinct Bieberbach groups $\Gamma$, according as the finite quotient $\Phi = \Gamma/T$ has order 1 or 2. In dimension 2, the 17 Bieberbach groups can be tabulated as follows. (Compare Burkhardt or Hilbert and Cohn-Vossen.)

Table of Bieberbach groups in the plane

| group $\Phi$ | of order | number of embeddings in GL(2, **Z**) | number of Bieberbach groups |
|---|---|---|---|
| cyclic | 1 | 1 | 1 |
| cyclic | 2 | 1 | 1 |
| cyclic | 3 | 1 | 1 |
| cyclic | 4 | 1 | 1 |
| cyclic | 6 | 1 | 1 |
| dihedral | 2 | 2 | 3 |
| dihedral | 4 | 2 | 4 |
| dihedral | 6 | 2 | 2 |
| dihedral | 8 | 1 | 2 |
| dihedral | 12 | 1 | 1 |
| total | | 13 | 17 |

The finite quotient $\Phi$ is cyclic, consisting only of rotations, or dihedral, containing reflections also. (Of course the dihedral group of order 2 is isomorphic to the cyclic group of order 2, but these two groups are listed separately in this table, since they are not conjugate as subgroups of GL(2, **R**).)

The classification of Bieberbach groups in dimension 3 is of great practical importance in the study of crystals. This classification was carried out independently by Fedorov and by Schoenfliess around 1890. (For a more modern presentation see Burkhardt.)

Table of orientation preserving groups in 3-space

| group $\Phi$ | of order | number of embeddings in SL(3, **Z**) | number of Bieberbach groups |
|---|---|---|---|
| cyclic | 1 | 1 | 1 |
| cyclic | 2 | 2 | 3 |
| cyclic | 3 | 2 | 3 (+1) |
| cyclic | 4 | 2 | 5 (+1) |
| cyclic | 6 | 1 | 4 (+2) |
| dihedral | 4 | 4 | 9 |
| dihedral | 6 | 3 | 5 (+2) |
| dihedral | 8 | 2 | 8 (+2) |
| dihedral | 12 | 1 | 4 (+2) |
| tetrahedral | 12 | 3 | 5 |
| octahedral | 24 | 3 | 7 (+1) |
| total | | 24 | 54 (+11) |

Here the quantities in parentheses on the right indicate the number of additional Bieberbach groups which we would obtain if we wish to distinguish between two groups of motions in 3-space which are mirror images of each other. In addition there are 165 Bieberbach groups in 3-space which contain orientation reversing elements. I will not attempt to tabulate these.

Finally, in dimension 4, a recent computation by Brown, Bülow, Neubüser, Wondratschek and Zassenhaus shows that there are 4783 distinct Bieberbach groups. (Compare Dade.)

It would be natural to ask whether Bieberbach's theorems would still hold if we forget the metric on Euclidean space and simply look for groups of one-to-one affine transformations of $\mathbf{R}^n$ which are discrete and have compact fundamental domain. A simple counter-example has been given by L. Auslander. Let $N_{\mathbf{R}}$ denote the group consisting of all $3 \times 3$ real matrices of the form

$$\begin{bmatrix} 1 & * & * \\ 0 & 1 & * \\ 0 & 0 & 1 \end{bmatrix},$$

and let $N_{\mathbf{Z}}$ denote the subgroup consisting of integer matrices of the same form. Then $N_{\mathbf{Z}}$ acts by left multiplication on $N_{\mathbf{R}}$. Identifying $N_{\mathbf{R}}$ with the coordinate space $\mathbf{R}^3$ in the obvious way, each element of $N_{\mathbf{Z}}$ gives rise to a one-to-one affine transformation of 3-space. This action is discrete, has no fixed points, and has the unit cube as fundamental domain. Yet the group $N_{\mathbf{Z}}$ does not contain three linearly independent translations. Furthermore, for $k = 1, 2, 3, \ldots$ the subgroup $N_{k\mathbf{Z}}$ consisting of matrices in $N_{\mathbf{Z}}$ whose off diagonal entries are divisible by $k$ also acts freely on $N_{\mathbf{R}}$ with compact quotient, and no two of these groups are isomorphic. It is interesting to note that each quotient space $N_{k\mathbf{Z}} \backslash N_{\mathbf{R}}$ is a smooth compact manifold with flat affine connection having fundamental group $N_{k\mathbf{Z}}$.

From the point of view of differential geometry, the most interesting Bieberbach groups are those which are torsionfree, i.e., contain no non-trivial elements of finite order. For it is easy to check that the action of a Bieberbach group $\Gamma$ on $\mathbf{R}^n$ is free (i.e., no non-trivial element has a fixed point) if and only if $\Gamma$ is torsionfree. For each torsionfree $\Gamma$ the quotient $\Gamma \backslash \mathbf{R}^n$ is a compact flat riemannian manifold with fundamental group $\Gamma$. Conversely, every compact flat Riemannian manifold is covered by $\mathbf{R}^n$, and hence is isometric to $\Gamma \backslash \mathbf{R}^n$ for some torsionfree Bieberbach group $\Gamma$. Since $\Gamma$ contains a free abelian subgroup of finite index, it follows that every such manifold is covered by a flat torus.

The simplest torsionfree Bieberbach group (other than $\mathbf{Z}^n$) is the Klein bottle group, that is the split extension

$$1 \longrightarrow Z \longrightarrow \Gamma \longrightarrow Z \longrightarrow 1$$

with non-trivial action. In dimension three there are ten distinct examples. (See Wolf, p. 111.) Six of these are orientable, the quotient $\Phi = \Gamma / T$ being either cyclic of order 1, 2, 3, 4, 6, or dihedral of order 4.

The collection consisting of all torsionfree Bieberbach groups is noteworthy since these groups are easily characterized, and all of them are fundamental groups of compact aspherical manifolds. (A manifold is aspherical if its universal covering space is contractible.) From this point of view, an important generalization has recently been given by F. E. A. Johnson.

Let $\Gamma$ be any torsionfree group which can be obtained as an iterated group extension

$$\Gamma = \Gamma_0 \supset \Gamma_1 \supset \cdots \supset \Gamma_k = 1$$

where each $\Gamma_{i+1}$ is a normal subgroup of $\Gamma_i$ and each quotient $\Gamma_i/\Gamma_{i+1}$ is either finite or cyclic. Then Johnson shows that $\Gamma$ is the fundamental group of a smooth, compact, aspherical manifold. He proves this by embedding $\Gamma$, as a discrete subgroup with compact quotient, in a Lie group $G$ having finitely many components, where the component $G_0$ of the identity is solvable. Choosing a maximal compact subgroup $K$, it follows that $\Gamma$ acts freely on the coset space $G/K$, which is topologically Euclidean space.

Now we turn to the second portion of Hilbert's eighteenth problem.

**§2. "Whether polyhedra also exist which do not appear as fundamental regions of groups of motions, by means of which nevertheless by a suitable juxtaposition of congruent copies a complete filling up of all [Euclidean] space is possible?"**

A rather complicated 3-dimensional counter-example to this question was constructed by K. Reinhardt in 1928. A much simpler 2-dimensional counter-example was constructed by Heesch in 1935, and is illustrated in Figure 1. The reader with a taste for plane geometry should find it an amusing and not difficult exercise to prove the following two statements:

Figure 1.

(1) The plane can be covered without overlap by congruent copies of this polygon $H$.

(2) Nevertheless $H$ is not the fundamental domain for any Bieberbach group.

A somewhat simpler example of a polygon with the same two properties is illustrated in Figure 2. This example has the advantage of being star-shaped. (I do not know whether a convex example exists.) However, the proof for this second polygon is slightly more difficult.

Figure 2.

For rather startling related examples, we refer to Robinson's paper "Undecidability and nonperiodicity for tilings of the plane." (See also Stein.)

**§3. "How can one arrange most densely in space an infinite number of equal solids of given form, e.g., spheres with given radii . . . , that is, how can one so fit them together that the ratio of the filled to the unfilled space may be as great as possible?"**

For 2-dimensional disks this problem has been solved by Thue and Fejes Tóth, who showed that the expected hexagonal (or honeycomb) packing of circular disks in the plane is the densest possible. However, the corresponding problem in 3 dimensions remains unsolved. This is a scandalous situation since the (presumably) correct answer has been known since the time of Gauss. (Compare Hilbert and Cohn-Vossen.) All that is missing is a proof.

Here are some definitions that will help to clarify this problem. Given positive real numbers $a_1, \ldots, a_n$, and $d$, let $N_d(a_1, \ldots, a_n)$ be the maximum number of $n$-dimensional balls of diameter $d$ which can be packed, without overlapping, into the $n$-dimensional "suitcase" consisting of all $(x_1, \ldots, x_n)$ with $0 \le x_i \le a_i$ for $i = 1, \ldots, n$. Clearly $N_d$ is superadditive as a function of any one of its $n$ variables if the other $n - 1$ variables are kept fixed. For example

$$N_d(a_1' + a_1'', a_2, \ldots, a_n) \ge N_d(a_1', a_2, \ldots, a_n) + N_d(a_1'', a_2, \ldots, a_n).$$

An elementary argument, which will be left to the reader, then shows that the ratio

$$N_d(a_1, \ldots, a_n)/a_1 \cdots a_n$$

tends to a finite non-zero limit as $a_1, \ldots, a_n \to \infty$. This limit measures the number of balls per unit volume, and is called the maximum center-density for balls of diameter $d$ in $n$-space. Using the identity $N_d(a_1, \ldots, a_n) = N_1(a_1/d, \ldots, a_n/d)$, we see that this maximum center-density is proportional to $d^{-n}$. Hence we can denote it by $c_n/d^n$ where $c_n$ is the maximum center-density for unit diameter balls in $\mathbf{R}^n$. Note that the $n$-th root ${c_n}^{1/n}$ is equal to the largest possible diameter for balls which can be packed with center-density 1 in $n$-space.

With these preliminaries we can quote the values $c_1 = 1$, $c_2 = \sqrt{4/3}$, and

$$c_3 \ge \sqrt{2} = 1.41421\ldots.$$

It is believed that equality holds here, but this has never been proved.

A beautiful upper bound for $c_n$ has been constructed by C. A. Rodgers, based on earlier work by Blichfeldt. The Rogers upper bound is precisely equal to $c_n$ for $n = 1$, 2, and takes the form

$$c_3 \le 6\sqrt{2}\left(\frac{3}{\pi}\cos^{-1}(1/3) - 1\right) = 1.48899\ldots$$

for $n = 3$. In dimensions 8 and 24 there are known packings whose center-density comes fairly close to the Rogers upper bound, but the discrepancy is rather wide for most dimensions.

The situation becomes somewhat more manageable if we restrict the situation by considering only arrangements of balls in $n$-space such that the center points form a lattice. By a lattice we mean a discrete additive subgroup $L$ of the vector space $\mathbf{R}^n$. The determinant $\det(L)$ of such a lattice $L$, with basis $v_1, \ldots, v_n$, will mean the real number

$$\text{volume}(\mathbf{R}^n/L) = |\det(v_1, \ldots, v_n)| = \sqrt{\det[v_i \cdot v_j]}.$$

Clearly the reciprocal $1/\det(L)$ can be identified with the density of lattice points per unit volume in $\mathbf{R}^n$. Let

$$\min(L) = \min_{v \in L-0} \|v\|$$

be the minimum distance between distinct lattice points. Then we can center a ball of diameter $\min(L)$ at each lattice point without overlap. Using the inequality

$$\text{density} \leq c_n/(\text{diameter})^n$$

we obtain

$$1/\det(L) \leq c_n/\min(L)^n$$

or in other words

$$\min(L)^n/\det(L) \leq c_n.$$

Let $c_n^{\text{lattice}}$ denote the maximum, among all lattices in $\mathbf{R}^n$, of this ratio $\min(L)^n/\det(L)$. Equivalently, $c_n^{\text{lattice}}$ can be defined as the maximum center-density for lattice packings of $\mathbf{R}^n$ by balls of diameter 1. Evidently $c_n^{\text{lattice}} \leq c_n$.

Korkin and Zolotarev computed this maximum lattice center-density for $n \leq 5$, and Blichfeldt later extended their work to the case $n \leq 8$. For any dimension $n$, it follows from the work of Korkin and Zolotarev that this maximum $c_n^{\text{lattice}}$ is actually attained for some lattice $L_n$ in $\mathbf{R}^n$, and is the square root of a rational number.

For $n \leq 8$ the densest possible lattice can be constructed out of an

appropriate Dynkin diagram as follows

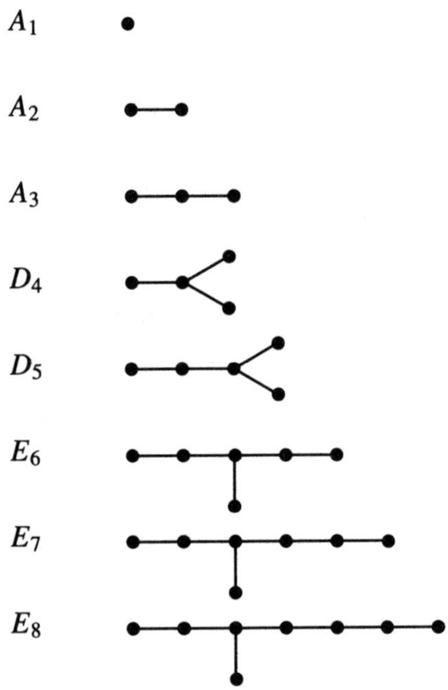


In each of these diagrams a dot stands for a basis vector $v$ with $v \cdot v = 2$. The inner product of any two of these basis vectors is defined to be either $-1$ or 0 according as they are or are not joined by a line segment. Thus each such diagram with $n$ vertices determines a symmetric $n \times n$ inner product matrix with 2's along the diagonal. It can be verified that each of these eight matrices is positive definite, and hence coincides with the inner product matrix $[v_i \cdot v_j]$ for suitably chosen vectors $v_1, \ldots, v_n \in \mathbf{R}^n$, spanning a lattice $L_n$ with $\min(L_n) = \sqrt{2}$. The determinant of this inner product matrix is 2, 3, 4, 4, 4, 3, 2, or 1 respectively, so the determinant of $L_n$ is equal to $\sqrt{2}$, $\sqrt{3}$, 2, 2, 2, $\sqrt{3}$, $\sqrt{2}$, or 1 respectively. Setting $c_n^{\text{lattice}} = \min(L_n)^n / \det(L_n)$ we obtain the following table.

| $n$ | 1 | 2 | 3 | 4 | 5 | 6 | 7 | 8 |
|---|---|---|---|---|---|---|---|---|
| $c_n^{\text{lattice}}$ | 1 | $\sqrt{4/3}$ | $\sqrt{2}$ | 2 | $2\sqrt{2}$ | $8/\sqrt{3}$ | 8 | 16 |

For dimensions $n \geq 9$ the maximal center-density $c_n^{\text{lattice}}$ is not known. For estimates when $n \leq 24$, see the work of Leech and Sloane. For large values of $n$, the best one has are rather crude and non-constructive lower bounds. For example, Minkowski obtained a lower bound which is very slightly stronger than the following:

$$c_n^{\text{lattice}} > \frac{2}{\text{volume of ball of radius 1}} = 2(n/2)!/\pi^{n/2}.$$

This has been sharpened by Rogers and others. Perhaps the situation can best be put in perspective by taking the $n$-th root of both sides. Then the quantity $\left(c_n^{\text{lattice}}\right)^{1/n}$, which can be identified with the largest possible value of $\min(L)$ for a unimodular lattice in $n$-space, has the order of magnitude of $\sqrt{n}$. In fact, the Minkowski or Rogers lower bounds reduce to

$$\liminf \left(c_n^{\text{lattice}}\right)^{1/n} / \sqrt{n} \geq 1/\sqrt{2\pi e}$$

as $n \to \infty$, while the Blichfeldt or Rogers upper bounds reduce to

$$\limsup {c_n}^{1/n} / \sqrt{n} \leq 1/\sqrt{\pi e}.$$

Thus for large $n$ the numbers

$${c_n}^{1/n} \geq \left(c_n^{\text{lattice}}\right)^{1/n}$$

are known to within a factor of approximately $\sqrt{2}$. (Compare Rogers, Sloane, Milnor and Husemoller.)

To conclude, here are three open problems.

One needs some technique for actually computing $c_n$, the maximum center-density for unit diameter balls in $n$-space, for small values of $n$. In particular one would like to prove that $c_3 = \sqrt{2}$.

In principle, one knows how to compute $c_n^{\text{lattice}}$, the maximum center-density for lattice packings of unit diameter balls. The procedure used by Korkin and Zolotarev can be reformulated as follows. Consider the convex polyhedron consisting of all symmetric $n \times n$ matrices $[a_{ij}]$ which satisfy the inequality $\sum a_{ij} k_i k_j \geq 1$ for every non-zero $n$-tuple $(k_1, \ldots, k_n)$ of integers. Then the determinant function attains its minimum at some vertex, which we may take to lie in a carefully chosen compact subpolyhderon. (Compare Watson.) With modern computing and programming techniques, one should be able to actually find this minimum for small values of $n \geq 9$. One reward of such a computation might be a proof that $c_n > c_n^{\text{lattice}}$ in some cases. For Leech and Sloane have given examples of sphere packings in dimensions 10, 11, and 13 which are denser than any known lattice packings in these dimensions.

Finally, there is the problem of determining the asymptotic behavior of the numbers $c_n \geq c_n^{\text{lattice}}$ (or at least of their $n$-th roots) as $n \to \infty$. In particular, do the ratios

$$c_n{}^{1/n}/\sqrt{n} \geq \left(c_n^{\text{lattice}}\right)^{1/n}/\sqrt{n}$$

tend to limits as $n \to \infty$? As noted earlier, these limits would have to lie between $1/\sqrt{2\pi e}$ and $1/\sqrt{\pi e}$.

## Bibliography

L. Auslander,, *Examples of locally affine spaces*, Annals of Math. **64** (1956), 255–259.

______, *An account of the theory of crystallographic groups*, Proc. Amer. Math. Soc. **16** (1956), 1230–1236.

L. Bieberbach, *Über die Bewegungsgruppen des n-dimensionalen euklidischen Raumes mit einem endlichen Fundamentalbereich*, Gött. Nachr. (1910), 75–84.

______, *Über die Bewegunsgruppen der Euklidischen Räume* II, *Die Gruppen mit einem endlichen Fundamentalbereich*, Math. Ann. **72** (1912), 400–412.

______, *Über die Minkowskische Reduktion der positiven quadratischen Formen und die endlichen Gruppen linearer ganzzahliger Substitutionen*, Gött. Nachr. (1912), 207–216.

H. F. Blichfeldt, *The minimum values of quadratic forms in six, seven, and eight variables*, Math. Zeits. **39** (1935), 1–15.

H. Brown, R. Bülow, J. Neubüser, H. Wondratschek and H. Zassenhaus, "Crystallographic Groups of Four-Dimensional Space," Wiley, 1978.

J. J. Burkhardt, "Die Bewegungsgruppen der Kristallographie," Basel, 1947.

E. C. Dade, *The maximal finite groups of* $4 \times 4$ *integral matrices*, Ill. J. Math. **9** (1965), 99–122.

E. S. Fedorov, "Symmetry of Crystals," (translated from 1949 Russian edition), Amer. Crystallographic Assoc., New York, 1971.

L. Fejes Tóth (= L. Fejes), *Über einem geometrischen Satz*, Math. Zeits. **46** (1940), 79–83.

G. Frobenius, *Über die unzerlegbaren diskreten Bewgungsgruppen*, Sitzungsb. Akad. Wiss. Berlin (1911), 654–665.

C. F. Gauss, *Untersuchungen über die Eigenschaften der positiven ternären quadratischen Formen von L. A. Seeber*, Werke **2**, 188–196, Göttingen 1876.

H. Heesch, *Aufbau der Ebene aus kongruenten Bereichen*, Gött. Nachr. (1935), 115–117.

______, "Reguläres Parkettierungsproblem," Arbeitsg. Forschung Nordrhein-Westf. **172**, Westdeutscher Verlag, 1968.

D. Hilbert and S. Cohn-Vossen, "Anschauliche Geometrie," Springer, 1932. (English transl., Chelsea 1952).

F. E. A. Johnson, *On a conjecture of C. T. C. Wall,* (preprint 1974); F. E. A. Johnson and L. Auslander, J. Lond. Math. Soc. **14** (1976) 331–332.

C. Jordan, *Mémoire sur l'équivalence des formes,* J. de l'École Polytech. **29** #48 (1880), 111–150. (See p. 133.)

A. Korkine and G. Zolotareff, *Sur les formes quadratiques positives,* Math. Ann. **11** (1877), 242–292.

J. Leech, *Notes on sphere packings,* Canad. J. Math. **19** (1967), 251–267.

______ and N. J. A. Sloane, *Sphere packings and error correcting codes,* Canad. J. Math. **23** (1971), 718–745.

J. Milnor and D. Husemoller, "Symmetric Bilinear Forms," Ergebnisse **73**, Springer, 1973.

H. Minkowski, *Diskontinuitätsbereich für arithmetische Äquivalenz,* J. reine agnew. Math. **129** (1905), 220–274. (See pp. 240, 269.)

M. Newman, "Integral Matrices," Academic Press, 1972. (See p. 175.)

K. Reinhardt, *Zur Zerlegung der euklidschen Räume in kongruente Polytope,* Sitzungsb. Akad. Wiss. Berlin (1928), 150–155.

R. M. Robinson, *Undecidability and nonperiodicity for tilings of the plane,* Inventiones math. **12** (1971), 177–209.

C. A. Rogers, *The packing of equal spheres,* Proc. London Math. Soc. **8** (1958), 609–620.

______, "Packing and Covering," Cambridge U. Press, 1964.

A. Schoenfliess, "Kristallsysteme und Kristallstruktur," Leipzig, 1891.

N. J. A. Sloane, *Sphere packings constructed from BCH and Justesen codes,* Mathematika **19** (1972), 183–190.

S. K. Stein, *Algebraic Tiling,* Amer. Math. Monthly **81** (1974), 445–462.

A. Thue, *Über die dichteste Zusamenstellung von Kongruenten Kreisen in einer Ebene,* Skr. Vidensk-Selsk, Christ. **1** (1910), 1–9.

G. L. Watson, "Integral Quadratic Forms," Cambridge U. Press, 1960.

J. Wolf, "Spaces of Constant Curvature," 5th ed., Publish or Perish, Inc., Houston, Texas, 1984.

H. Zassenhaus, *Neuer Beweis der Endlichkeit der Klassenzahl bei unimodularer Äquivalenz endlicher ganzzahliger Substitutionsgruppen,* Abh. Math. Sem. Hamburg **12** (1938), 276–288.

______, *Über einen Algorithmus zur Bestimmung der Raumgruppen,* Commentarii **21** (1948), 117–141.

# HOW TO COMPUTE VOLUME IN HYPERBOLIC SPACE[1]

## Preface

The late seventies were an exciting time in topology. The theory of 3-dimensional manifolds, which had been an isolated backwater of mathematics, suddenly began to occupy a much more prominent role. In particular, Thurston had begun to formulate his geometrization conjecture suggesting that 3-manifold theory was intimately connected with several very classical forms of geometry.[2] The most interesting and important of these is the '***hyperbolic***' or '***Lobachevsky***' geometry. If a manifold of any dimension can be realized as a hyperbolic manifold, complete with curvature $-1$, then according to the Rigidity Theorem of Mostow, Margulis and Prasad, its ***volume*** is a topological invariant.[3] Given two such manifolds, if one could prove that the ratio $\mathrm{volume}(M_1)/\,\mathrm{volume}(M_2)$ is irrational, then it would follow that no finite covering manifold of $M_1$ could be homeomorphic[4] to a finite covering manifold of $M_2$.

I was fascinated by Thurston's ideas, and corresponded with him regularly at this time. (We were both in Princeton, but seldom actually saw each other.) The following paper is based on a handwritten manuscript "Notes on Hyperbolic Volume" from 1978, which attempted to understand how one can actually compute hyperbolic volume. Related material was circulated as part of the Thurston lecture notes.[5] However, this 1978 manuscript also

---

[1] Not previously published.

[2] See W. Thurston, *Three-dimensional manifolds, Kleinian groups and hyperbolic geometry*, Bull. Amer. Math. Soc. **6** (1982), 357–381; as well as *Hyperbolic structures on 3-manifolds, I, deformation of acyclic manifolds*, Annals of Math. **124** (1986), 203–246; and "Three-Dimensional Topology and Geometry, I", Princeton U. Press, to appear. See also C. McMullen, *Riemann surfaces and geometrization of 3-manifolds*, Bull. Amer. Math. Soc. **27** (1992), 207–216.

[3] G. D. Mostow, *Strong Rigidity of Locally Symmetric Spaces*, Ann. Math. Studies **78** Princeton U. Press 1973; G. Margulis, *Isometry of closed manifolds of constant negative curvature with the same fundamental group*, Sov. Math. Doklady **11** (1970), 722–723; and G. Prasad, *Strong rigidity of* $\mathbf{Q}$*-rank 1 lattices*, Invent. Math. **21** (1973), 255–286. (Compare M. Gromov, *Volume and bounded cohomology*, Pub. Math. IHES **56** (1983), 5–99.)

[4] Compare Milnor and Thurston, *Characteristic numbers of 3-manifolds*, Enseign. Math. **23** (1977), 249–254; and see A. Borel, *Commensurability classes and volumes of hyperbolic 3-manifolds*, Ann. Sc. Norm. Sup. Pisa Cl. Sci. **8** (1981), 1–33 or "Collected Papers" **3**, 617–649, Springer 1983.

[5] "The Geometry and Topology of Three-Manifolds", Lecture Notes, Princeton University 1979 (available through The Geometry Center, University of Minnesota, Minneapolis).

included a discussion of groups generated by reflections in $H^3$, and a discussion of higher dimensional power series computations. It has been revised for clarity and mildly updated.[1] For example I have added a statement of the Haagerup-Munkholm Theorem which appeared a few years later, and added an Appendix.

## §1. Some Standard Models for Hyperbolic Space

The ***upper half space model*** for the hyperbolic space $H^{n+1}$ consists of all $(n+1)$-tuples $(x_1, \ldots, x_n, y)$ with $y > 0$, provided with the Riemannian metric

$$ds^2 = \left(dx_1{}^2 + \cdots + dx_n{}^2 + dy^2\right)/y^2. \tag{1}$$

Its $(n+1)$-dimensional volume element is

$$dV_{n+1} = dx_1 \cdots dx_n \, dy/y^{n+1} .$$

The hyperbolic straight lines in this model, or in other words the geodesics, consist of

a) all Euclidean semi-circles in this half-space which meet the boundary hyperplane $y = 0$ orthogonally, and
b) all vertical half-lines $x_1 = \text{constant}, \ldots, x_n = \text{constant}$.

By the ***sphere of points at infinity*** in this model we mean the set $\mathbf{R}^n \cup \mathbf{v}_\infty$ consisting of the Euclidean hyperplane $y = 0$, compactified by adding an infinite point $\mathbf{v}_\infty$. Thus every line in $H^{n+1}$ joins two distinct points on $\mathbf{R}^n \cup \mathbf{v}_\infty$, and is uniquely determined by these two points. The hyperplanes of $H^{n+1}$ in this model consist of

a′) all Euclidean hemispheres $y = +\sqrt{c^2 - (x_1 - a_1)^2 - \cdots - (x_n - a_n)^2}$ which meet $y = 0$ orthogonally, and
b′) all vertical half-planes, specified by a linear relation between the $x_i$.

The standard unit hemisphere

$$y = \sqrt{1 - x_1{}^2 - \cdots - x_n{}^2}$$

[1] For further references on hyperbolic volume, see *Hyperbolic geometry: The first 150 years*, this volume, pp. 245–260, as well as *The Schläfli differential equality*, this volume, pp. 281–295.

provides a typical example. In this way we obtain the ***unit hemisphere model*** for the hyperbolic space $H^n$, consisting of this unit hemisphere with the metric (1). The geodesics in this model are the intersections of the hemisphere with vertical 2-planes.

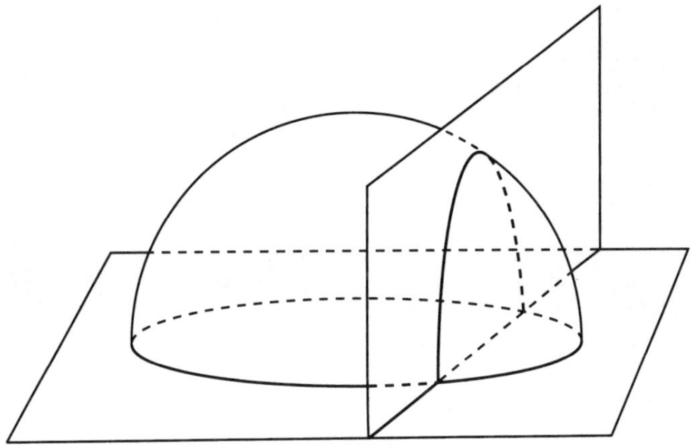

Closely related is the ***projective unit disk model*** for $H^n$, obtained simply by ignoring the $y$-coordinate, or by projecting vertically to $y = 0$. Thus we consider the unit disk

$$x_1{}^2 + \cdots + x_n{}^2 < 1$$

with the metric

$$ds^2 = \left(dx_1{}^2 + \cdots + dx_n{}^2 + \left(d\sqrt{1-r^2}\,\right)^2\right)/(1-r^2), \tag{1'}$$

where $r = \sqrt{x_1{}^2 + \cdots + x_n{}^2}$. Geodesics in this model are just straight line segments joining two points on the sphere $x_1{}^2 + \cdots + x_n{}^2 = 1$, which is thought of as the ***sphere of points at infinity*** of $H^n$. Note however that angles are distorted in this model. (Compare the Appendix.)

As a first example of a volume computation using the upper half space model, consider a region defined by an inequality of the form

$$y \geq f(x_1, \ldots, x_n)$$

where $(x_1, \ldots, x_n)$ varies over some bounded region $\mathcal{R}$ of $\mathbf{R}^n$. If

$$G = \left\{\left(x_1, \ldots, x_n, f(x_1, \ldots, x_n)\right) : (x_1, \ldots, x_n) \in \mathcal{R}\right\}$$

is the graph of $f$, we will describe this region as the ***infinite cone*** which is obtained by joining $G$ to the point $\mathbf{v}_\infty$.

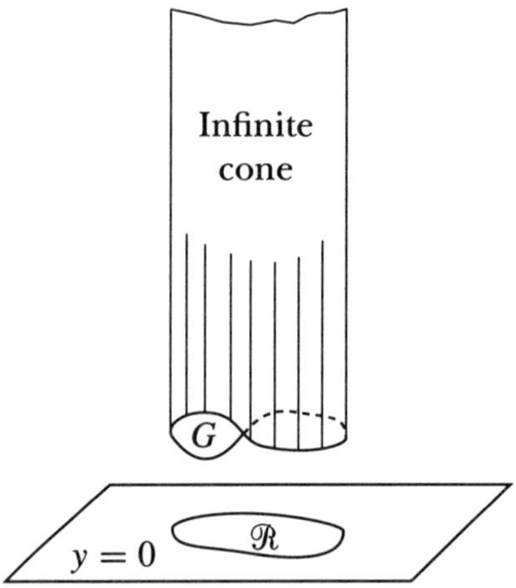


Picture in upper half-space model

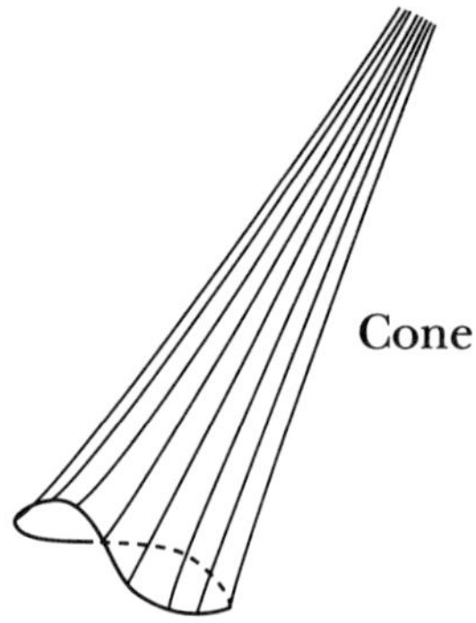


Intuitive picture

Integrating the volume element $dV_{n+1} = dx_1 \cdots dx_n\, dy/y^{n+1}$ over this cone $y > f(\mathbf{x})$, $\mathbf{x} \in \mathcal{R}$, we obtain the formula

$$V_{n+1}(\text{Cone}) = \frac{1}{n} \int_{\mathcal{R}} \frac{dx_1 \cdots dx_n}{f(x_1, \ldots, x_n)^n}. \tag{2}$$

As an easy special case, if $f(\mathbf{x})$ is identically equal to 1, then this volume is proportional to the $n$-dimensional Euclidean volume

$$V_{n+1}(\text{Cone}) = \frac{1}{n} \int_{\mathcal{R}} dx_1 \cdots dx_n.$$

(The locus $y = 1$ is described as a ***horosphere*** in hyperbolic space. By definition, a horosphere is a locus, isometric to Euclidean $n$-space, which forms an orthogonal trajectory to the family consisting of all geodesics which terminate at a common point on the sphere at infinity.)

Now suppose in particular that our hypersurface $y = f(x_1, \ldots, x_n)$ is the standard hyperplane $y = \sqrt{1 - x_1{}^2 - \cdots - x_n{}^2}$. Then we obtain the following:

**Lemma.** *The volume of the infinite cone obtained by joining a compact region* $\mathfrak{R}$ *in the hyperplane* $y = \sqrt{1 - x_1{}^2 - \cdots - x_n{}^2} > 0$ *to* $\mathbf{v}_\infty$ *is given by the formula*

$$V_{n+1} = \frac{1}{n} \int_{\mathfrak{R}} \frac{dx_1 \cdots dx_n}{(1 - x_1{}^2 - \cdots - x_n{}^2)^{n/2}}. \tag{3}$$

## §2. The Two Dimensional Case: Area in the Hyperbolic Plane

For example consider a triangle in $H^2$ with one vertex at infinity, with

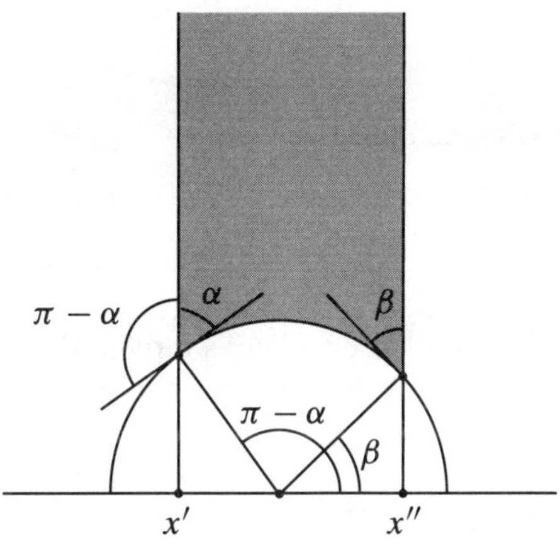


Picture in upper half-plane model

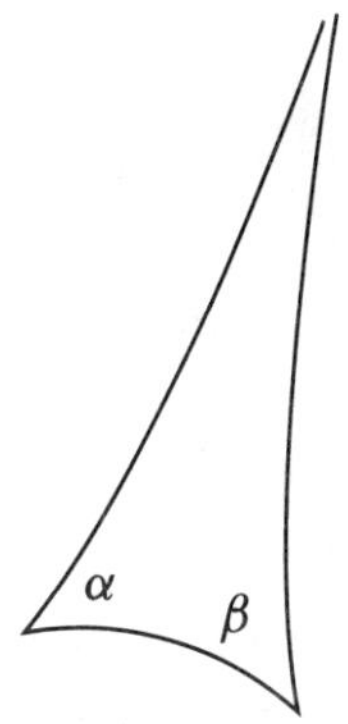


Intuitive picture

angles $\alpha$, $\beta$ as illustrated. Then Formula (3) takes the form

$$\text{area} = V_2 = \int_{x'}^{x''} \frac{dx}{\sqrt{1-x^2}} = \arccos(x') - \arccos(x'') = (\pi - \alpha) - \beta.$$

The largest possible area which can be obtained in this way is

$$V_2^{\max} = \int_{-1}^{1} \frac{dx}{\sqrt{1-x^2}} = \pi.$$

This corresponds to the case of an ***ideal*** hyperbolic 2-simplex, that is a 2-simplex with distinct vertices all located on the 1-sphere at infinity. *Thus the area of an ideal hyperbolic 2-simplex is equal to* $\pi$.

The formula

$$\text{area}(T) = \pi - \alpha - \beta - \gamma$$

for a bounded triangle $T$ with angles $\alpha$, $\beta$, $\gamma$ is then easily obtained by extending one edge to infinity and setting $\text{area}(T) = \text{area}(T \cup T') - \text{area}(T')$.

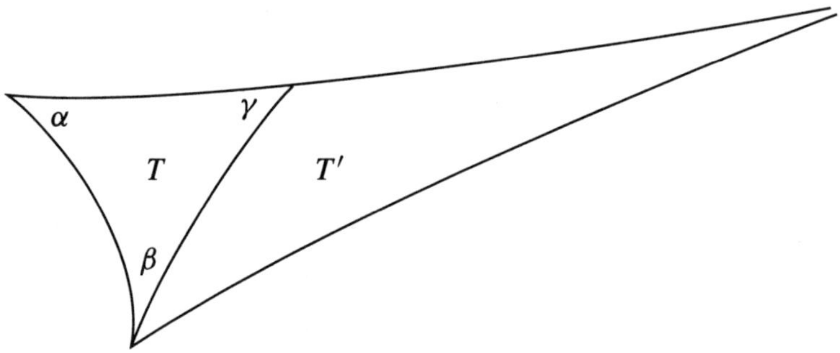


### §3. The Three Dimensional Case

Consider the right triangle $T$ in the unit 2-disk consisting of all $(x, y)$ with:

$$0 \leq x \leq b, \quad 0 \leq y \leq x \tan \alpha.$$

Coning to $\mathbf{v}_\infty$, as in (3) above, we obtain a hyperbolic 3-simplex $\Sigma$ in the upper half space model, with one vertex at infinity, consisting of all $(x, y, z)$ with $(x, y)$ in $T$ and $z > \sqrt{1 - x^2 - y^2}$.

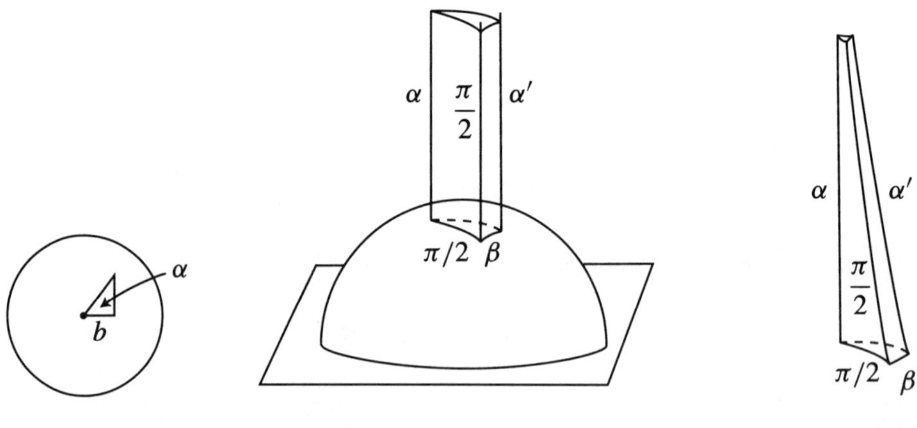


Projection to plane Upper half-space picture Intuitive picture

Figure 1. The right triangle $T$ and two views of the 3-simplex $\Sigma$ with edges labeled by their dihedral angles. Here $\alpha' = \pi/2 - \alpha$ and $b = \cos\beta$. (The dihedral angles along the dotted edges are also $\pi/2$.)

Applying (3) we have

$$V_3(\Sigma) = \frac{1}{2}\int_T \frac{dx\,dy}{1-x^2-y^2}.$$

Using the identity

$$\int_0^c \frac{dy}{a^2-y^2} = \frac{1}{2a}\log\Big(\frac{a+c}{a-c}\Big),$$

we have

$$(4) \qquad \int_0^{x\tan\alpha} \frac{dy}{1-x^2-y^2} = \frac{1}{2\sqrt{1-x^2}}\log\left(\frac{\sqrt{1-x^2}+x\tan\alpha}{\sqrt{1-x^2}-x\tan\alpha}\right).$$

Now we must integrate this expression (4) from $x=0$ to $x=b$ with respect to $dx$. Using the substitution $x=\cos\theta$ for $\beta \le \theta \le \pi/2$, where $\cos\beta = b$, $\sin\theta = \sqrt{1-x^2}$, and $d\theta = -dx/\sqrt{1-x^2}$, we obtain the preliminary formula

$$(5) \qquad V_3(\Sigma) = \frac{1}{4}\int_\beta^{\pi/2} \log\left(\frac{\sin(\theta+\alpha)}{\sin(\theta-\alpha)}\right) d\theta.$$

(Lobachevsky, p. 56. Note that $\sin(\theta\pm\alpha) > 0$ throughout the range of integration.)

It will be convenient to express this result in terms of the following version of the log sine integral. Let

$$(6) \qquad Л(\theta) = -\int_0^\theta \log|2\sin u|\,du.$$

I will call Л the ***Lobachevsky function***.[1] Evidently the indefinite integral of $\log\big(\sin(\theta+\alpha)/\sin(\theta-\alpha)\big)\,d\theta$ is equal to $Л(\theta-\alpha)-Л(\theta+\alpha)$, and the definite integral (5) is given by

$$(7) \qquad V_3(\Sigma) = \tfrac{1}{4}\big(Л(\pi/2-\alpha) - Л(\pi/2+\alpha) - Л(\beta-\alpha) + Л(\beta+\alpha)\big).$$

[1] This is not quite accurate historically. Lobachevsky's volume formulas were actually expressed in terms of the function

$$\int_0^\theta \log|\sec u|du = \theta\log 2 + Л(\theta+\pi/2),$$

which is much more awkward to work with.

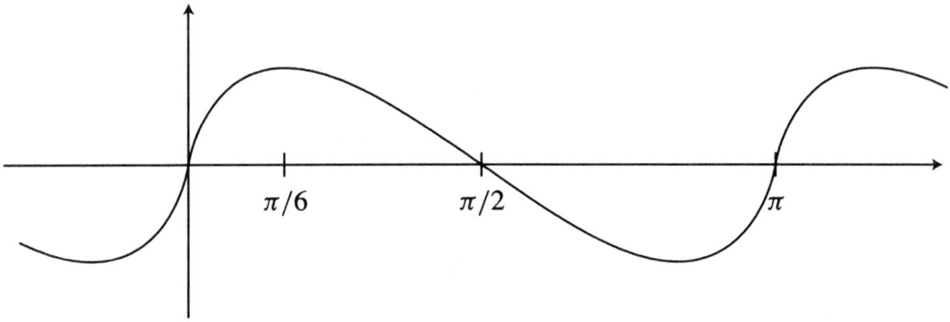


Figure 2. Graph of the Lobachevsky function Л.

We will return to this expression after developing some properties of the function Л.

It follows immediately from the defining equation (6) that the Lobachevsky function Л is smooth and analytic except when $\theta$ is a multiple of $\pi$. Evidently Л is an odd function. Note that the second derivative $d^2 Л/d\theta^2$ is equal to $-\cot\theta$.

Closely related is the ***dilogarithm function***

$$\psi(z) = \sum_{n=1}^{\infty} \frac{z^n}{n^2} = -\int_0^z \log(1-w)\, dw/w,$$

studied by Abel. (Here $|z| \le 1$, $|w| \le 1$.) In fact substituting

$$w = e^{2i\theta}, \quad dw/w = 2i\,d\theta,$$
$$\log(1-w) = \log\big((e^{-i\theta} - e^{i\theta})e^{i\theta}\big) = \log(2\sin\theta) + i(\theta - \pi/2)$$

for $0 < \theta < \pi$, and recalling that

$$\psi(1) = \sum \frac{1}{n^2} = \pi^2/6,$$

we obtain the formula

$$\psi(e^{2i\theta}) = 2i\,Л(\theta) + \pi^2/6 - \theta(\pi - \theta)$$

for $0 < \theta < \pi$. Taking the imaginary part of both sides, it follows easily that

(8) $$Л(\theta) = \frac{1}{2}\sum_{n=1}^{\infty} \frac{\sin(2n\theta)}{n^2}.$$

(Compare [Clausen].) In particular, $Л(0) = Л(\pi/2) = Л(\pi) = 0$. Since the derivative

$$\frac{dЛ(\theta)}{d\theta} = -\log|2\sin\theta|$$

is periodic of period $\pi$, it follows that the function $Л(\theta)$ itself is periodic of period $\pi$. In particular the formula (8), which was proved above only for $0 < \theta < \pi$, must actually be true for all $\theta$. Since Л is continuous and periodic, it has a finite maximum, namely $Л(\pi/6) = 0.5074708\ldots$.

For computational purposes, it is much better to work with the power series

$$Л(\theta) = \theta\left(1 - \log(2\theta) + \frac{\theta^2}{18} + \frac{\theta^4}{900} + \frac{\theta^6}{19845} + \cdots\right)$$

which converges for $|\theta| < \pi$, and hence converges reasonably well for $|\theta| < \pi/2$. The coefficient of the $\theta^{2n}$ term is

$$\frac{B_n}{2n}\frac{2^{2n}}{(2n+1)!}$$

for $n \geq 1$, where $B_1 = 1/6$, $B_2 = 1/30$, ... are Bernoulli numbers.

Integrating the equation $\log|2\sin 2u| = \log|2\sin u| + \log|2\cos u|$, we obtain the basic identity

$$(9_2) \qquad \tfrac{1}{2}Л(2\theta) = Л(\theta) + Л(\theta + \pi/2) = Л(\theta) - Л(\pi/2 - \theta).$$

As an example, taking $\theta = \pi/6$, it follows that

$$\tfrac{3}{2}Л(\pi/3) = Л(\pi/6).$$

More generally, the trigonometric identity

$$2\sin nu = \prod_{j=0}^{n-1} 2\sin(u + j\pi/n)$$

leads to the relation

$$(9_n) \qquad Л(n\theta)/n = \sum_{j \bmod n} Л(\theta + j\pi/n).$$

Compare [Bloch, pp. 6–7].

**Problem.** If we consider only angles $\theta$ for which $\theta/\pi$ is rational, is every **Q**-linear relation between the numbers $Л(\theta)$ a consequence of the relations $(9_n)$, together with the trivial relations

$$Л(-\theta) = -Л(\theta), \quad Л(\theta + \pi) = Л(\theta)?$$

I have searched for other linear relations by means of a crude computer program, but failed to find any. More generally, one might even ask the same question for angles for which $e^{i\theta}$ is algebraic.

Now let us return to the volume computation which was begun above. Since the function Л is odd and periodic, we can restate Formula (7) as follows.

**Assertion.** *For the hyperbolic 3-simplex $\Sigma = \Sigma(\alpha, \alpha', \beta)$ illustrated in the Figure on page 194 we have*

$$V_3\big(\Sigma(\alpha, \alpha', \beta)\big) = \tfrac{1}{4}\big(Л(\alpha + \beta) + Л(\alpha - \beta) + 2Л(\alpha')\big). \tag{10}$$

(Here the notation $\Sigma(\alpha, \alpha', \beta)$ means that the dihedral angles along three consecutive edges are $\alpha$, $\alpha'$ and $\beta$, while the remaining dihedral angles are $\pi/2$. See the discussion of "orthosimplexes" below.) If we specialize to the case $\beta = \alpha$, then the simplex will have two vertices at infinity, and the formula becomes: $V_3 = \big(Л(2\alpha) + 2Л(\alpha')\big)/4$. Making use of $(9_2)$, this reduces to

$$V_3\big(\Sigma(\alpha, \alpha', \alpha)\big) = \tfrac{1}{2}Л(\alpha). \tag{11}$$

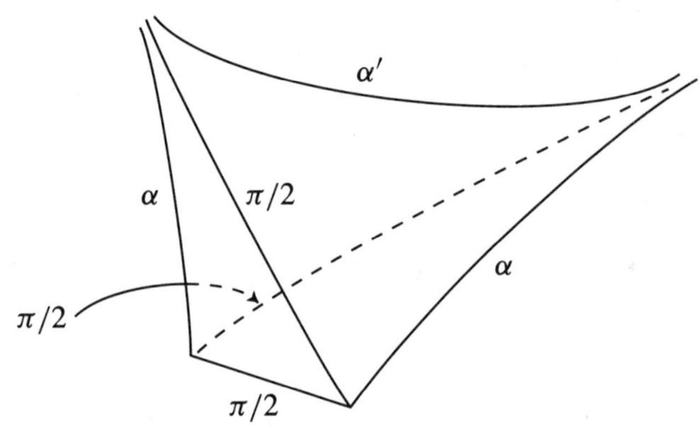


Picture of $\Sigma(\alpha, \alpha', \alpha)$.

Now consider a triangle which is inscribed in the unit circle, as shown below. Again coning to $\mathbf{v}_\infty$, we obtain a hyperbolic 3-simplex with all four

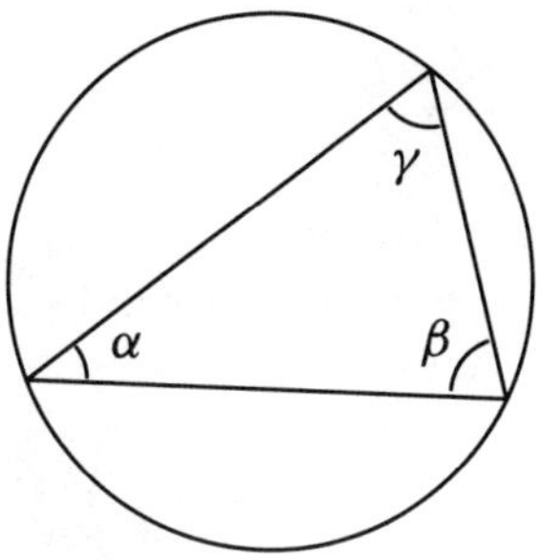


Projection to plane

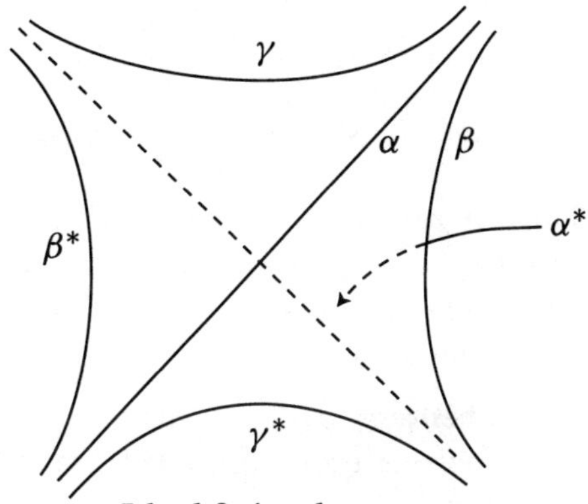


Ideal 3-simplex,
with dihedral angles as indicated

vertices on the sphere at infinity. The sum of the dihedral angles around the vertex $\mathbf{v}_\infty$ is $\alpha + \beta + \gamma = \pi$. There are analogous relations

$$\alpha^* + \beta^* + \gamma = \pi , \quad \alpha^* + \beta + \gamma^* = \pi, \quad \alpha + \beta^* + \gamma^* = \pi$$

associated with the other three vertices, and it follows easily that

$$\alpha = \alpha^*, \quad \beta = \beta^*, \quad \gamma = \gamma^*.$$

*Thus the opposite dihedral angles of an ideal hyperbolic* 3*-simplex are equal.* To compute volume, assuming for the moment that $\alpha, \beta, \gamma < \pi/2$, we drop perpendiculars from the center point of the disk to the three edges, thus

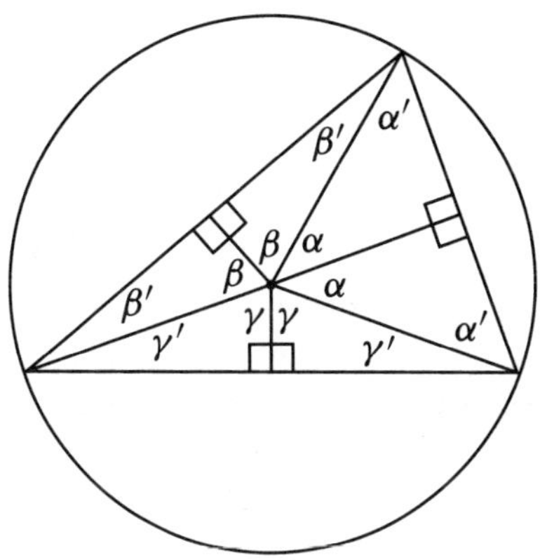

splitting the given triangle into six right triangles. Here

$$\alpha' + \beta' = \gamma, \quad \alpha' + \gamma' = \beta, \quad \beta' + \gamma' = \alpha,$$

and it follows that

$$\alpha' = \pi/2 - \alpha, \quad \beta' = \pi/2 - \beta, \quad \gamma' = \pi/2 - \gamma.$$

Thus we obtain two copies each of the 3-simplexes $\Sigma(\alpha, \alpha', \alpha)$, $\Sigma(\beta, \beta', \beta)$ and $\Sigma(\gamma, \gamma', \gamma)$. Computing the volumes of these six simplexes by Formula (11) and adding, we obtain the following statement, at least when the angles are acute.

**Theorem.** *The volume of an ideal hyperbolic* 3*-simplex with dihedral angles* $\alpha, \alpha, \beta, \beta, \gamma, \gamma$ *is given by*

$$V_3 = Л(\alpha) + Л(\beta) + Л(\gamma). \tag{12}$$

The proof when one of the angles is obtuse is analogous (or one can use analytic continuation).

The function $Л(\alpha)+Л(\beta)+Л(\gamma)$ is continuous on the locus consisting of all $\alpha, \beta, \gamma > 0$ with $\alpha+\beta+\gamma = \pi$, being smooth and positive in the interior, but zero on the boundary. It has just one critical point, characterized by the equations

$$\frac{dЛ(\alpha)}{d\alpha} = \frac{dЛ(\beta)}{d\beta} = \frac{dЛ(\gamma)}{d\gamma},$$

hence $\sin\alpha = \sin\beta = \sin\gamma$, hence $\alpha = \beta = \gamma = \pi/3$. This proves the following.

**Corollary** (Coxeter). *The maximum volume of a hyperbolic* 3*-simplex is*

$$V_3^{\max} = 3Л(\pi/3) = 1.014941606\ldots. \tag{13}$$

Note that $3Л(\pi/3) = 2Л(\pi/6))$, according to $(9_2)$.

By definition, an $n$-dimensional simplex will be called an ***orthosimplex*** if its faces can be numbered as $F_0, \ldots, F_n$ so that $F_i \perp F_j$ whenever $|i-j| \geq 2$. We will denote such an orthosimplex by $\Sigma(\theta_1, \ldots, \theta_n)$, where $\theta_i$ is the dihedral angle between $F_{i-1}$ and $F_i$. Here are some results about 3-dimensional orthosimplexes, quoted without proof from Lobachevsky, as described in [Coxeter].

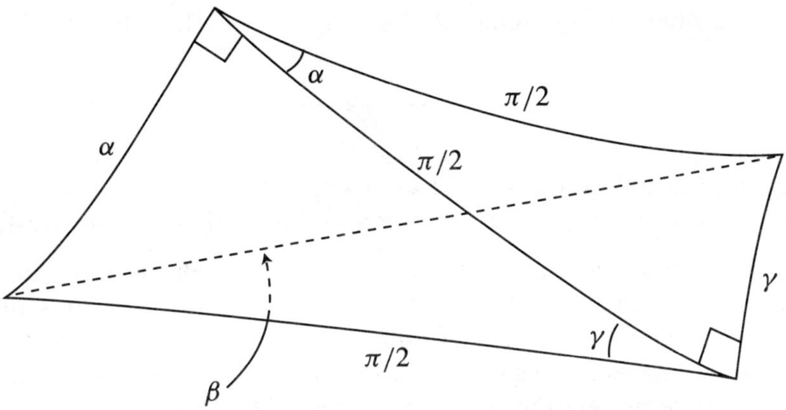


A 3-dimensional orthosimplex $\Sigma(\alpha, \beta, \gamma)$.

*The orthosimplex* $\Sigma(\alpha, \beta, \gamma)$ *exists*[1] *in hyperbolic space if and only if*

$$0 < \alpha, \beta, \gamma < \pi/2, \quad (\sin\alpha)(\sin\gamma) < \cos\beta, \quad \textit{and} \quad \alpha+\beta \geq \pi/2,\ \beta+\gamma \geq \pi/2.$$

Here equality holds on the right if and only if an associated vertex lies on the sphere at infinity. Define an auxiliary angle $\delta$ by the equation

$$\sin\delta = \frac{\cos\alpha\cos\gamma}{\sqrt{\cos^2\alpha + \cos^2\beta + \cos^2\gamma - 1}} \quad \text{with} \quad 0 < \delta < \frac{\pi}{2}.$$

Fixing $\delta$, introduce the abbreviation $F(\theta) = F(-\theta) = Л(\delta+\theta) + Л(\delta-\theta)$. As usual, let $\alpha' = \pi/2 - \alpha$, $\gamma' = \pi/2 - \gamma$. With these notations, Lobachevsky's formula can be written as

$$V_3\big(\Sigma(\alpha, \beta, \gamma)\big) = \tfrac{1}{4}\big(F(0) - F(\alpha') + F(\beta) - F(\gamma')\big). \tag{14}$$

If $\Sigma(\alpha, \beta, \gamma)$ has a vertex at infinity, say if $\beta = \alpha'$, then $\delta$ is equal to $\alpha'$ also, and Formula (14) reduces to $V_3 = \frac{1}{4}(F(0) - F(\gamma'))$, equivalent to (10) above.

One particularly interesting case occurs when

$$\alpha = \pi/p, \quad \beta = \pi/q, \quad \gamma = \pi/r,$$

so that reflections in the faces generate a properly discontinuous group $\Gamma$ with fundamental domain $\Sigma(\alpha, \beta, \gamma)$. (Compare the discussion of Poincaré's Theorem in [Beardon], [Maskit].) Let me write this simplex briefly

[1] See *The Schläfli differential equality*, in this volume, for more general existence criteria.

as $\Sigma_{p,q,r}$. The corresponding group $\Gamma = \Gamma_{p,q,r}$ has a presentation which can be described by the following diagram. Here each vertex stands for a

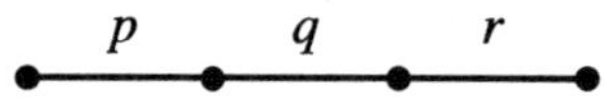


generating involution, the symbol $\overset{p}{\bullet\!\!-\!\!\bullet}$ stands for the relation $(xy)^p = 1$ (where $x$ and $y$ are the generators corresponding to these two vertices), and where $\overset{2}{\bullet\!\!-\!\!\bullet}$ is understood, so that $xy = yx$, when no edge is drawn in.

There are ten cases, as tabulated below. Here $V_3^{\max} = 2Л(\pi/6)$ is the volume of the maximal hyperbolic 3-simplex. The superscript $\infty$ indicates that a corresponding vertex lies on the sphere at infinity. The column L.C.M. lists the least common multiple of the orders of finite subgroups of $\Gamma_{p,q,r}$, while the column least $\#(\Gamma/\Pi)$ lists the smallest index of a subgroup $\Pi \subset \Gamma$ which is torsion-free, and hence equal to the fundamental group of a hyperbolic manifold $H^3/\Pi$. (I don't know whether this subgroup is essentially unique.)

**Table of hyperbolic reflection groups $\Gamma_{p,q,r}$**

| $p, q, r$ | $V_3(\Sigma_{p,q,r})$ | L.C.M. | least $\#(\Gamma/\Pi)$ | $V_3(H^3/\Pi)$ |
|---|---|---|---|---|
| $3, 5, 3$ | .0390503 | 120 | | ? |
| $4, 3, 5$ | .035885 | 240 | | ? |
| $5, 3, 5$ | .0933255 | 120 | 120 | 11.199064 $^{(a)}$ |
| $3, 3, 6^\infty$ | $.0422892 = V_3^{\max}/24$ | 24 | 24 | $1.0149416 = V_3^{\max}$ $^{(b)}$ |
| $4, 3, 6^\infty$ | $.1057231 = (5/48)V_3^{\max}$ | 48 | 48 | $5.074708 = 5V_3^{\max}$ |
| $3, 4, 4^\infty$ | $.0763305 = (1/6)Л(\pi/4)$ | 48 | 48 | $3.66386 = 8Л(\pi/4)$ $^{(c)}$ |
| $5, 3, 6^\infty$ | .1715017 | 120 | | ? |
| $3^\infty, 6, 3^\infty$ | $.1691569 = V_3^{\max}/6$ | 12 | 12 | $2.029883 = 2V_3^{\max}$ $^{(d)}$ |
| $6^\infty, 3, 6^\infty$ | $.2537354 = V_3^{\max}/4$ | 24 | | ? |
| $4^\infty, 4, 4^\infty$ | $.2289914 = (1/2)Л(\pi/4)$ | 16 | 16 | $3.66386 = 8Л(\pi/4)$ $^{(e)}$ |

$^{(a)}$ Weber-Seifert manifold $^{(b)}$ Gieseking manifold (non-orientable) $^{(c)}$ Non-orientable

$^{(d)}$ $S^3 \setminus$  $^{(e)}$ Non-orientable

Closely related are the eight regular polyhedra in $H^3$ which have dihedral angles of the form $2\pi/r$, so that they form one tile of a regular tiling of $H^3$. If the faces of such a regular $N$-hedron are regular $p$-gons, with $q$ edges meeting at each vertex, then it splits into $2Np = 4/(p^{-1}+q^{-1}-2^{-1})$ copies of $\Sigma_{p,q,r}$ and the automorphism group of the associated tiling is precisely $\Gamma_{p,q,r}$. Here is a table. (The superscript $\infty$ again indicates vertices at infinity. The two starred volumes have ratio 2:5.)

**Table of regular polyhedra in $H^3$**

| | $N$ | $2Np$ | $p$ | $q$ | $r$ | volume |
|---|---|---|---|---|---|---|
| Tetrahedron | 4 | 24 | 3 | 3 | $6^\infty$ | $1.07494 = V_3^{\max} = 2Л(\pi/6)$ |
| Cube | 6 | 48 | 4 | 3 | 5 | 1.72248* |
| | | | | | $6^\infty$ | $5.07471 = 5V_3^{\max}$ |
| Octahedron | 8 | 48 | 3 | 4 | $4^\infty$ | $3.66386 = 8Л(\pi/4)$ |
| Dodecahedron | 12 | 120 | 5 | 3 | 4 | 4.30621* |
| | | | | | 5 | 11.19906 |
| | | | | | $6^\infty$ | 20.58020 |
| Icosohedron | 20 | 120 | 3 | 5 | 3 | 4.68603 |

For $N = 4, 6, 8, 12$, let $P^N$ denote the regular $N$-hedron with vertices at infinity. Coxeter notes that the maximal tetrahedron $P^4$ splits into four copies of $\Sigma_{6,3,6}$ or six copies of $\Sigma_{3,6,3}$. The ideal cube $P^6$ splits into five $P^4$'s, while the ideal octahedron $P^8$ splits into sixteen $\Sigma_{4,4,4}$'s. It follows that the groups $\Gamma_{3,4,4}$, $\Gamma_{4,4,4}$ are commensurable, as are the groups $\Gamma_{3,3,6}$, $\Gamma_{4,3,6}$, $\Gamma_{3,6,3}$, $\Gamma_{6,3,6}$. It may be conjectured that there are exactly six distinct commensurability classes of hyperbolic reflection groups $\Gamma_{p,q,r}$. There are also a few reflection groups associated with more general simplexes in $H^3$. A preliminary computation shows that those with all vertices finite are described by the following diagrams. I do not have any further information about these.

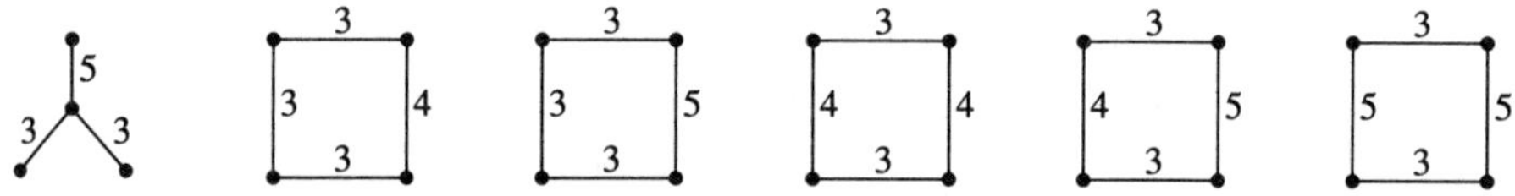


## §4. Higher dimensional cases

Let me first describe a general method for expressing a simplex volume as a power series, and then illustrate the method for the special case of a regular simplex with all vertices at infinity. (For other algorithms for volume computation, see [Aomoto], [Böhm].)

Using the projective unit disk model, any simplex $\Delta^*$ in the hyperbolic space $H^n$ corresponds to a rectilinear simplex $\Delta \subset D^n$. It is not difficult to compute that the hyperbolic volume element can be expressed in the projective model as

$$dV_n = dx_1 \cdots dx_n / (1 - {x_1}^2 - \cdots - {x_n}^2)^\beta, \tag{15}$$

where $\beta = (n+1)/2$. (See the Appendix.) Hence we can express the hyperbolic volume as

$$V_n(\Delta^*) = I_n\left(\frac{n+1}{2}, \Delta\right) \tag{16}$$

where $I_n(\beta, \Delta)$ stands for the $n$-fold integral

$$I_n(\beta, \Delta) = \int \cdots \int_\Delta \frac{dx_1 \cdots dx_n}{(1 - \sum {x_i}^2)^\beta}. \tag{17}$$

(It is interesting to compare Formula (3), which says in this notation that the volume of an infinite cone in the upper half space model, obtained by joining a simplex $\Delta'$ in the unit hemisphere to the point $\mathbf{v}_\infty$, is given by

$$V_{n+1}(\mathrm{Cone}(\Delta')) = I_n\left(\frac{n}{2}, \Delta\right) \Big/ n, \tag{18}$$

where $\Delta \subset D^n$ is the image of $\Delta' \subset S^n$ under projection.) To evaluate $I_n(\beta, \Delta)$ one can carry out the following two steps in either order.

a) Replace the integrand by a power series, using the expansion

$$(1-r^2)^{-\beta} = 1 + \beta r^2 + \frac{\beta(\beta+1)}{2} r^4 + \cdots ,$$

which converges throughout the unit disk. It will be convenient to write this binomial expansion briefly as

$$(1-r^2)^{-\beta} = \sum_{k=0}^{\infty} \binom{\beta+k-1}{k} r^{2k}.$$

b) Make a linear change of variable by setting $x = (x_1, \ldots, x_n) \in \Delta$ equal to $\xi_0 \mathbf{w}_0 + \cdots + \xi_n \mathbf{w}_n$ where $\mathbf{w}_0, \ldots, \mathbf{w}_n$ are the vertices of $\Delta$ and $\xi_0, \ldots, \xi_n$ are the ***barycentric coordinates***, with $\xi_\alpha \geq 0$, $\sum \xi_\alpha = 1$. Note that

$$dx_1 \cdots dx_n = J \, d\xi_1 \cdots d\xi_n$$

where the Jacobian

$$J = \det(\mathbf{w}_1 - \mathbf{w}_0, \ldots, \mathbf{w}_n - \mathbf{w}_0) = \sqrt{\det\left[(\mathbf{w}_i - \mathbf{w}_0) \cdot (\mathbf{w}_j - \mathbf{w}_0)\right]}$$

can be interpreted as $n!$ times the Euclidean volume $I_n(0, \Delta)$. Thus we obtain

$$I_n(\beta, \Delta) = J \sum_{k=0}^{\infty} \binom{\beta+k-1}{k} \int \cdots \int_{\Delta_0} r^{2k} d\xi_1 \cdots d\xi_n,$$

where $\Delta_0$ is the ***standard $n$-simplex*** consisting of all $\xi_i \geq 0$ with $\sum \xi_i = 1$, and where $r^2 = {x_1}^2 + \cdots + {x_n}^2$ is to be expressed as a polynomial in the $\xi_i$. (The order of summation and integration can be interchanged since all terms are positive.)

Finally, any polynomial function of the $\xi_i$ can be integrated, using the identity:

$$(19) \qquad \int \cdots \int_{\Delta_0} {\xi_0}^{i_0} {\xi_1}^{i_1} \cdots {\xi_n}^{i_n} \, d\xi_1 \cdots d\xi_n = \frac{i_0! \, i_1! \cdots i_n!}{(i_0 + \cdots + i_n + n)!}$$

where $\xi_0$ stands for $1 - \xi_1 - \cdots - \xi_n$. To prove this formula, it is convenient to pass to the more general statement that the integral of

$$\left(a - \xi_1 - \cdots - \xi_n\right)^{i_0} {\xi_1}^{i_1} \cdots {\xi_n}^{i_n}$$

over the region $\xi_\alpha \geq 0$, $\xi_1 + \cdots + \xi_n \leq a$ is equal to $a^{i_0 + \cdots + i_n + n}$ multiplied by the expression above. In this form there is a straightforward proof by induction.

To illustrate this method, let $\Delta$ be the regular Euclidean $n$-simplex inscribed in $D^n$. Then $\Delta^*$ is the regular hyperbolic $n$-simplex with all vertices at infinity and the associated Cone$(\Delta')$ is the regular $(n+1)$-simplex with all vertices at infinity. This special case is of interest because of the following.

**Theorem of Haagerup and Munkholm.** *The volume $V_n$ of this infinite regular n-simplex is the maximum possible volume of a hyperbolic n-simplex.*

This has been proved above for $n = 2$ and $n = 3$ (Coxeter). Thurston had earlier given a proof for $n = 4$ based on the identity

$$\text{volume} = \frac{\pi}{3}\left(4\pi - \sum(\text{dihedral angles})\right)$$

for a 4-simplex with all vertices at infinity. (Compare page 231 of *Euler characteristic and finitely additive Steiner measures,* in this volume.)

For this regular simplex $\Delta \subset D^n$ we have

$$\mathbf{w}_i \cdot \mathbf{w}_j = \begin{cases} 1 & \text{for } i = j \\ -1/n & \text{for } i \neq j, \end{cases}$$

and it follows easily that the Jacobian $\sqrt{\det\left[(\mathbf{w}_i - \mathbf{w}_0) \cdot (\mathbf{w}_j - \mathbf{w}_0)\right]}$ is given by

$$J = \sqrt{n+1}^{\,n+1} / \sqrt{n}^{\,n}.$$

Furthermore, setting $\rho^2 = \xi_0{}^2 + \cdots + \xi_n{}^2$, we have

$$1 - r^2 = \left(\frac{n+1}{n}\right)(1 - \rho^2),$$

hence

$$I_n(\beta, \Delta) = \frac{\sqrt{n+1}^{\,n+1}}{\sqrt{n}^{\,n}} \left(\frac{n}{n+1}\right)^{\beta} \sum_{k=0}^{\infty} \binom{\beta + k - 1}{k} \int \cdots \int_{\Delta_0} \rho^{2k}\, d\xi_1 \cdots d\xi_n.$$

Expanding $\rho^{2k}$ by the multinomial theorem and using (19) above, this yields

$$\text{(20)} \quad I_n(\beta, \Delta) = \frac{\sqrt{n+1}^{\,n+1}}{\sqrt{n}^{\,n}} \left(\frac{n}{n+1}\right)^{\beta} \sum_{k=0}^{\infty} \frac{\beta(\beta+1)\cdots(\beta+k-1)}{(n+2k)!} A_{n,k}$$

where $A_{n,k}$ denotes the positive integer

$$A_{n,k} = \sum_{\substack{i_0 + \cdots + i_n = k \\ i_\alpha \geq 0}} \frac{(2i_0)!}{i_0!} \cdots \frac{(2i_n)!}{i_n!}.$$

According to (16) and (18) we are interested in the special cases $\beta = (n+1)/2$ and $\beta = n/2$. In these cases, note that the initial coefficient

$$\frac{\sqrt{n+1}^{\,n+1}}{\sqrt{n}^{\,n}}\left(\frac{n}{n+1}\right)^{\beta}$$

simplifies to $\sqrt{n}$ or to $\sqrt{n+1}$ respectively. This yields the two alternative formulas

$$\begin{aligned} V_n^{\max} &= \frac{\sqrt{n}}{n-1}\sum_{k=0}^{\infty}\frac{\beta(\beta+1)\cdots(\beta+k-1)}{(n-1+2k)!}A_{n-1,k} \qquad \text{with} \quad \beta=(n-1)/2, \\ &= \sqrt{n}\sum_{k=0}^{\infty}\frac{\beta(\beta+1)\cdots(\beta+k-1)}{(n+2k)!}A_{n,k} \qquad \text{with} \quad \beta=(n+1)/2 \end{aligned}$$

for the volume $V_n^{\max}$ of the infinite regular $n$-simplex. Here are some explicit numerical estimates:

$$\begin{aligned} V_2^{\max} &= 3.1415926\ldots = \pi \\ V_3^{\max} &= 1.0149416\ldots = 2Л(\pi/6) \\ V_4^{\max} &= \ .2688956\cdots = \frac{\pi}{3}\big(4\pi - 10\arccos(1/3)\big) \\ V_5^{\max} &\cong .057564 \\ V_6^{\max} &\cong .010239 \\ V_7^{\max} &\cong .0015528 \\ V_8^{\max} &\cong .00020497 \\ V_9^{\max} &\cong .000023936 \\ V_{10}^{\max} &\cong .0000025052 \end{aligned}$$

One can show that[1]

$$V_n^{\max} \sim e\sqrt{n}/n! \quad \text{as} \quad n\to\infty.$$

Here is a moderately explicit volume formula for one important class of simplexes. Recall that a (Euclidean or non-Euclidean) $n$-dimensional ***orthosimplex*** is an $n$-simplex whose faces $F_0, \ldots, F_n$ can be numbered so that $F_i \perp F_j$ for $|i-j| \ge 2$. Assuming that the initial vertex $\mathbf{v}_0$ is finite,

[1] Proof available on request.

we can choose the projective unit disk model so that $\mathbf{v}_0$ corresponds to the origin. Then the corresponding Euclidean simplex $\Delta \subset D^n$ will also be an orthosimplex. In other words the edges $\mathbf{v}_0\mathbf{v}_1$, $\mathbf{v}_1\mathbf{v}_2$, ... , $\mathbf{v}_{n-1}\mathbf{v}_n$ will be mutually orthogonal. Denoting the Euclidean lengths of these edges by $a_1, \ldots, a_n$, we can describe $\Delta$ (after rotating coordinates) as the set of all $(x_1, \ldots, x_n) = (a_1\xi_1, \ldots, a_n\xi_n)$ with $1 \geq \xi_1 \geq \cdots \geq \xi_n \geq 0$. Now expanding the integral

$$I_n((n+1)/2, \Delta) = V_n(\Delta^*) \quad \text{or} \quad I_n(n/2, \Delta) = n \cdot V_{n+1}(\text{Cone}\Delta')$$

as above, and using the identity

$$\int \cdots \int_{1\geq\xi_1\geq\cdots\geq\xi_n\geq 0} \xi_1^{i_1} \cdots \xi_n^{i_n}\, d\xi_1 \cdots d\xi_n$$
$$= (i_1 + \cdots + i_n + n)^{-1}(i_2 + \cdots + i_n + n - 1)^{-1} \cdots (i_n + 1)^{-1},$$

one finds the formula

(21)
$$I_n(\beta, \Delta) = a_1 \cdots a_n \sum_{i_1,\ldots,i_n=0}^{\infty} \frac{\beta(\beta+1)\cdots(\beta+k-1)\, a_1^{2i_1} \cdots a_n^{2i_n}}{i_1! \cdots i_n!\, (2i_1 + \cdots + 2i_n + n) \cdots (2i_n + 1)},$$

where $k = i_1 + \cdots + i_n$.

To really make use of this formula, one must express the Euclidean lengths $a_1, \ldots, a_n$ in terms of the hyperbolic geometry of $\Delta^*$. For example, if $\ell_1$, ... , $\ell_n$ denote the hyperbolic lengths of the corresponding edges of $\Delta^*$, then it is not difficult to show that

$$\tanh \ell_1 = a_1, \ \tanh \ell_2 = \frac{a_2}{\sqrt{1-a_1^2}}, \ \ldots, \ \tanh \ell_n = \frac{a_n}{\sqrt{1-a_1^2-\cdots-a_{n-1}^2}}.$$

Compare the Appendix.

**Appendix. Non-Euclidean Trigonometry and the Projective Disk Model**

A simplex in hyperbolic space can be specified either by its matrix of dihedral angles $\angle(F_i, F_j)$ or by its matrix of edge lengths $\mathrm{dist}(\mathbf{v}_i, \mathbf{v}_j)$. In the 2-dimensional case it is easy to pass from one description to the other by means of the Law of Cosines. In fact there are two Cosine Laws for a hyperbolic triangle, one to compute angles in terms of sides, and one to compute sides in terms of angles. There are two corresponding Cosine Laws for a spherical triangle, and there is also a Law of Sines in each case. For a hyperbolic or spherical triangle with angles $\alpha$, $\beta$, $\gamma$ and edge lengths $A$, $B$, $C$, as illustrated, these laws can be stated as follows.

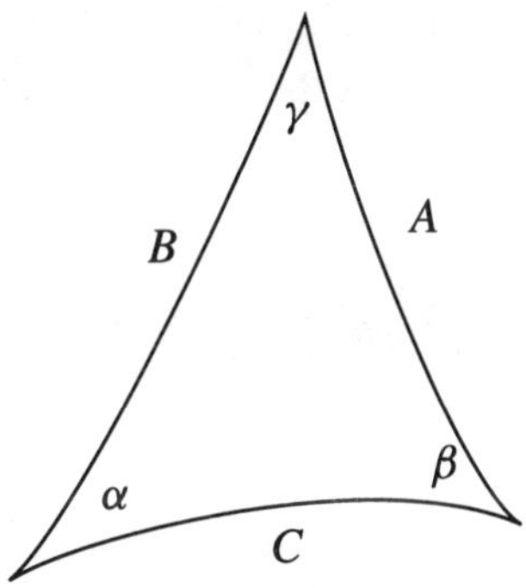


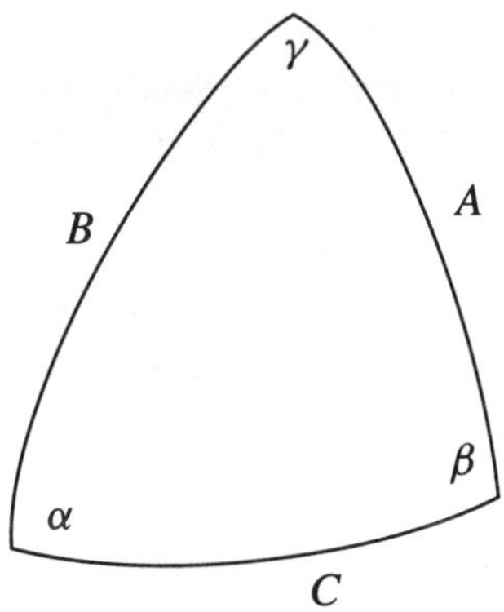


$$\cos\alpha = \frac{-\cosh A + \cosh B \cosh C}{\sinh B \sinh C} \qquad \cos\alpha = \frac{\cos A - \cos B \cos C}{\sin B \sin C}$$

$$\cosh A = \frac{\cos\alpha + \cos\beta\cos\gamma}{\sin\beta\sin\gamma} \qquad \cos A = \frac{\cos\alpha + \cos\beta\cos\gamma}{\sin\beta\sin\gamma}$$

$$\frac{\sinh A}{\sin\alpha} = \frac{\sinh B}{\sin\beta} = \frac{\sinh C}{\sin\gamma} \qquad \frac{\sin A}{\sin\alpha} = \frac{\sin B}{\sin\beta} = \frac{\sin C}{\sin\gamma}$$

Proofs can be found in [Beardon]. For the spherical case, a proof can be found in any text on spherical trigonometry. To justify the statements in the hyperbolic case, we could also insert a factor of $\sqrt{K}$ before the edge lengths $A$, $B$, $C$ in the spherical formulas, where $K$ is Gaussian curvature, in order to obtain expressions which are true for a sphere of any radius $r = 1/\sqrt{K}$. Substituting $K = -1$ and using the identities

$$\cos iA = \cosh A \qquad \sin iA = i \sinh A,$$

we obtain the analogous hyperbolic formulas. (Compare [Coxeter].)

In order to obtain relations between edge length and dihedral angle for an $n$-dimensional non-Euclidean simplex, we proceed inductively as follows. Given the edge lengths, the Law of Cosines determines the angles between incident edges. Now intersect with a small sphere around any vertex. This yields a spherical simplex of dimension $n-1$ with known edge lengths. By induction, we know how to compute its dihedral angles. But these are evidently equal to dihedral angles of the original $n$-simplex.

**Remark.** In the Euclidean case, a similar inductive algorithm computes angles as functions of edge lengths. But of course in the Euclidean case the angles do not determine the size of the simplex.

If we use the projective unit disk model, there is an additional problem since it is not immediately clear how to compute lengths, angles, or volumes. Let us express each point $\mathbf{x}$ of the disk $D^n$ in polar coordinates as a product $\mathbf{x} = r\mathbf{u}$ with $0 \le r < 1$, where $\mathbf{u}$ is a unit vector. Then the Euclidean metric can be written as

$$\|d\mathbf{x}\|^2 = dr^2 + r^2\|d\mathbf{u}\|^2,$$

hence the hyperbolic metric (1′) can be written as

$$ds^2 = \frac{\|d\mathbf{x}\|^2 + \left(d\sqrt{1-r^2}\right)^2}{1-r^2} = \frac{dr^2}{(1-r^2)^2} + \frac{r^2\|d\mathbf{u}\|^2}{1-r^2}.$$

Thus Euclidean distances are squeezed by a factor of $1-r^2$ in radial directions, but only by a factor of $\sqrt{1-r^2}$ in the orthogonal directions. From this we see easily that the hyperbolic volume element $dV_n$ is the product of the Euclidean volume element $dx_1 \cdots dx_n$ multiplied by a factor of $1/(1-r^2)$ for the radial direction and $1/\sqrt{1-r^2}^{\,n-1}$ for the orthogonal directions. Thus

$$dV_n = dx_1 \cdots dx_n/(1-r^2)^{(n+1)/2},$$

as stated in (15).

Distance in the radial direction is given by

$$\int \frac{dr}{1-r^2} = \tanh^{-1} r.$$

In other words, a sphere of hyperbolic radius $\ell$ about the origin corresponds to a sphere of Euclidean radius $r = \tanh \ell$. (Note that $\tanh \ell$ converges to $+1$ as $\ell \to +\infty$.) More generally, along any line $L$ which intersects the unit sphere in points **p** and **q** with midpoint $\mathbf{c} = (\mathbf{p}+\mathbf{q})/2$, it is not hard to check that the Euclidean distance $\|\mathbf{x}-\mathbf{c}\|$ from **x** to **c** and the hyperbolic distance

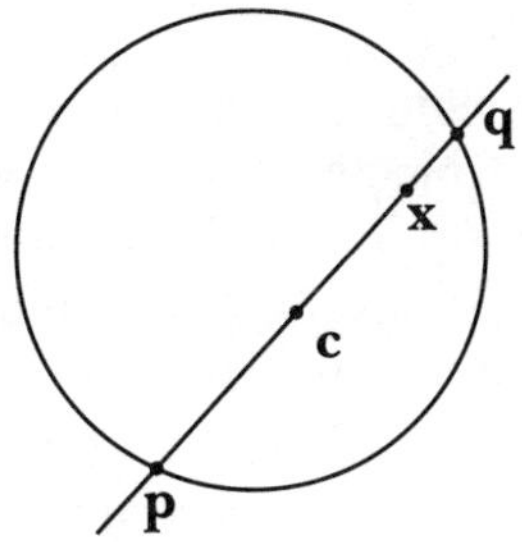


$\mathrm{dist}(\mathbf{x},\mathbf{c})$ are related by the equation

$$\frac{\|\mathbf{x}-\mathbf{c}\|}{\|\mathbf{q}-\mathbf{c}\|} = \tanh(\mathrm{dist}(\mathbf{x},\mathbf{c})).$$

More generally, for any two points **x** and **y** along this line, the hyperbolic distance can be computed from a cross-ratio

$$\mathrm{dist}(\mathbf{x},\mathbf{y}) = \frac{1}{2}\left|\log\frac{(\mathbf{q}-\mathbf{x})\cdot(\mathbf{y}-\mathbf{p})}{(\mathbf{q}-\mathbf{y})\cdot(\mathbf{x}-\mathbf{p})}\right|.$$

The hyperbolic angle $\theta$ between two intersecting lines $L$ and $L'$ can also be computed by means of a cross-ratio, as follows. Form the tangent lines to the unit circle at its intersection points with $L'$. Let $\mathbf{p}^*$ and $\mathbf{q}^*$ be the intersection points of these tangent lines with $L$. Then

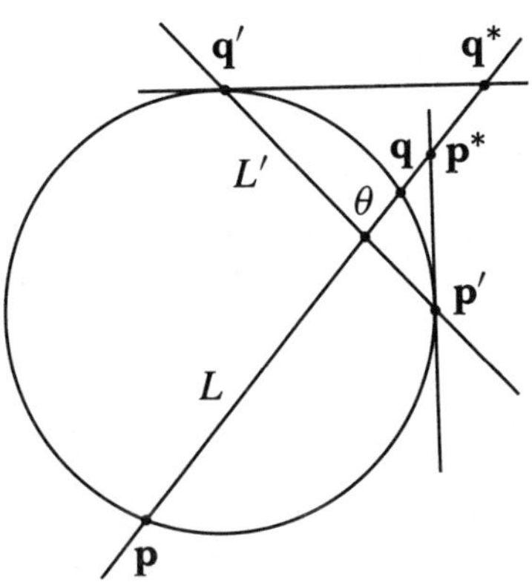


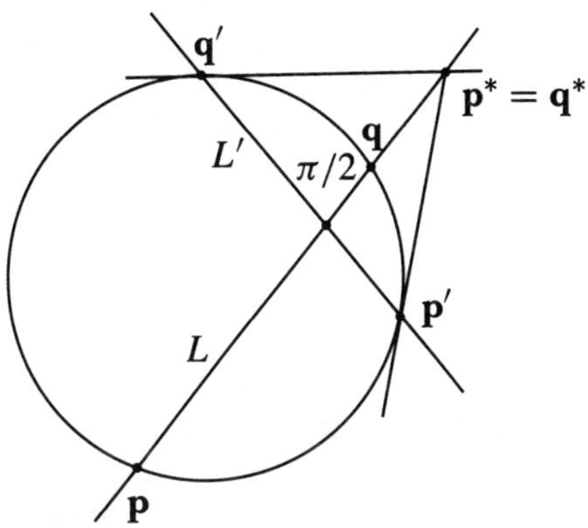

$$\frac{(\mathbf{p}-\mathbf{q})\cdot(\mathbf{p}^*-\mathbf{q}^*)}{(\mathbf{p}-\mathbf{p}^*)\cdot(\mathbf{q}-\mathbf{q}^*)} = \frac{-4\cos\theta}{(1-\cos\theta)^2}.$$

In particular, we have $\cos\theta = 0$, or in other words $L \perp L'$, if and only if $\mathbf{p}^* = \mathbf{q}^*$, as in the right hand picture above. These statements can be proved by noting that any projective transformation which carries the unit disk onto itself corresponds to a hyperbolic isometry. By means of such a projective transformation, we can place the intersection point $L \cap L'$ at the origin. The computation is then straightforward.

If we identify $D^2$ with the unit disk of complex numbers, then we can also compute this angle by a cross-ratio of complex numbers:

$$\frac{(\mathbf{p}-\mathbf{p}')\,(\mathbf{q}-\mathbf{q}')}{(\mathbf{p}-\mathbf{q})\,(\mathbf{p}'-\mathbf{q}')} = \frac{\cos\theta - 1}{2}.$$

The proof is not difficult.

## Bibliography


K. Aomoto, *Analytic Structure of Schläfli function,* Nagoya Math. J. **68** (1977), 1–16.

A. Beardon, "The Geometry of Discrete Groups," Springer, 1983.

S. Bloch, *Applications of the dilogarithm function in algebraic K-theory and algebraic geometry,* in "Proc. Intern. Symp. Alg. Geom. (Kyoto 1977)," Kyoto Univ, 1978, pp. 103–114.

J. Böhm, *Zu Coxeter's Integrationsmethode in gekrümmten Räumen,* Math. Nachr. **27** (1964), 179–214.

T. Clausen, *Über die Function* $\sin\phi + (1/2^2)sin2\phi + (1/3^2)\sin 3\phi + etc.$, J. reine angew. Math. **8** (1832), 298–300.

H. S. M. Coxeter, "The Functions of Schläfli and Lobatschefsky, Twelve Geometric Essays," South Ill. Press, 1968 (or Quart. J. Math **6** (1935)).

H. Gieseking, *Analytische Untersuchungen über topologische Gruppen,* Thesis, Münster 1912 (as described in [Magnus]).

U. Haagerup and H. Munkholm, *Simplices of maximal volume in hyperbolic n-space,* Acta Math. **147** (1981), 1–11.

G. Humbert, *Sur la mesure des Classes d'Hermite de discriminant donné dans un corps quadratique imaginaire, et sur certains volumes non euclideans,* Comptes Rendues (Paris) **169** (1919), 448–454.

N. I. Lobachevsky, "On the foundations of geometry," Kazan, 1829–30, translated as 'Zwei geometrische Abhandlungen,' Teubner, 1898.

B. Maskit, "Kleinian Groups," Springer, 1988.

W. Magnus, "Noneuclidean tessellations and their groups," Academic Press, 1974.

C. Weber & H. Seifert, *Die beiden Dodekaederräume,* Math. Zeits. **37** (1933), 237–253.

# EULER CHARACTERISTIC AND FINITELY ADDITIVE STEINER MEASURES[1]

## §1. Introduction

This note will study geometrically defined finitely additive set functions $\mu_k$ and $\alpha_k$ which generalize the $k$-dimensional Euclidean volume (§2) and the $k$-dimensional solid angle (§3). There is a discussion of open problems in §4. To give the flavor of the subject, for a compact region $A \subset \mathbf{R}^3$ with smooth boundary $\partial A$, we have

$$\begin{aligned}
\mu_3(A) &= \text{volume}(A),\\
\mu_2(A) &= \text{area}(\partial A)/2,\\
\mu_1(A) &= \frac{1}{2\pi}\int_{\partial A}(k_1 + k_2)\, d(\text{area})\\
\mu_0(A) &= \chi(A) = \frac{1}{4\pi}\int_{\partial A} k_1 k_2\, d(\text{area}),
\end{aligned}$$

where $k_1$ and $k_2$ are the principal curvatures. (These expressions were introduced by J. Steiner in 1840. Compare [Hadwiger], as well as (12) and (19) below.) The measure $\mu_k$ is defined for much more general sets, and satisfies the product formula

$$\mu_k(A \times B) = \sum_{i+j=k} \mu_i(A)\, \mu_j(B).$$

The angular measure $\alpha_k(C)$, where $C \subset \mathbf{R}^n$ is a polyhedral cone, has analogous properties.

## §2. The Steiner Measure $\mu_k$ in Euclidean Space

A compact set $A$ in the Euclidean space $\mathbf{R}^n$ will be called ***piecewise smooth*** if it can be triangulated[2] into simplexes which are $C^1$-smoothly embedded

[1] Not previously published. The original version of this note, with title 'Euler characteristic, volume, and angle', was written by hand at the University of Warwick in July 1978. Revised and corrected in Stony Brook, May 1993.

[2] This is a rather strong hypothesis. However, even stronger hypotheses will be needed for the proofs of several fundamental results.

into $\mathbf{R}^n$. For any such $A \subset \mathbf{R}^n$, let $\mu_0(A)$ denote the Euler characteristic $\chi(A)$. For $k = 1, 2, \ldots, n$ set

$$\mu_k(A) = c \int \chi(A \cap P_k)\, d(P_k), \tag{1}$$

to be integrated over the space of dimension $k(n+1-k)$ consisting of all planes $P_k$ of codimension $k$ in $\mathbf{R}^n$. Thus, $\mu_k(A)$ is evaluated by taking the Euler characteristic of the intersection with a generic plane of dimension $n-k$, and then integrating over all such planes. The normalizing constant $c = c_{k,n}$ is to be chosen so that

$$\mu_k(I^k) = 1,$$

where $I^k = [0, 1]^k$ is the standard unit cube in $\mathbf{R}^k \subset \mathbf{R}^n$. It will be convenient to set $\mu_k(A) = 0$ for $k > n$ and for $k < 0$.

Here we are implicitly making use of the following well known fact. (Compare [Santaló, 1953 or 1976].)

(2) **Lemma.** *The space consisting of all codimension k planes $P_k \subset \mathbf{R}^n$ has one and up to a multiplicative constant only one absolutely continuous measure, denoted here by $d(P_k)$, which is invariant under the action of the Euclidean group.*

**Proof.** Uniqueness is clear, since the Euclidean group acts transitively. To prove existence, we define the integral $\int \phi(P_k)\, d(P_k)$ of a measurable function of $P_k$ in two steps as follows:

**First Step.** Choose some fixed plane $P_k^0$ through the origin and integrate over all parallel planes. Each of these can be expressed uniquely as $P_k^0 + \mathbf{x}$ where $\mathbf{x}$ varies over the orthogonal complement of $P_k^0$. Thus we can integrate with respect to the usual $k$-dimensional Euclidean measure in this orthogonal $k$-plane.

**Second Step.** Now average over all codimension $k$ planes through the origin. Equivalently, we could form the image $\rho(P_k^0)$, where $\rho$ ranges over the group $SO(n)$ of rotations, and then average with respect to the Haar measure on $SO(n)$. Euclidean invariance is easily checked. □

**Definition.** This number $\mu_k(A)$ will be called the (Euclidean) ***k-dimensional Steiner measure*** of $A$. However, it should be emphasized that $\mu_k$ is not a measure in the usual sense, that is, it does not seem to extend to a countably additive set function, except when $k = n$. Furthermore, it can take both positive and negative values, at least when $k < n-1$. Here are five basic properties.

**(3) Additivity.**

$$\mu_k(A \cup B) = \mu_k(A) + \mu_k(B) - \mu_k(A \cap B)$$

*whenever both sides are defined.*

**(4) Euclidean Invariance.** *The measure $\mu_k(A)$ is invariant under Euclidean motions of $A$.*

**(5) Independence of Ambient Dimension.** *If $A \subset \mathbf{R}^m \subset \mathbf{R}^n$, then $\mu_k(A)$ is the same whether computed in $\mathbf{R}^m$ or $\mathbf{R}^n$.*

**(6) Scaling Law.** *If all distances are multiplied by a constant $t > 0$, then $\mu_k$ is multiplied by $t^k$. That is*

$$\mu_k(tA) = t^k \mu_k(A)$$

*where $tA$ is the set of all products $t\mathbf{x}$ with $\mathbf{x} \in A$.*

**(7) Top Dimensional Case.** *If $A$ is piecewise smooth of dimension $d$, then $\mu_d(A)$ is the ordinary $d$-dimensional Riemannian volume of $A$, while $\mu_k(A) = 0$ for $k > d$.*

The proofs are all straightforward. □

The interpretation of $\mu_k(A)$ when $0 < k < \dim(A)$ is more difficult. (See (12) and (38) below. For examples, see (14) and (17).) We will first prove the following.

**(8) Boundary Theorem.** *If $M^d$ is a $C^1$-smooth $d$-dimensional manifold with boundary, then*

$$\mu_k(\partial M^d) = \begin{cases} 2\mu_k(M^d) & \text{if the difference } d-k \text{ is odd,} \\ 0 & \text{if } d-k \text{ is even.} \end{cases}$$

Combining (7) and (8), we see that $\mu_{d-1}(M^d)$ is just half of the $(d-1)$-dimensional Riemannian volume of the boundary $\partial M^d$. If $M^d$ is a closed manifold, so that $\partial M^d = \emptyset$, then it follows that $\mu_k(M^d) = 0$ whenever $d-k$ is odd. However, this is certainly not true when $d-k$ is even.

The proof of (8) can be sketched as follows. First use Sard's Theorem to guarantee that almost every $P_k$ intersects $M^d$ transversally. Thus the intersection $Q = M^d \cap P_k$ is a $(d-k)$-dimensional manifold with boundary.

Now $\chi(Q, \partial Q) = \chi(Q) - \chi(\partial Q)$ is equal to $(-1)^{d-k}\chi(Q)$ by Poincaré duality. Integrating with respect to $d(P_k)$, we obtain the required equality. □

**Some Notation.** Let $\delta_n = \mu_n(D^n)$ be the $n$-dimensional volume of the closed unit disk in $\mathbf{R}^n$. A straightforward double integration shows that

$$\delta_n = \frac{2\pi}{n}\,\delta_{n-2},$$

with $\delta_0 = 1$, $\delta_1 = 2$, $\delta_2 = \pi$, $\delta_3 = 4\pi/3, \ldots$ . If we introduce the idiosyncratic notation

$$\text{(9)}\quad u! = \Gamma(u+1) = \int_0^\infty e^{-x} x^u \, dx$$

$$\left( = \int_0^\infty e^{-x}\, dx^u = (u-1)!\,u \quad \text{if } u > 0 \right)$$

for the gamma function, then this can be written simply as

$$\text{(10)}\qquad \delta_n = \frac{\pi^{n/2}}{(n/2)!}.$$

Here

$$\left(\frac{2m-1}{2}\right)! = \sqrt{\pi}\,\frac{1}{2}\,\frac{3}{2}\cdots\frac{2m-1}{2} = \sqrt{\pi}\,\frac{(2m)!}{4^m m!}.$$

We will also need the easily verified formula

$$\text{(11)}\qquad \mu_k(S^k) = 2\pi\,\delta_{k-1} = (k+1)\delta_{k+1}$$

for the $k$-dimensional volume of the unit sphere $S^k \subset \mathbf{R}^{k+1}$.

Now suppose that $K \subset \mathbf{R}^n$ is compact and convex. Then the $\mu_k(K)$ can be characterized in a quite different way as follows. Let $N_r(K)$ be the neighborhood of radius $r$, consisting of all points with distance $\le r$ from $K$.

**(12) Theorem.** *The n-dimensional volume of this neighborhood can be expressed as a polynomial function of* $r$. *In fact* $\mu_n(N_r(K))$ *is equal to*

$$\mu_n(K) + \mu_{n-1}(K)\cdot\delta_1 r + \mu_{n-2}(K)\cdot\delta_2 r^2 + \cdots + \mu_0(K)\cdot\delta_n r^n.$$

(Compare [Hadwiger], as well as [Minkowski, p. 207].) The proof will depend on the following lemma. For each hyperplane $H$ through the origin let

$$K_H \subset H$$

be the image of $K$ under orthogonal projection to $H$.

**(13) Cauchy Lemma.** *For $k < n$, the average of $\mu_k(K_H)$ as $H$ varies over all hyperplanes through the origin in $\mathbf{R}^n$ is equal to some constant multiple of $\mu_k(K)$.*

**Proof.** Let $\pi_H$ be the orthogonal projection from $\mathbf{R}^n$ onto $H$. For every plane $P_k \subset H$ of codimension $k$ in $H$, the pre-image $\pi_H^{-1}(P_k)$ has codimension $k$ in $\mathbf{R}^n$ and satisfies $\chi(K \cap \pi_H^{-1}(P_k)) = \chi(K_H(\cap P_k) \in \{0, 1\}$. Integrating over all pairs $H \supset P_k$ we obtain the required statement. □

**Remarks.** Since each unit vector $\mathbf{u} \in \mathbf{R}^n$ has a hyperplane $H_\mathbf{u}$ as orthogonal complement, we could equally well take the average of $\mu_k(K_{H_\mathbf{u}})$ as $\mathbf{u}$ varies over the unit sphere $S^{n-1}$. The precise constant in (13) can be described as the ratio $\mu_k(D^{n-1})/\mu_k(D^n)$. Compare (14) below.

**Proof of (12).** For $n = 1$ the convex set $K$ is a line segment of length $\mu_1(K) \geq 0$ and the required formula reads $\mu_1(N_r(K)) = \mu_1(K) + 2r$, which is certainly true. We proceed by induction on $n$. If we project $K$ to a hyperplane $H$, then

$$\mu_{n-1}(N_r(K_H) \cap H) = \sum_{i=0}^{n-1} \mu_{n-1-i}(K_H \cap H) \cdot \delta_i\, r^i$$

by the induction hypothesis. Note that $N_r(K_H) \cap H$ is equal to the image $(N_r(K))_H$ under projection. Hence by (13),

$$\mu_{n-1}(N_r(K)) = \sum_{i=0}^{n-1} c_i\, \mu_{n-i-1}(K)\, r^i,$$

for suitable constants $c_i$ which depend only on $i$ and $n$. But $\mu_{n-1}(N_r(K))$ is equal to $\frac{1}{2}\mu_{n-1}(\partial N_r(K))$ by (8). Integrating from 0 to $r$, this yields

$$\begin{aligned}\mu_n(N_r(K)) &= \mu_n(K) + \int_0^r \mu_{n-1}\big(\partial N_s(K)\big)\, ds \\ &= \mu_n(K) + \sum_{i=0}^{n} c_i'\, \mu_{n+1-i}(K)\, r^{i+1},\end{aligned}$$

where $c_i' = 2c_i/(i+1)$. This has the required form. The precise values of the coefficients $c_i'$ can be evaluated by looking at special cases. Suppose for example that we take $K$ to be the $(n-k)$-dimensional unit disk $D^{n-k} \subset \mathbf{R}^{n-k} \subset \mathbf{R}^n$. Then $N_r(D^{n-k})$ can be described as the union of a product

$D^{n-k} \times rD^k$ which has $n$-volume exactly $\delta_{n-k}\,\delta_k \cdot r^k$ and a neighborhood of the sphere $S^{n-k-1} = \partial D^{n-k}$ which has $n$-volume bounded by some constant times $r^{k+1}$ as $r \to 0$. Therefore

$$\mu_n(N_r(D^{n-k})) = \mu_{n-k}(D^{n-k}) \cdot \delta_k\, r^k + O(r^{k+1}) \qquad \text{as} \quad r \to 0.$$

It follows that the coefficient of $\mu_{n-k}(K) \cdot r^k$ in the formula for $\mu_n(N_r(K))$ must be equal to $\delta_k$. □

Here is an illustrative example. Suppose that $K$ is the unit disk $D^n$. Then $N_r(D^n)$ is the disk of radius $1+r$, with volume $\delta_n(1+r)^n$. Expanding by the binomial theorem and comparing coefficients in (12), we see that

**(14)** $$\mu_k(D^n) = \binom{n}{k}\frac{\delta_n}{\delta_{n-k}}.$$

Now let $t$ be a real variable. We can combine all of the Steiner measures $\mu_k(K)$ into one polynomial

$$\mu(K,t) = \mu_0(K) + t\mu_1(K) + \cdots + t^n \mu_n(K).$$

Let $r(\mathbf{x}) \geq 0$ be the distance from $\mathbf{x}$ to the closest point of $K$.

**(15) Corollary.** *If $K$ is convex, then for $t > 0$ the polynomial $\mu(K,t)$ can be expressed as an integral*

$$\mu(K,t) = t^n \int_{\mathbf{R}^n} e^{-\pi r(\mathbf{x})^2 t^2}\, d(\mathbf{x}).$$

*Here $d(\mathbf{x})$ stands for the usual $n$-dimensional Euclidean volume element.*

**Proof.** The integral of $e^{-\pi r^2 t^2}$ over the region $K$ yields the constant $\mu_n(K)$. The integral over the complement of $K$ can be written as

$$\int_{r=0}^{\infty} e^{-\pi r^2 t^2}\, d\Big(\sum_{k=1}^{n} \mu_{n-k}(K)\delta_k r^k\Big) = \sum_{k=1}^{n} \mu_{n-k}(K)\,\delta_k \int_0^{\infty} e^{-\pi r^2 t^2}\, dr^k$$

by (12). Using the substitution $u = \pi r^2 t^2$, this becomes

$$\sum_{k=1}^{n} \mu_{n-k}(K)\,\delta_k \int_0^{\infty} e^{-u} d\,u^{k/2}/(\pi t^2)^{k/2} = \sum_{k=1}^{n} \delta_k\, \mu_{n-k}(K)(k/2)!/(\pi t^2)^{k/2},$$

using (9). Since $\delta_k = \pi^{k/2}/(k/2)!$ by (10), this equals $\sum_1^n \mu_{n-k}(K)/t^k$, as required. □

**(16) Product Theorem.** *If $A \subset \mathbf{R}^m$ and $B \subset \mathbf{R}^n$ are finite unions of compact, convex sets, then*

$$\mu(A \times B, t) = \mu(A, t)\mu(B, t),$$

*or in other words*

$$\mu_k(A \times B) = \sum_{i+j=k} \mu_i(A)\mu_j(B).$$

**Proof.** If $A$ and $B$ are convex, then the assertion follows by inspection, since the integral (15) is set up so as to convert cartesian products to products. The general case follows by additivity (3). □

It may be conjectured that this product formula holds for much more general sets $A$.

**Examples.** For the unit interval $I = [0, 1]$, we have $\mu(I, t) = 1 + t$. Therefore for the unit cube $I^n$ we have $\mu(I^n, t) = (1+t)^n$, or in other words

$$\mu_k(I^n) = \binom{n}{k}. \tag{17}$$

More generally, if $I_1, \ldots, I_n$ are intervals of lengths $\lambda_1, \ldots, \lambda_n$, then $\mu_k(I_1 \times \cdots \times I_n)$ is equal to the $k^{\text{th}}$ elementary symmetric function of the $\lambda_i$. On the other hand, if $A$ is a figure eight with total length $\lambda_1$ and $B$ is an interval of length $\lambda_2$, then

$$\mu(A \times B, t) = (-1 + \lambda_1 t)(1 + \lambda_2 t).$$

Thus $\mu_1(A \times B) = \lambda_1 - \lambda_2$, which can be either positive or negative.

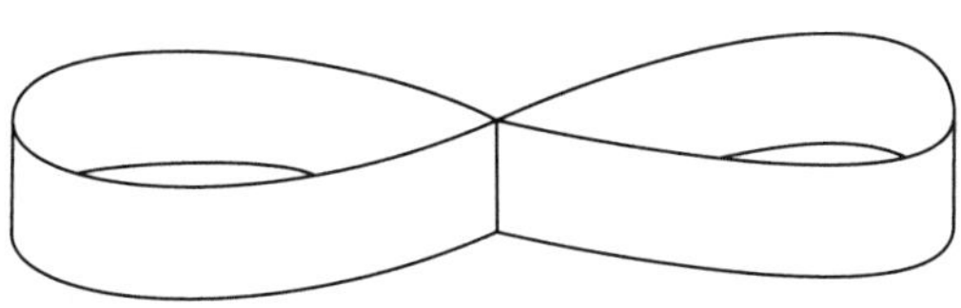

Theorem (12) can also be generalized so as to apply to finite unions of compact convex sets. Let $N_r(\mathbf{x}) = \mathbf{x} + r\, D^n$ be the $n$-disk of radius $r$ centered at $\mathbf{x}$.

**(18) Corollary.** *If $A$ is a finite union of compact, convex sets, then the integral of the function $\mathbf{x} \mapsto \chi(A \cap N_r(\mathbf{x}))$ over $\mathbf{R}^n$ is given by*

$$\int_{\mathbf{R}^n} \chi\big(A \cap N_r(\mathbf{x})\big)\, d(\mathbf{x}) = \mu_n(A) + \mu_{n-1}(A) \cdot \delta_1\, r + \cdots + \mu_0(A) \cdot \delta_n\, r^n.$$

In fact if $A$ is convex then the left side is equal to $\mu_n(N_r(A))$, so this follows from (12). The general case follows since both sides are additive. □

Again, it is conjectured that this formula holds for more general sets $A$. One interesting case occurs when $A$ is a region bounded by a $C^2$-smooth hypersurface in $\mathbf{R}^n$. Let $k_1, \ldots, k_{n-1}$ be the principal curvatures along $\partial A$, choosing signs so that $k_i > 0$ in the convex case, and let

$$\kappa_0 = 1, \quad \kappa_1 = \sum k_i, \quad \kappa_2 = \sum_{i<j} k_i k_j, \quad \cdots \quad \kappa_{n-1} = \prod k_i$$

be the elementary symmetric functions of the $k_i$. Then

$$\mu_{n-1-h}(A) = \frac{1}{\mu_h(S^h)} \int_{\partial A} \kappa_h(\mathbf{x})\, d(\mathbf{x}), \tag{19}$$

where now $d(\mathbf{x})$ denotes the $(n-1)$-dimensional Riemannian volume element along the boundary.

As an example, if $h = n-1$ then $\kappa_{n-1}$ is the Gaussian curvature, and the resulting formula

$$\chi(A) = \frac{1}{\mu_{n-1}(S^{n-1})} \int_{\partial A} \kappa_{n-1}(\mathbf{x})\, d(\mathbf{x})$$

is just an expression of the Gauss-Bonnet Theorem. At the other extreme, for $h = 0$ since $\kappa_0 = 1$, formula (19) reduces to the formula (8): $\mu_{n-1}(A) = \mu_{n-1}(\partial A)/2$. If $A$ is the unit disk $D^n$, then (19) reduces easily to (14).

**Proof of (19).** In the convex case, if we choose some local parameterization $\mathbf{x} = \mathbf{x}(u_1, \ldots, u_{n-1})$ for $\partial A$, and let $\mathbf{n}(u_1, \ldots, , u_{n-1})$ be the outward unit normal vector, then we can use

$$\mathbf{x}(u_1, \ldots, u_{n-1}) + u_n \cdot \mathbf{n}(u_1, \ldots, u_{n-1})$$

as parameterization for the neighborhood $N_r(A) \setminus A$. Using these parameters to compute the volume of $N_r(A)$ and applying (12), we find the required expression (19). In the non-convex case, if we could apply (18), then restricting to small values of $r$ the same proof would apply. However, since (18) has not been proved in this generality, a more complicated proof, based directly on the definition (1) and the Gauss-Bonnet formula, is needed. Details will be omitted. □

## §3. Spherical and Angular Steiner Measures

Now let $A$ be a compact piecewise smooth subset of the unit sphere $S^{n-1} \subset \mathbf{R}^n$. For $0 \le k < n$ define the ***k-dimensional spherical Steiner measure*** $\sigma_k(A)$ to be

**(20)** $$\sigma_k(A) = \text{Average}\,\chi(A \cap \Sigma_k),$$

to be averaged over all great spheres $\Sigma_k \subset S^{n-1}$ of codimension $k$. We set $\sigma_k(A) = 0$ for $k \ge n$. (The definition for $k < 0$ will be given presently.)

Note that $\sigma_0(A)$ is just the Euler characteristic $\chi(A)$. These measures $\sigma_k$ satisfy the analogues of Properties (3), (4), (5), (7). That is, they are additive, invariant under rotation, and independent of the ambient dimension. The normalization is such that $\sigma_k(S^k) = 2$ (and in fact $\sigma_{k-2}(S^k) = \sigma_{k-4}(S^k) = \cdots = 2$ also). If $H^k \subset S^k$ denotes a $k$-dimensional hemisphere, note that

$$\sigma_0(H^k) = \sigma_1(H^k) = \cdots = \sigma_k(H^k) = 1.$$

If $A$ is a set of dimension $\le k$, then $\sigma_k(A)$ coincides with the $k$-dimensional Riemannian volume, or in other words with the Euclidean Steiner measure $\mu_k(A)$, up to a normalizing factor of $\mu_k(H^k) = \pi\delta_{k-1}$. Theorem (8) also generalizes: For a $d$-dimensional manifold with boundary $M^d \subset S^{n-1}$ we have

**(21)** $$\sigma_k(M^d) = \tfrac{1}{2}\sigma_k(\partial M^d),$$

provided that the difference $d - k$ is odd. In particular, it follows easily that

**(22)** $$\sigma_k(M^d) = \mu_k(M^d)/\mu_k(H^d) = \mu_k(M^d)/(\pi\delta_{k-1})$$

whenever $k \ge d - 1$.

**(23) Remark.** It might seem natural to conjecture that (22) holds for all dimensions; however, this would be false. The spherical measure $\sigma_1(S^3)$ is equal to 2, while it follows from (14) that $\mu_1(S^3)/\mu_1(H^1) = (3\pi)/\pi = 3$. Perhaps a more reasonable conjecture would be that $\mu_k(A)$ can be expressed as a linear combination of $\sigma_k(A), \sigma_{k+2}(A), \sigma_{k+4}(A), \ldots$ whenever $A \subset S^{n-1}$.

Suppose that we think of $S^{n-1}$ as the equator of an $n$-dimensional sphere. Then every compact set $A \subset S^{n-1}$ has a ***suspension*** $A * S^0 \subset S^n$, which can be described as the union of all semi-circles which join the north pole to the south pole through some point of $A$.

**(24) Suspension Theorem.** *The spherical measures of a suspension are given by*

$$\sigma_k(A * S^0) = \sigma_{k-1}(A) \qquad \text{for } k > 0.$$

*However the Euler characteristic $\sigma_0 = \chi$ satisfies*

$$\sigma_0(A * S^0) = 2 - \sigma_0(A).$$

The proof is not difficult. □

It will be convenient to define $\sigma_k$ for $k < 0$ in such a way that $\sigma_k(A * S^0)$ will be equal to $\sigma_{k-1}(A)$ for all $k$. In fact we set

**(25)** $$\sigma_{-1} = 2 - \chi, \quad \sigma_{-2} = \chi, \quad \sigma_{-3} = 2 - \chi, \quad \sigma_{-4} = \chi, \quad \ldots.$$

It follows easily that $\sigma_k(A * S^0) = \sigma_{k-1}(A)$ for all values of $k$. With this convention, note that $\mu_k(H^n) = 1$ for all $k \le n$, and that

$$\mu_n(S^n) = \mu_{n-2}(S^n) = \mu_{n-4}(S^n) = \cdots = 2,$$

although $\mu_k(S^n) = 0$ whenever $n - k$ is odd.

Now let $C$ be a ***polyhedral cone*** in $\mathbf{R}^n$, that is, a finite union of finite intersections of closed half-spaces bounded by hyperplanes through the origin. Let

**(26)** $$\hat{\sigma}_k(C) = \sigma_k(C \cap S^{n-1}).$$

These quantities can be used as measures of angle. However, it turns out that certain linear combinations of the $\hat{\sigma}_k$ have properties which are more convenient. We define the ***angular Steiner measure*** $\alpha_k(C)$ by the formula

**(27)** $$\alpha_k(C) = \tfrac{1}{2}\big(\hat{\sigma}_{k-1}(C) - \hat{\sigma}_{k+1}(C)\big),$$

using the convention (25). It follows that $\alpha_k(C) = 0$ for $k < 0$ and for $k > n$. We can also write

$$\hat{\sigma}_k(C) = 2\big(\alpha_{k+1}(C) + \alpha_{k+3}(C) + \alpha_{k+5}(C) + \cdots\big).$$

Note that

**(28)** $$\alpha_0(C) + \alpha_1(C) + \cdots + \alpha_n(C) = 1$$

for all $C$. In the special case $k = 0$, note that

$$\alpha_0(C) = 1 - \tfrac{1}{2}\big(\hat{\sigma}_0(C) + \hat{\sigma}_1(C)\big).$$

Evidently these measures are additive, invariant under rotation, independent of the ambient dimension, and satisfy a boundary formula analogous to (8) or (21).

The definition (27) may look somewhat strange, but we will show that it has useful properties. As a first example, note the simple formula

**(29)**
$$\alpha_k(\mathbf{R}^n) = \begin{cases} 1 & \text{for } k = n, \\ 0 & \text{for } k \neq n. \end{cases}$$

As a further example, note that the suspension operation $A \mapsto A * S^0$ for a subset $A \subset S^{n-1}$ corresponds to the product operation $C \mapsto C \times \mathbf{R}$ for a polyhedral cone $C \subset \mathbf{R}^n$, where $C \times \mathbf{R} \subset \mathbf{R}^{n+1}$. Using these expressions $\alpha_k$, the Suspension Theorem (24) takes the simple form

**(30)**
$$\alpha_k(C \times \mathbf{R}) = \alpha_{k-1}(C).$$

In particular, $\alpha_0(C \times \mathbf{R}) = 0$. The proof is easily supplied.

Note that $\alpha_n$ is the usual $n$-dimensional solid angle, suitably normalized:

**(31)**
$$\alpha_n(C) = \mu_n(C \cap D^n)/\mu_n(D^n) \qquad \text{for a cone } C \subset \mathbf{R}^n,$$

where $\mu_n$ is Lebesgue measure and $D^n$ is the unit disk. An alternative form of this equation which is often useful is

**(32)**
$$\alpha_n(C) = \int_C e^{-\pi\|\mathbf{x}\|^2}\, d(\mathbf{x}).$$

(Compare [Kneser].) One easy consequence of this formula is the identity

**(33)**
$$\alpha_{n+m}(C_1 \times C_2) = \alpha_n(C_1)\alpha_m(C_2)$$

which holds for cones $C_1 \subset \mathbf{R}^n$ and $C_2 \subset \mathbf{R}^m$. (Compare (43) below.)

Now let $C \subset \mathbf{R}^n$ be a *convex* cone. The ***dual cone*** $C^* \subset \mathbf{R}^n$ is defined to be the set of all $\mathbf{x} \in \mathbf{R}^n$ such that $\mathbf{x} \cdot \mathbf{y} \leq 0$ for all $\mathbf{y} \in C$. (Compare [Gale].) As an example, if $V_k \cong \mathbf{R}^{n-k}$ is a vector space of codimension $k$ contained in $\mathbf{R}^n$, then $V_k^*$ is the orthogonal vector space of dimension $k$.

**(34) Duality Theorem.** *For any convex cone $C \subset \mathbf{R}^n$, the angular measures of the dual cone are given by*

$$\alpha_k(C^*) = \alpha_{n-k}(C).$$

To prove this statement, we first prove a corresponding statement for $\hat{\sigma}_k$.

**(35) Lemma.** *If $\mathbf{C} \subset \mathbf{R}^n$ is any convex cone which is not a vector space, so that $\mathbf{C} \ncong \mathbf{R}^m$ for any $m$, then for almost every codimension $k$ plane $V_k$ through the origin we have*

$$C \cap V_k \neq \{0\} \iff C^* \cap V_k^* = \{0\}.$$

**Proof outline.** First suppose that $C$ has full dimension $n$. If $V_k$ intersects the interior of $C$, then an easy argument shows that $C^* \cap V_k^* = \{0\}$. On the other hand, if $C \cap V_k = \{0\}$, let $\widehat{C} \subset V_k^*$ be the image of $C$ under orthogonal projection from $\mathbf{R}^n$ to the orthogonal complement of $V_k$. Evidently the origin must be a boundary point of $\widehat{C}$. Hence $\widehat{C}$ is contained in a half-space in $V_k^*$, and the unit normal vector to this half-space is the required non-zero vector in $C^* \cap V_k^*$. This completes the proof when $\dim(C) = n$, since the collection of planes $V_k$ which intersect $C \setminus \{0\}$ but are disjoint from its interior forms a set of measure zero. Finally, suppose that $C$ has dimension less than $n$. Then $C$ is contained in a hyperplane $H$, and $C^*$ splits as an orthogonal product $(H \cap C^*) \times H^*$. The proof in this case is easily completed by induction on the ambient dimension $n$. □

**Proof of (34).** It follows immediately from (35) that

$$\hat{\sigma}_k(C) + \hat{\sigma}_{n-k}(C^*) = 1 \qquad \text{for } 0 \leq k \leq n.$$

whenever $C$ is convex but not a vector space. Evidently (34) follows easily, still assuming that $C$ is not a vector space. On the other hand if $C \cong \mathbf{R}^m$ is a vector space, then the required statement follows from (29). □

As an example, it follows that $\alpha_0(C)$ can be interpreted as the $n$-dimensional solid angle of the dual cone, with the usual normalization. For a convex cone in $\mathbf{R}^3$, we can interpret $\alpha_2(C)$ as the sum of the angles between consecutive edges, divided by $4\pi$, while $\alpha_1(C)$ is the corresponding expression for the dual cone. (See the Figure on the next page.)

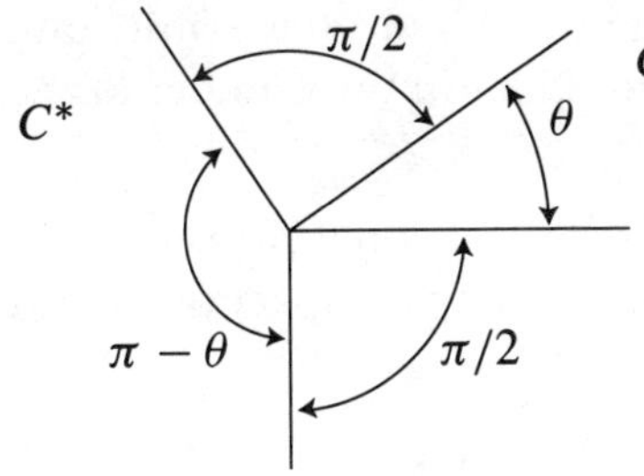


$$2\pi\alpha_2(C) = \theta$$
$$2\pi\alpha_1(C) = \pi$$
$$2\pi\alpha_0(C) = \pi - \theta.$$

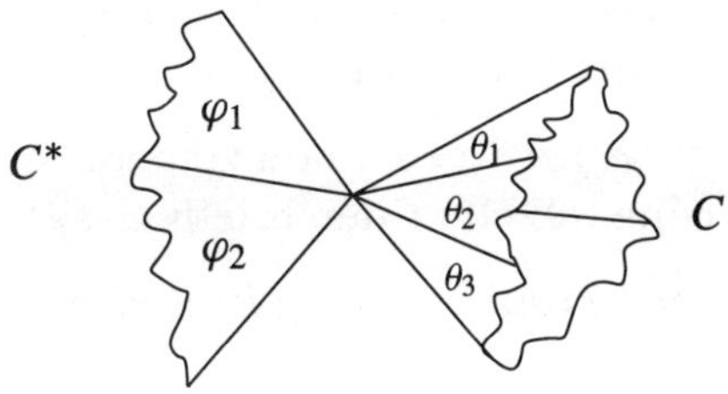


$$4\pi\alpha_3(C) = \text{solid angle}(C) = 2\pi - \sum\varphi_i$$
$$4\pi\alpha_2(C) = \sum\theta_i$$
$$4\pi\alpha_1(C) = \sum\varphi_i$$
$$4\pi\alpha_0(C) = \text{solid angle}(C^*) = 2\pi - \sum\theta_i$$

**(36) Lemma.** *The average, over all hyperplanes $H$ through the origin, of $\alpha_k(C \cap H)$ is equal to $\alpha_{k+1}(A)$ when $k \geq 1$, but is equal to $\alpha_0(A) + \alpha_1(A)$ when $k = 0$.*

**Proof.** This follows easily from the definitions when $k > 0$, and the case $k = 0$ then follows from (28). □

If A is a piecewise smooth subset of a smooth manifold, define the ***tangent cone*** $T_x A$ at a point $x \in A$ to be the polyhedral cone consisting of all tangent vectors $p'(0)$ to smooth paths $p\colon [0, \epsilon) \to A$ with $p(0) = x$.

**(37) Gauss-Bonnet Theorem for Polyhedra.** *If $P$ is a compact polyhedron in Euclidean space, then*

$$\chi(P) = \sum_{\mathbf{v}} \alpha_0(T_{\mathbf{v}} P),$$

*to be summed over all vertices $\mathbf{v}$ of $P$.*

Compare [Banchoff]. Alternatively, we could sum over all points of $P$. For if $\mathbf{x}$ is not a vertex then it follows easily from (34) that $\alpha_0(T_{\mathbf{x}} P) = 0$.

**Proof.** If P is convex, then the dual cones $(T_{\mathbf{v}} P)^*$ cover $\mathbf{R}^n$ without any $n$-dimensional overlap. Hence

$$\sum_{\mathbf{v}} \alpha_n((T_{\mathbf{v}} P)^*) = \alpha_n(\mathbf{R}^n) = 1 = \chi(P),$$

where $\alpha_n((T_{\mathbf{v}}P)^*) = \alpha_0(T_{\mathbf{v}}P)$ by (34). Since any polyhedron is a finite union of convex sets, the general case follow by induction on the number of these convex sets, using additivity. □

More generally, we can also express the $k$-dimensional measure $\mu_k(P)$ as a similar sum. It will be convenient to express any compact polyhedron $P \subset \mathbf{R}^n$ as a disjoint union

$$P = \bigcup_F \overset{\circ}{F},$$

where $F$ ranges over all faces of $P$ and where $\overset{\circ}{F}$ denotes the "interior" of $F$, that is, the set of all points of $F$ which do not belong to lower dimensional faces. Note that any two points $\mathbf{x}$ and $\mathbf{y}$ of $\overset{\circ}{F}$ have a common tangent cone, $T_{\mathbf{x}}(P) = T_{\mathbf{y}}(P)$. This common tangent cone will be written briefly as $T_F(P)$.

**(38) Theorem.** *If $P \subset \mathbf{R}^n$ is a compact polyhedron, then $\mu_k(P)$ is given by the formula*

$$\mu_k(P) = \sum_F \mu_k(F)\,\alpha_k(T_F P),$$

*to be summed over all k-dimensional faces $F \subset P$.*

Note that $\mu_k(F)$ is just ordinary $k$-dimensional volume. If $P$ is convex, then $\alpha_k(T_F P)$ is just the ordinary $(n-k)$-dimensional solid angle of the dual cone, suitably normalized,

$$\alpha_k(T_F P) = \alpha_{n-k}((T_F P)^*) = \mu_{n-k}((T_F P)^* \cap D^n)/\delta_{n-k},$$

since the dual cone clearly has dimension precisely $n-k$. For example, with $k = 1$ and $n = 3$, if an edge $E$ has length $\lambda$ and dihedral angle $\theta$, then

$$\alpha_1(T_E P) = \alpha_2((T_E P)^*) = \frac{\pi - \theta}{2\pi},$$

so the contribution of $E$ to $\mu_1(P)$ is $(\pi - \theta)\lambda/2\pi$. (See the first Figure on the next page.)

**Proof of (38).** By additivity, it suffices to check this formula when $P$ is convex. Let $\rho\colon \mathbf{R}^n \to P$ be the natural retraction, which carries each point of $\mathbf{R}^n$ to the closest point of $P$. Then for each face $F$ we have

$$\rho^{-1}(\overset{\circ}{F}) \cong \overset{\circ}{F} \times (T_F P)^*.$$

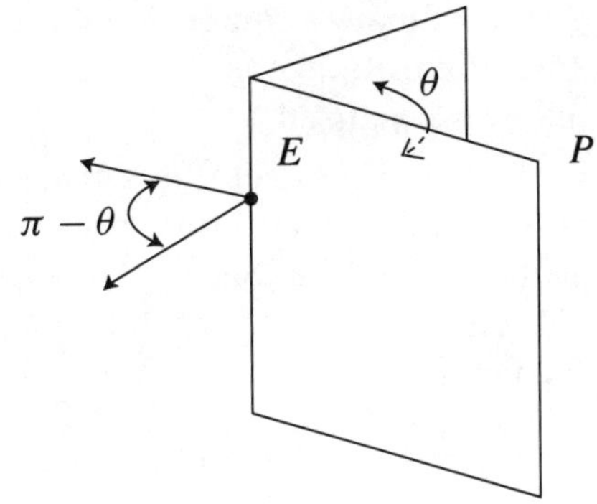


If $\dim(F) = k$ so that $\dim(T_F P)^* = n - k$, then the intersection of $(T_F P)^*$ with a disk of radius $r$ has $(n-k)$-dimensional volume

$$r^{n-k} \cdot \mu_{n-k}\big((T_F P)^* \cap D^n\big) = r^{n-k}\delta_{n-k}\,\alpha_{n-k}\big((T_F P)^*\big) = r^{n-k}\,\delta_{n-k}\,\alpha_k(T_F P).$$

Hence the $n$-dimensional volume of $N_r(P)$ equals

$$\sum_{k=0}^{n} \sum_{\dim F = k} r^{n-k}\delta_{n-k}\,\alpha_k(T_F P)\mu_k(F).$$

Together with (12), this completes the proof. □

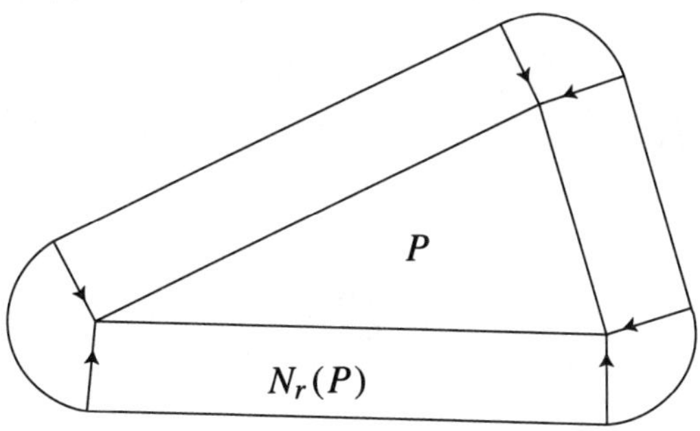


There is a completely analogous formula for the angular measure $\alpha_k(C)$.

**(39) Theorem.** *For any polyhedral cone $C \subset \mathbf{R}^n$, we have*

$$\alpha_k(C) = \sum_F \alpha_k(F)\,\alpha_k(T_F C),$$

*to be summed over all k-dimensional faces $F \subset C$.*

Note that $\alpha_k(F)$ is just an ordinary $k$-dimensional solid angle, and that in the convex case $\alpha_k(T_F C)$ is the ordinary $(n-k)$-dimensional solid angle of the dual cone, both suitably normalized.

Before giving the proof, let us interpret the statement of (39) both geometrically and analytically, and derive an important corollary. Let $C \subset \mathbf{R}^n$ be a convex polyhedral cone. First note that there is a duality between faces of $C$ and faces of the dual cone $C^*$. In fact to each face $F \subset C$ of dimension $k$ there corresponds a dual face

$$F' = (T_F C)^* \subset C^*$$

of dimension $n-k$. Let $\rho\colon \mathbf{R}^n \to C$ be the canonical retraction which carries each point of $R^n$ to the closest point of $C$. Since $C$ is the disjoint union of the subsets $\overset{\circ}{F}$, it follows that $\mathbf{R}^n$ is the disjoint union of the subsets $\rho^{-1}(\overset{\circ}{F})$. The closure of $\rho^{-1}(\overset{\circ}{F})$ is itself a polyhedral cone; in fact

$$\text{closure}(\rho^{-1}\overset{\circ}{F}) \cong F \times (T_F C)^* = F \times F'$$

(where $F$ and $F'$ are perpendicular to each other). Thus we have proved the following.

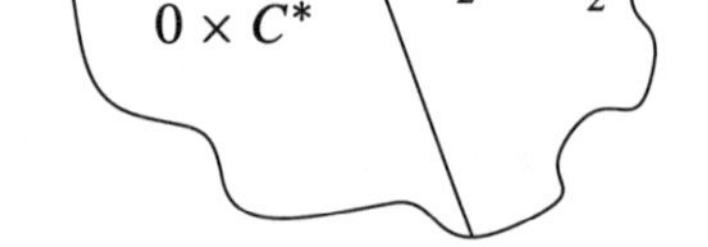

Picture for $n = 2$

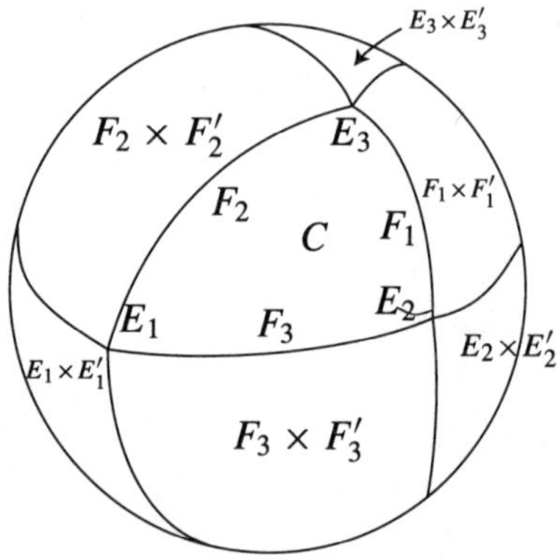

Picture for $n = 3$, intersected with the unit sphere

**(40) Lemma.** *Any convex polyhedral cone $C \subset \mathbf{R}^n$ determines a partition of $\mathbf{R}^n$ into non-overlapping convex cones $F \times F'$. Here $F$ ranges over all faces of $C$ and $F'$ denotes the dual face of $C^*$, with*

$$\dim(F) + \dim(F') = n.$$

*Thus*

$$\sum_F \alpha_n(F \times F') = \alpha_n(\mathbf{R}^n) = 1.$$

On the other hand, if $F$ has dimension $k$, then by (33) and (34) we have

$$\alpha_n(F \times F') = \alpha_k(F)\alpha_{n-k}(F') = \alpha_k(F)\alpha_{n-k}((T_F C)^*) = \alpha_k(F)\alpha_k(T_F C).$$

Thus in the convex case we can restate (39) as follows.

**(41) Corollary.** *For a convex polyhedral cone $C$, the angular measure $\alpha_k(C) = \alpha_{n-k}(C^*)$ is equal to the sum, over all faces $F$ of dimension $k$, of the product $\alpha_k(F)\alpha_{n-k}(F')$.*

Evidently this formulation gives a direct geometric meaning to the identity $\sum_k \alpha_k(C) = 1$, and also to the duality statement $\alpha_k(C) = \alpha_{n-k}(C^*)$.

We can express (41) analytically as follows. Let $r(\mathbf{x}) = \|\rho(\mathbf{x}) - \mathbf{x}\|$ be the distance of $\mathbf{x}$ from $\mathbf{C}$. Note that

$$\|\mathbf{x}\|^2 = \|\rho(\mathbf{x})\|^2 + r(x)^2.$$

**(42) Corollary.** *For $t > 0$, the polynomial $\alpha(C, t) = \sum_k t^k \alpha_k(C)$ can be expressed as an integral*

$$\alpha(C, t) = t^n \int_{\mathbf{R}^n} e^{-\pi(\|\rho(\mathbf{x})\|^2 + t^2 r(\mathbf{x})^2)}\, d(\mathbf{x}).$$

**Proof.** This follows easily from (32) and (41). □

**(43) Product Theorem.** *The angular measures for any cartesian product of polyhedral cones are given by*

$$\alpha_k(C \times C') = \sum_{i+j=k} \alpha_i(C)\,\alpha_j(C').$$

This follows since the integral (42) over a product is the product of the integrals. □

Here is an example. If $P^n \subset \mathbf{R}^n$ is the ***positive cone***, consisting of all $(x_1, \ldots, x_n)$ with $x_i \geq 0$, then it follows from (43) that $\alpha_k(P^n) = \binom{n}{k}/2^n$.

**Proof of (39).** We may assume that $C$ is convex. If $C$ contains a line $L$ through the origin, or equivalently if $C^*$ is contained in some hyperplane $L^*$, then $C$ is isomorphic to an orthogonal product $(C \cap L^*) \times L$, and the conclusion follows by induction on the dimension $n$, making use of (30). Thus we may assume that $C$ does not contain any line through the origin. For $k = 0$, since $\alpha_0(\mathbf{R}^0) = 1$ and $T_0(C) = C$, the statement is a tautology. For $k = 1$, since the dual cone $C^*$ is assumed to be $n$-dimensional, we have

$$\alpha_1(C) = \alpha_{n-1}(C^*) = \tfrac{1}{2}\alpha_{n-1}(\partial C^*).$$

The $(n-1)$-dimensional faces of $C^*$ can be identified with the dual cones $(T_E C)^*$, where $E$ ranges over the 1-dimensional edges of $C$. Each half-line $E$ satisfies $\alpha_1(E) = \frac{1}{2}\alpha_1(\mathbf{R}^1) = \frac{1}{2}$, so

$$\sum_{\text{edges } E} \alpha_1(E)\alpha_1(T_E C) = \tfrac{1}{2}\sum \alpha_{n-1}\big((T_E C)^*\big) = \alpha_{n-1}(C^*),$$

which proves the required statement.

For $k > 1$ we proceed by induction. Applying (36) $k-1$ times, we see that $\alpha_k(C)$ is the average, over all planes $\Pi_{k-1}$ of codimension $k-1$ through the origin, of

$$\alpha_1(C \cap \Pi_{k-1}) = \sum_E \tfrac{1}{2}\alpha_1(T_E(C \cap \Pi_{k-1})),$$

to be summed over all edges $E$ of $C \cap \Pi_{k-1}$. The probability that $\Pi_{k-1}$ intersects a given $k$-dimensional face $F \subset C$ non-trivially is equal to $2\,\alpha_k(F)$ (since we assume that $F \not\cong \mathbf{R}^k$). Setting $E = F \cap \Pi_{k-1}$, we must then average over all planes $\Pi_{k-1}$ containing $E$, using an appropriate conditional probability measure. The result will be an expression of the form

$$\alpha_k(C) = \sum_{\dim F = k} \alpha_k(F)\Phi(T_F C)),$$

where $\Phi$ is some unknown function. However, we can evaluate this unknown function by the simple expedient of applying this formula to the convex cone $T_F C \subset \mathbf{R}^n$, which has a unique $k$-dimensional face, isomorphic to $\mathbf{R}^k$. Thus

$$\alpha_k(T_F C) = \alpha_k(\mathbf{R}^k)\Phi(T_F C),$$

so the unknown function $\Phi$ is just $\alpha_k$. □

Here is a version of the Gauss-Bonnet Theorem which applies to non-Euclidean polyhedra. (Compare [Allendoerfer and Weil] as well as [Cheeger, Müller and Schrader].)

**(44) Corollary.** *If $P$ is a compact polyhedron in a space of constant curvature $K$, then*

$$\chi(P) = \sum_m \frac{(2m)!}{m!}\left(\frac{K}{4\pi}\right)^m \sum_{\dim F = 2m} \mu_{2m}(F)\alpha_{2m}(T_F P).$$

*Here $F$ ranges over all even dimensional faces of $P$, and $\mu_{2m}(F)$ is the ordinary Riemannian volume.*

**Proof.** For $K = 1$, with $P \subset S^{n-1}$, let $C \subset \mathbf{R}^n$ be the associated polyhedral cone consisting of all half-lines from the origin through $P$. Then

$$\chi(P) = \hat{\sigma}_0(C) = 2\big(\alpha_1(C) + \alpha_3(C) + \alpha_5(C) + \cdots\big).$$

Applying (39) and (22), we easily obtain the required formula (44). Now the case $K > 0$ follows by a scale change. For $K = 0$ the statement is known by (37), and the case $K < 0$ follows by analytic continuation. □

**Example.** For a 4-simplex $P$ in Lobachevsky space, with $K = -1$, the formula takes the form

$$\chi(P) = 1 = \sum_{\text{vertices } \mathbf{v}} \alpha_0(T_{\mathbf{v}}P) - \sum_{\text{triangles } F} \frac{\text{area}(F)}{2\pi}\alpha_2(T_F P) + \frac{\text{vol}_4(P)}{4\pi^2/3},$$

where $\text{vol}_4$ is 4-dimensional volume, and $\alpha_2(T_F P) = (\pi - \text{dihedral angle})/2\pi$. In the limit, as all vertices tend to the 3-sphere at infinity, the five tangent cones $T_{\mathbf{v}}P$ tend to half-lines with $\alpha_0 = \frac{1}{2}$, and the ten 2-dimensional faces tend to ideal triangles with area $\pi$. This yields Thurston's formula

**(45)** $$\text{vol}_4(P) = \frac{\pi}{3}\big(4\pi - \textstyle\sum(\text{dihedral angles})\big)$$

for the volume of an ideal 4-simplex.

For further information and references concerning convex cones, see [McMullen].

## §4. Open Problems

This exposition leaves a number of large gaps. Here are a few basic problems or conjectures which remain unanswered:

Which compact sets $A \subset \mathbf{R}^n$ have well defined measures $\mu_k(A)$?

In what sense do sets $A$ and $B$ have to be close to each other, in order to guarantee that $\mu_k(A)$ is close to $\mu_k(B)$?

In what generality are formulas such as (16) and (18) valid? It is conjectured that these formulas are true at least for arbitrary piecewise smooth subsets of $\mathbf{R}^n$.

In the piecewise smooth case, it is conjectured that $\mu_k(A)$ is an intrinsic invariant of $A$, considered as a union of Riemannian manifolds with corners, independent of the particular embedding in Euclidean space. Furthermore, for any Riemannian manifold with corners, it is conjectured that one can compute $\mu_k(A)$ by integrating suitable curvatures over the interior and the various faces. (Compare [Allendoerfer and Weil].)

I am indebted to Graeme Segal for pointing out the following important reference. Suppose that $M^p$ is a compact manifold without boundary, smoothly embedded in $\mathbf{R}^n$. Then according to [Weyl], the volume $\mu_n(N_r(M^p))$ of the neighborhood of radius $r$ can be expressed as a polynomial in $r$, provided that $r \geq 0$ is sufficiently small. (See also [Gray].) Furthermore, the coefficients of this polynomial are explicitly computed intrinsic invariants of the Riemannian manifold $M^p$, independent of its embedding into $\mathbf{R}^n$. If we write this polynomial, in analogy with (12), as

**(46)** $$\mu_n(N_r(M^p)) = \sum_i \hat{\mu}_{n-i}(M^p) \cdot \delta_i \, r^i$$

for small values of $r$, then Weyl's formula for the coefficients can be stated more precisely as follows. The coefficient $\hat{\mu}_k(M^p)$ is zero unless $k$ is of the form $p - 2h$. (Compare (8).) For $k = p - 2h$ we have

**(47)** $$\hat{\mu}_{p-2h}(M^p) = \frac{1}{(4\pi)^h} \int_{M^p} \boldsymbol{K}_h(\mathbf{x}) \, d(\mathbf{x}).$$

Here $d(\mathbf{x})$ is the $p$-dimensional Riemannian volume element, and

$$\boldsymbol{K}_h = \frac{1}{2^h h!} \sum \delta^{i_1 \cdots i_{2h}}_{j_1 \cdots j_{2h}} R^{j_1 j_2}_{i_1 i_2} \cdots R^{j_{2h-1} j_{2h}}_{i_{2h-1} i_{2h}}$$

is a curvature invariant which is obtained by contracting the $h$-fold product of the Riemann tensor with itself. For example $\boldsymbol{K}_0$ is identically equal to one, and $\boldsymbol{K}_1$ is the usual scalar curvature of $M^p$, taking the constant value $p(p-1)/a^2$ on a $p$-dimensional sphere of radius $a$. In this expression, $R^{j_1 j_2}_{i_1 i_2}$ stands for the version $\sum g^{j_1 k_1} g^{j_2 k_2} R_{i_1 i_2 k_1 k_2}$ of the Riemann tensor, while the symbol $\delta^{i_1 \cdots i_{2h}}_{j_1 \cdots j_{2h}}$ is defined to be $+1$ or $-1$ if $(i_1, \ldots, i_{2h})$ forms an even or odd permutation of the distinct integers $(j_1, \ldots, j_{2h})$, and is defined to be zero otherwise. The normalizing constant is chosen so that the curvatures $K_h$ at a point $(\mathbf{x}, \mathbf{x}')$ of a Cartesian product $M \times M'$ are given by

$$K_h(\mathbf{x}, \mathbf{x}') = \sum_{j+k=h} K_j(\mathbf{x}) K_k(\mathbf{x}').$$

It seems natural to conjecture that formula (18), or formula (12) for small values of $r$, is indeed valid for every smooth closed manifold in $\mathbf{R}^n$, so that the coefficient $\hat{\mu}_k(M^p)$ in (46) and (47) can simply be identified with $\mu_k(M^p)$.

Here is a different question: do $\{\mu_0, \dots, \mu_n, 1\}$ form a basis for the vector space consisting of all real valued set functions, defined for compact polyhedra in $\mathbf{R}^n$ (or perhaps even for compact piecewise smooth subsets of $\mathbf{R}^n$), which are additive, invariant under the Euclidean group, and whose value on a simplex depends continuously on the simplex? For convex sets in 3-space, this is proved in [Hadwiger].

Most of these questions have analogues for the spherical Steiner measures $\sigma_i$ (compare [Weyl]), or equivalently for the angular measures $\alpha_i$. For example, do $\{\alpha_0, \dots, \alpha_n\}$ form a basis for the vector space consisting of all real valued set functions, defined for polyhedral cones in $\mathbf{R}^n$, which are additive, rotation invariant, and continuous in a suitable topology?

## Bibliography

C. Allendoerfer and A. Weil, *The Gauss-Bonnet theorem for Riemannian polyhedra,* Trans. Amer. Math. Soc. **53** (1943), 101–129.

T. Banchoff, *Critical points and curvature for embedded polyhedra,* J. Diff. Geom. **1** (1967), 245–256.

J. Cheeger, W. Müller and R. Schrader, *On the curvature of piecewise flat spaces,* Comm. Math. Phys. **92** (1984), 405–454.

D. Gale, "The Theory of Linear Economic Models," reprinted, U. Chicago Press, 1989.

A. Gray, "Tubes," Addison-Wesley, 1990.

H. Hadwiger, "Altes und Neues über konvexe Körper," Birhäuser, 1955.

H. Kneser, *Der Simplexinhalt in der nichteuklidische Geometrie,* Deutsche Math. **1** (1936), 337–340.

N. Kuiper, *Der Satz von Gauss-Bonnet für Abbildungen in $E^N$ und damit verwandte Probleme,* Jber. Deutsch. Math. Verein. **69** (1967), 77–88.

P. McMullen, *Non-linear angle sum relations for polyhedral cones and polytopes,* Math. Proc. Cambr. Phil. Soc. **78** (1975) 247–261.

H. Minkowski, *Theorie der konvexen Körper,* in "Gesamm. Abh. **2**," Leipzig 1911, reprinted by Chelsea 1967, pp. 131–229.

L. A. Santaló, "Introduction to Integral geometry," Hermann, 1953.

———, "Integral Geometry and Geometric Probability," Addison-Wesley 1976 (Encycl. of Math. and its Appl., v. **1**).

J. Steiner, *Über parallele Flächen*, Mber. preuss. Akad. Wiss. 1840, 114–118. (Gesamm. Werke **2** 173-176).

H. Weyl, *On the volume of tubes*, Amer. J. Math. **61** (1939), 461–472.

# ANALYTIC PROOFS OF THE "HAIRY BALL THEOREM" AND THE BROUWER FIXED POINT THEOREM

*Reprinted from*

The American Mathematical Monthly
Volume 85 (1978), pp. 521–524

The theorem that every continuous vector field on the 2-sphere must have a zero is a basic topological result which is not completely easy to prove. This note presents an extremely curious proof, which seems to have nothing to do with topology, and little to do with anything else. The Brouwer fixed point theorem is an easy consequence. However, this is less interesting since[1] there are several fairly elementary proofs in this case.

The original text has been reset in TEX, and one footnote added.

[1] See for example M. Hirsch, *A proof of the nonretractability of a cell onto its boundary*, Proc. Amer. Math. Soc. **14** (1963), 364–365.

# ANALYTIC PROOFS OF THE "HAIRY BALL THEOREM" AND THE BROUWER FIXED POINT THEOREM

JOHN MILNOR

This note will present strange but quite elementary proofs of two classical theorems of topology, based on a volume computation in Euclidean space and the observation that the function $(1+t^2)^{n/2}$ is not a polynomial when $n$ is odd. The argument was suggested by the methods of Asimov [**1**]. Familiar proofs of these theorems all use either combinatorial arguments, homology theory, differential forms, or methods from geometric topology. Compare [**2**], [**3**], [**4**], [**6**], [**7**].

Here is a preliminary version of the first one.

THEOREM 1. *An even-dimensional sphere does not posses any continuously differentiable field of unit tangent vectors.*

By definition, the *sphere* $S^{n-1}$ is the set of all vectors $\mathbf{u} = (u_1, \ldots, u_n)$ in the Euclidean space $\mathbf{R}^n$ such that the Euclidean length $\|\mathbf{u}\|$ equals 1. A vector $\mathbf{v}(\mathbf{u})$ in $\mathbf{R}^n$ is *tangent* to $S^{n-1}$ at $\mathbf{u}$ if the Euclidean inner produce $\mathbf{u} \cdot \mathbf{v}(\mathbf{u})$ is zero.

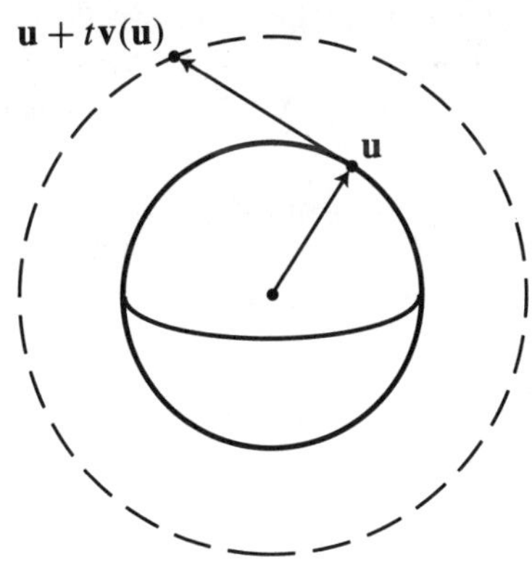


Fig. 1.

The hypothesis that the dimension $n-1$ is even is essential. For if $n-1$ is odd then the formula

$$\mathbf{v}(u_1, \ldots, u_n) = (u_2, -u_1, \ldots, u_n, -u_{n-1})$$

John Milnor received his Ph.D. from Princeton and taught there until 1967. After two years at M.I.T., he joined the Institute for Advanced Study; he has also been a visiting professor at Berkeley and at UCLA. He was awarded a Fields Medal in 1962 and the National Medal of Science in 1966; he is a member of the National Academy of Sciences. His principal research has been in the topology of manifolds.—*Editors* defines a differentiable field of unit tangent vectors on $S^{n-1}$.

The proof of Theorem 1 will depend on two lemmas. The first involves a volume computation. Let $A$ be a compact region in $\mathbf{R}^n$, and let $\mathbf{x} \mapsto \mathbf{v}(\mathbf{x})$ be a continuously differentiable vector field which is defined throughout a neighborhood of $A$. The values $\mathbf{v}(\mathbf{x})$ can be arbitrary vectors in $\mathbf{R}^n$. For each real number $t$, consider the function

$$\mathbf{f}_t(x) = \mathbf{x} + t\mathbf{v}(\mathbf{x})$$

which is defined for all $\mathbf{x}$ in $A$.

LEMMA 1. *If the parameter $t$ is sufficiently small, then this mapping $\mathbf{f}_t$ is one-to-one and transforms the region $A$ onto a nearby region $\mathbf{f}_t(A)$ whose volume can be expressed as a polynomial function of $t$.*

*Proof.* Since $A$ is compact, and since the function $\mathbf{x} \mapsto \mathbf{v}(\mathbf{x})$ is continuously differentiable, there exists a *Lipschitz constant* $c$ so that

$$\|\mathbf{v}(\mathbf{x}) - \mathbf{v}(\mathbf{y})\| \leq c\|\mathbf{x} - \mathbf{y}\| \tag{$*$}$$

for all $\mathbf{x}$ and $\mathbf{y}$ in $A$. [This is proved as follows. First consider the special cases where $A$ is a cube with edges parallel to the coordinate axes. Passing from $\mathbf{x}$ to $\mathbf{y}$ in $n$ steps by changing one coordinate at a time, and applying the Mean Value Theorem of differential calculus, one sees that $|v_i(\mathbf{x}) - v_i(\mathbf{y})| \leq \sum_j c_{ij}|x_j - y_j|$, where

$$c_{ij} = \sup_A |\partial v_i/\partial x_j|.$$

Therefore

$$\|\mathbf{v}(\mathbf{x}) - \mathbf{v}(\mathbf{y})\| \leq \sum_i |v_i(\mathbf{x}) - v_i(\mathbf{y})| \leq \sum_{i,j} c_{ij}|x_j - y_j| \leq \sum_{i,j} c_{ij}\|\mathbf{x} - \mathbf{y}\|,$$

as required. Now an arbitrary compact set $A$ in $\mathbf{R}^n$ can be covered by finitely many open cubes $I_\alpha$, chosen so that a Lipschitz condition ($*$) holds whenever $\mathbf{x}$ and $\mathbf{y}$ belong to the same cube. But if $\mathbf{x}$ and $\mathbf{y}$ in $A$ *do not* belong to any common cube $I_\alpha$, then the distance $\|\mathbf{x} - \mathbf{y}\|$ is bounded away from zero. In fact this expression $\|\mathbf{x} - \mathbf{y}\|$ can be thought of as a continuous and nowhere zero function on the compact set $A \times A - \bigcup I_\alpha \times I_\alpha$. It is now easy to choose a constant $c$ so that the Lipschitz condition ($*$) holds uniformly for all $\mathbf{x}$ and $\mathbf{y}$ in $A$.]

Choose any $t$ with $|t| < c^{-1}$. Then $\mathbf{f}_t$ is one-to-one; for if $\mathbf{f}_t(\mathbf{x}) = \mathbf{f}_t(\mathbf{y})$ then $\mathbf{x} - \mathbf{y} = t(\mathbf{v}(\mathbf{y}) - \mathbf{v}(\mathbf{x}))$, hence the inequality $\|\mathbf{x} - \mathbf{y}\| \leq |t|c\|\mathbf{x} - \mathbf{y}\|$ implies that $\mathbf{x} = \mathbf{y}$.

The matrix of first derivatives of $\mathbf{f}_t$ can be written as $I + t[\partial v_i/\partial x_j]$, where $I$ is the identity matrix. Hence its determinant is a polynomial function of $t$, of the form $1 + t\sigma_1(\mathbf{x}) + \cdots + t^n\sigma_n(\mathbf{x})$, the coefficients being continuous functions of $\mathbf{x}$. This determinant is strictly positive for $|t|$ sufficiently small. Integrating over $A$, we see that the volume of the image region can be expressed as a polynomial function of $t$,

$$\text{volume}\,\mathbf{f}_t(A) = a_0 + a_1 t + \cdots + a_n t^n,$$

with coefficients $a_k = \int \cdots \int_A \sigma_k(\mathbf{x})\,dx_1 \cdots dx_n$. ∎

Now suppose that the sphere $S^{n-1}$ has a continuously differentiable field $\mathbf{u} \mapsto \mathbf{v}(\mathbf{u})$ of unit tangent vectors. For any real number $t$, note that the vector $\mathbf{u} + t\mathbf{v}(\mathbf{u})$ has length $\sqrt{1+t^2}$.

LEMMA 2. *If the parameter $t$ is sufficiently small, then the transformation $\mathbf{u} \mapsto \mathbf{u} + t\mathbf{v}(\mathbf{u})$ maps the unit sphere in $\mathbf{R}^n$* **onto** *the sphere of radius $\sqrt{1+t^2}$.*

Assuming this Lemma for the moment, we can now prove Theorem 1. As region $A$ we take the region between two concentric spheres, defined by the inequalities $a \leq \|\mathbf{x}\| \leq b$. We extend the vector field $\mathbf{v}$ throughout this region by setting $\mathbf{v}(r\mathbf{u}) = r\mathbf{v}(\mathbf{u})$ for $a \leq r \leq b$. It follows that the mapping $\mathbf{f}_t(\mathbf{x}) = \mathbf{x} + t\mathbf{v}(\mathbf{x})$ is defined throughout the region $A$, and maps the sphere of radius $r$ *onto* the sphere of radius $r\sqrt{1+t^2}$ providing that $t$ is sufficiently small. (Note that $\mathbf{f}_t(r\mathbf{u}) = r\mathbf{f}_t(\mathbf{u})$.) Hence it maps $A$ onto the region between spheres of radius $a\sqrt{1+t^2}$ and $b\sqrt{1+t^2}$. Evidently

$$\text{volume}\,\mathbf{f}_t(A) = \left(\sqrt{(1+t^2})\right)^n \text{volume}(A).$$

Thus, if $n$ is odd, this volume is not a polynomial function of $t$. Comparing Lemma 1, we obtain a contradiction which proves Theorem 1.

Now we must prove Lemma 2. Here are two alternative arguments, the first based on the *Shrinking Lemma*,* and the second based on elementary point set topology.

*First Proof.* Let the region $A$ considered above be defined by the inequalities $1/2 \leq \|\mathbf{x}\| \leq 3/2$. Choose $t$ small enough so that $|t| < 1/3$ and $|t| < c^{-1}$, where $c$ is a Lipschitz constant for $\mathbf{v}$. Then for each fixed $\mathbf{u}_0$ in $S^{n-1}$ the auxiliary mapping

$$\mathbf{x} \mapsto \mathbf{u}_0 - t\mathbf{v}(\mathbf{x})$$

carries the complete metric space $A$ into itself (since $|t\mathbf{v}(\mathbf{x})| < 1/2$); and satisfies a Lipschitz condition with Lipschitz constant less than 1. Hence, by the Shrinking Lemma, this auxiliary map has a unique fixed point. In other words, the equation $\mathbf{f}_t(\mathbf{x}) = \mathbf{u}_0$ has a unique solution $\mathbf{x}$. Multiplying $\mathbf{x}$ and $\mathbf{u}_0$ by $\sqrt{1+t^2}$, Lemma 2 follows.

*Second Proof.* We may assume that $n \geq 2$. If $t$ is sufficiently small, then the matrix of first derivatives of $\mathbf{f}_t$ is non-singular throughout the compact region $A$. (Compare the proof of Lemma 1.) Using the Inverse Function Theorem, it follows that $\mathbf{f}_t$ maps

---

* Also called the *Contraction Lemma*. This is the statement that any self-map with Lipschitz constant strictly less than one, on a complete metric space, must have a unique fixed point. Compare [5].

open sets in the interior of $A$ to open sets. Hence the image $\mathbf{f}_t(S^{n-1})$ is a relatively open subset of the sphere of radius $\sqrt{1+t^2}$. But this image $\mathbf{f}_t(S^{n-1})$ is also compact and hence closed. Since $S^{n-1}$ is connected, an open and closed subset must be the entire sphere, and the conclusion follows. ■

A slightly sharper version of Theorem 1, which does not mention differentiability or unit vectors, follows as an immediate corollary.

THEOREM 1′. *An even dimensional sphere does not admit any* **continuous** *field of* **non-zero** *tangent vectors.*

*Proof.* Suppose that the sphere $S^{n-1}$ possesses a continuous field of non-zero tangent vectors $\mathbf{v}(\mathbf{u})$. Let $m > 0$ be the minimum of $\|\mathbf{v}(\mathbf{u})\|$. By the Weierstrass Approximation Theorem [5], there exists a polynomial mapping $\mathbf{p}$ from $S^{n-1}$ to $\mathbf{R}^n$ satisfying

$$\|\mathbf{p}(\mathbf{u}) - \mathbf{v}(\mathbf{u})\| < m/2$$

for all $\mathbf{u}$. Defining a differentiable vector field $\mathbf{w}(\mathbf{u})$ by the formula

$$\mathbf{w} = \mathbf{p} - (\mathbf{p} \cdot \mathbf{u})\mathbf{u}$$

for every $\mathbf{u}$, the computation $\mathbf{w} \cdot \mathbf{u} = 0$ shows that $\mathbf{w}(\mathbf{u})$ is tangent to $S^{n-1}$ at $\mathbf{u}$, while the computation

$$\|\mathbf{w} - \mathbf{p}\| = |\mathbf{p} \cdot \mathbf{u}| < m/2,$$

together with the triangle inequality, shows that $\mathbf{w} \neq \mathbf{0}$. Therefore, the quotient $\mathbf{w}(\mathbf{u})/\|\mathbf{w}(\mathbf{u})\|$ is an infinitely differentiable field of unit tangent vectors on the sphere $S^{n-1}$. If $n-1$ is even, this is impossible by Theorem 1. ■

Starting with Theorem 1′, it is quite easy to prove the Brouwer Fixed Point Theorem:

THEOREM 2. *Every continuous mapping* $\mathbf{f}$ *from the disk* $D^n$ *to itself possesses at least one fixed point.*

Here $D^n$ is defined to be the set of all vectors $\mathbf{x}$ in $\mathbf{R}^n$ with $\|\mathbf{x}\| \leq 1$.

*Proof.* If $\mathbf{f}(x) \neq \mathbf{x}$ for all $\mathbf{x}$ in $D^n$, then the formula $\mathbf{v}(\mathbf{x}) = \mathbf{x} - \mathbf{f}(\mathbf{x})$ would define a non-zero vector field $\mathbf{v}$ on $D^n$ which points *outward* everywhere on the boundary, in the sense that $\mathbf{u} \cdot \mathbf{v}(\mathbf{u}) > 0$ for every point $\mathbf{u}$ in $S^{n-1}$.

With a little care, we can modify this definition to obtain a non-zero vector field $\mathbf{w}$ on $D^n$ which points *directly outward* on the boundary, in the sense that $\mathbf{w}(\mathbf{u}) = \mathbf{u}$ for every $\mathbf{u}$ in $S^{n-1}$. For example, set

$$\mathbf{w}(\mathbf{x}) = \mathbf{x} - \mathbf{y}(1 - \mathbf{x} \cdot \mathbf{x})/(1 - \mathbf{x} \cdot \mathbf{y}),$$

where $\mathbf{y} = \mathbf{f}(\mathbf{x}) \neq \mathbf{x}$. Evidently $\mathbf{w}(\mathbf{x}) = \mathbf{x}$ whenever $\mathbf{x} \cdot \mathbf{x} = 1$. This expression depends continuously on $\mathbf{x}$, since the denominator never vanishes. It is clearly non-zero whenever

$\mathbf{x}$ and $\mathbf{y}$ are linearly independent; while if $\mathbf{x}$ and $\mathbf{y}$ are linearly dependent the identity $(\mathbf{x}\cdot\mathbf{x})\mathbf{y} = (\mathbf{x}\cdot\mathbf{y})\mathbf{x}$ implies that $\mathbf{w}(\mathbf{x}) = (\mathbf{x}-\mathbf{y})/(1-\mathbf{x}\cdot\mathbf{y}) \neq \mathbf{0}$.

Let us transplant this hypothetical vector field $\mathbf{w}(\mathbf{x})$ to the southern hemisphere of the unit sphere $S^n$ in $\mathbf{R}^{n+1}$. Identifying $\mathbf{R}^n$ with the hyperplane $x_{n+1} = 0$ which passes through the "equator" of $S^n$, we will use stereographic projection from the north pole $(0, \ldots, 0, 1)$ to map each point $\mathbf{x}$ of $D^n$ to a point $\mathbf{s}(\mathbf{x}) = \mathbf{u}$ of the southern hemisphere $u_{n+1} \leq 0$. The precise formula is

$$\mathbf{s}(\mathbf{x}) = (2x_1, \ldots, 2x_n, \mathbf{x}\cdot\mathbf{x} - 1)/(\mathbf{x}\cdot\mathbf{x} + 1).$$

Applying the derivative of the mapping $\mathbf{s}$ at $\mathbf{x}$ to the vector $\mathbf{w}(\mathbf{x})$, we obtain a corresponding tangent vector $\mathbf{W}(\mathbf{u})$ to $S^n$ at the image point $\mathbf{s}(\mathbf{x}) = \mathbf{u}$. (The vector $\mathbf{W}(\mathbf{u})$ can be described as the velocity vector $d\mathbf{s}(\mathbf{x}+t\mathbf{w}(\mathbf{x}))/dt$ of the spherical curve $t \mapsto \mathbf{s}(\mathbf{x}+t\mathbf{w}(\mathbf{x}))$, evaluated at $t = 0$.) In this way, we obtain a non-zero tangent vector field $\mathbf{W}$ on the southern hemisphere. At every point $\mathbf{u} = \mathbf{s}(\mathbf{u})$ of the equator, since $\mathbf{w}(\mathbf{u}) = \mathbf{u}$ points directly outward, computation shows that the corresponding vector $\mathbf{W}(\mathbf{u}) = (0, \ldots, 0, 1)$ points due north (i.e., away from the southern hemisphere).

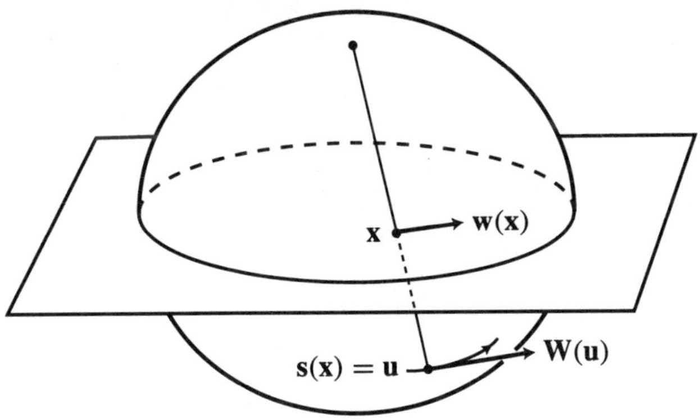


Fig. 2.

Similarly, using stereographic projection from the south pole, the vector field $-\mathbf{w}(\mathbf{x})$ corresponds to a vector field on the northern hemisphere which also points due north on the equator. Piecing these two vector fields together, we obtain a non-zero tangent vector field $\mathbf{W}$ which is defined and continuous everywhere on $S^n$. If $n$ is even, this is impossible by Theorem 1$'$.

This contradiction proves the Brouwer Fixed Point Theorem for even values of $n$. But this suffices to prove the Theorem for an odd value $n = 2k-1$ also. For any map $\mathbf{f}$ from $D^{2k-1}$ to itself without fixed point would give rise to a map $\mathbf{F}(x_1, \ldots, x_{2k}) = (\mathbf{f}(x_1, \ldots, x_{2k-1}, 0)$ from $D^{2k}$ to itself without fixed point. ■

## References

**1.** D. Asimov, Average Gaussian curvature of leaves of foliations, preprint, 1976. Bull. Amer. Math. Soc. 84 (1978) 131–133.

**2.** W. M. Boothby, On two classical theorems of algebraic topology, this MONTHLY, 78 (1971) 237–249.

**3.** P. Hilton and S. Wylie, Homology Theory, Cambridge University Press, New York, 1960, 218–219.

**4.** M. Hirsch, Differential Topology, Springer-Verlag, New York, 1976, 73, 134.

**5.** S. Lang, Analysis II, Addison-Wesley, Reading, Mass., 1969, 50, 121.

**6.** J. Milnor, Topology from the Differentiable Viewpoint, Univ. Press of Va. Charlottesville, 1965, 4, 30.

**7.** E. Spanier, Alegraic Topology, McGraw-Hill, New York, 1966, 151, 194.

INSTITUTE FOR ADVANCED STUDY, PRINCETON UUNIVERSITY, PRINCETON, NJ 87540.

# HYPERBOLIC GEOMETRY: THE FIRST 150 YEARS

*Reprinted from*

Bulletin of the American Mathematical Society
Volume 6 (1982), pp. 9–24

*and*

Proceedings of Symposia in Pure Mathematics 39
Poincaré Symposium, Part I
American Mathematical Society 1983, pp. 25–40

This survey presents some highlights of the history of Hyperbolic Geometry, from 1829 through 1981. (Some further references are given below. Some of the most recent contributions to the study of three dimensional hyperbolic volume are incorporated into the computer program '`snappea`', written by J. Weeks at the Geometry Center, University of Minnesota, Minneapolis.)

**Further References:**

C. Adams, M. Hildebrand and J. Weeks, *Hyperbolic invariants of knots and links,* Trans. Amer. Math. Soc. **326** (1991) 1–56.

A. B. Goncharov, *The classical polylogarithms, algebraic K-theory and* $\zeta_F(n)$, pp. 113–133 of "The Gelfand Mathematical Seminars, 1990–1992," (Corwin, Gelfand, Lepowsky, eds.), Birkhäuser 1993 (see §12).

U. Haagerup and H. Munkholm, *Simplices of maximal volume in hyperbolic n-space,* Acta Math. **147** (1981) 1–11.

B. Iversen, "Hyperbolic Geometry", Lond. Math. Soc. Student Texts **25**, 1992.

R. Meyerhoff, *Geometric invariants for 3-manifolds,* Math. Intelligencer **14** (1992) 37–53.

J. Milnor, *On polylogarithms, Hurwitz zeta functions, and the Kubert identities,* Enseign. Math. **29** (1983) 281–322 (see Appendix 3).

R. Riley, *Seven excellent knots,* pp. 81–151 of "Low Dimensional Topology (Bangor 1979)", Lond. Math. Soc. Lect. Notes **48**, Cambr. U. Press 1982.

G. Sforza, *Formula fondamentale per calcolo dei volumi poliedrici non euclidei,* Periodico di Matematica, Ser. 3, **6** (1909) 139–157.

D. E. Smith, "A Source Book in Mathematics", McGraw-Hill 1929 (contains excerpts from Saccheri, Lobachevsky, and Bolyai).

W. Thurston, *Hyperbolic geometry and 3-manifolds,* pp. 9–25 of "Low Dimensional Topology (Bangor 1979)", Lond. Math. Soc. Lect. Notes **48**, Cambr. U. Press 1982.

———, *Three dimensional manifolds, Kleinian groups and hyperbolic geometry,* Bull. A. M. S. **6** (1982) 357–381; or Proc. Symp. Pure Math. **39** (1983) 87–111.

———, "Three-Dimensional Geometry and Topology", to appear.

# HYPERBOLIC GEOMETRY: THE FIRST 150 YEARS

BY JOHN MILNOR

This will be a description of a few highlights in the early history of non-euclidean geometry, and a few miscellaneous recent developments. An Appendix describes some explicit formulas concerning volume in hyperbolic 3-space.

The mathematical literature on non-euclidean geometry begins in 1829 with publications by N. Lobachevsky in an obscure Russian journal. The infant subject grew very rapidly. Lobachevsky was a fanatically hard worker, who progressed quickly from student to professor to rector at his university of Kazan, on the Volga.

Already in 1829, Lobachevsky showed that there is a natural unit of distance in non-euclidean geometry, which can be characterized as follows. In the right triangle of Figure 1 with fixed edge $a$, as the opposite vertex $A$ moves infinitely far away, the angle $\theta$ will increase to a limit $\theta_0$ which is assumed to be strictly less than $\pi/2$. He showed that

$$a = -\log \tan(\theta_0/2)$$

if the unit of distance is suitably chosen. In particular,

$$a \approx (\pi/2) - \theta_0$$

if $a$ is very small. (In the interpretation introduced by Beltrami forty years later, this unit of distance is chosen so that curvature $\equiv -1$.)

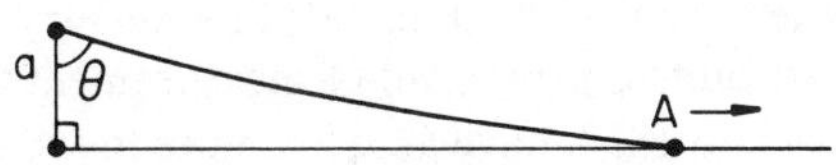


FIGURE 1. A right triangle in hyperbolic space

By early 1830, Lobachevsky was testing his "imaginary geometry" as a possible model for the real world. If the universe is non-euclidean in Lobachevsky's sense, then he showed that our solar system must be extremely small, in terms of this natural unit of distance. More precisely, taking the vertex $A$ in Figure 1 to be the star Sirius and taking the edge $a$ to be a suitably chosen radius of the Earth's orbit, he used the (unfortunately incorrect) estimate

$$\pi - 2\theta \cong 1.24 \text{ seconds of arc} \cong 6 \times 10^{-6} \text{ radians}$$

Presented to the Symposium on the Mathematical Heritage of Henri Poincaré, April 7–10, 1980; received by the editors March 10, 1981 and, in revised form, April 20, 1981.

1980 *Mathematics Subject Classification.* Primary 01A55, 01A60, 51M10; Secondary 57R15, 20H10.

for the parallax of Sirius to conclude that the diameter $2a$ of Earth's orbit is less than $6 \times 10^{-6}$. (The correct parallax of 0.37″ for Sirius would have given a sharper estimate. In fact, the first reliable measurements of parallax were made by Bessel several years later, in 1838.)

By late 1830, he was working out the volumes of pyramids, and other polyhedra in 3-dimensional non-euclidean space. Such computations are not easy, and can be quite interesting. (Compare the Appendix.)

A few years later, in 1832, J. Bolyai, a flamboyant officer in the Austro-Hungarian army, published an independent account of non-euclidean geometry. However, perhaps because he was discouraged by Gauss, he did not pursue the subject as far as Lobachevsky.

Although Lobachevsky and Bolyai were the first with the courage to publish, it should be noted that Gauss himself had been privately working on similar ideas for many years. In a letter to Taurinus in 1824 he wrote (roughly translated):[1]

"The assumption that the sum of the three angles [of a triangle] is smaller than 180° leads to a geometry which is quite different from our (euclidean) geometry, but which is in itself completely consistent. I have satisfactorily constructed this geometry for myself so that I can solve every problem, except for the determination of one constant, which cannot be ascertained a priori. The larger one chooses this constant, the closer one approximates euclidean geometry. . . . If non-euclidean geometry were the true geometry, and if this constant were comparable to distances which we can measure on earth or in the heavens, then it could be determined a posteriori. Hence I have sometimes in jest expressed the wish that euclidean geometry is not true. For then we would have an absolute a priori unit of measurement."

His words to Bessel, in 1830, have an even more modern ring:

" . . . we must admit with humility that, while number is purely a product of our mind, space has a reality outside of our mind, so that we cannot completely prescribe its laws a priori."

For the first forty years or so of its history, the field of non-euclidean geometry existed in a kind of limbo, divorced from the rest of mathematics, and without any firm foundation. However, Gauss' theory of curved surfaces [**1827**], and Riemann's theory of higher dimensional curved manifolds [**1868**] provided a way of integrating non-euclidean geometry into more respectable branches of mathematics. In fact Riemann briefly described the theory of manifolds of constant curvature $\alpha$, citing the metric

$$ds = \sqrt{dx_1^2 + \cdots + dx_n^2} \Big/ \left(1 + \frac{\alpha}{4}\left(x_1^2 + \cdots + x_n^2\right)\right)$$

as an example.

The turning point came in 1868, with the publication of two papers by E. Beltrami. In the first, Beltrami showed that two-dimensional non-euclidean geometry is nothing more nor less than the study of suitable surfaces of

[1]Gauss [**1900**, pp. 187, 201]. For other early workers in non-euclidean geometry, see for example Coxeter [**1942**].

constant negative curvature. He introduced the term *pseudosphere* of *radius R* for a surface of curvature $-1/R^2$. (In practice, he used the term "pseudosphere" only for complete, simply connected surfaces.) In this first paper, he was baffled by the 3-dimensional case. However, after encountering Riemann's inaugural address, which had been delivered in 1854 but published only after his death, in 1868, Beltrami published a second paper, on $n$-dimensional pseudospherical geometry. He started with what I will call the *hemisphere model*. Points of the $n$-dimensional non-euclidean geometry are identified with interior points of the hemisphere

$$y = \sqrt{a^2 - x_1^2 - \cdots - x_n^2}, \qquad y \geqslant 0,$$

in $(n + 1)$-dimensional euclidean space, provided with the Riemannian metric

$$ds = R\sqrt{dx_1^2 + \cdots + dx_n^2 + dy^2}\,/y.$$

He noted that this model is simply connected, and that the "limit space", consisting of boundary points with $y = 0$, is infinitely far from the interior in this metric. We will refer to such boundary points as *points at infinity*.

Projecting orthogonally down to the disk

$$x_1^2 + \cdots + x_n^2 \leqslant a^2,$$

he showed that each pseudospherical geodesic maps precisely to a euclidean straight line segment. Thus he obtained the *projective disk model*, later popularized by Klein.

On the other hand, projecting the hemisphere stereographically onto a disk, he obtained the *conformal disk model*, with the metric

$$ds = \sqrt{d\xi_1^2 + \cdots + d\xi_n^2}\,/\left(1 - (\xi_1^2 + \cdots + \xi_n^2)/4R^2\right)$$

which had been noted already by Riemann.

Finally, performing an inversion in a boundary point of this disk, he obtained the *half-space model*, with coordinates $x_1, \ldots, x_{n-1}$ and $y \geqslant 0$, and with metric $R(dx_1^2 + \cdots + dx_{n-1}^2 + dy^2)^{1/2}/y$. He pointed out that this half-space metric had been used already by Liouville in the 2-dimensional case.

Klein **[1871]** reinterpreted Beltrami's projective disk model in terms of projective geometry. Following Cayley **[1859]**, he took as his starting point the expression

$$\frac{1}{2}\log\frac{|q - a|\,|b - p|}{|p - a|\,|b - q|}$$

for the non-euclidean distance between two points $p$, $q$, as illustrated in Figure 2. (The factor $1/2$ is inserted so that curvature will be $-1$.) Here $|q - a|$ denotes the euclidean distance from $a$ to $q$. In this paper he introduced the term *hyperbolic geometry* for the non-euclidean geometry of Lobachevsky and Bolyai.

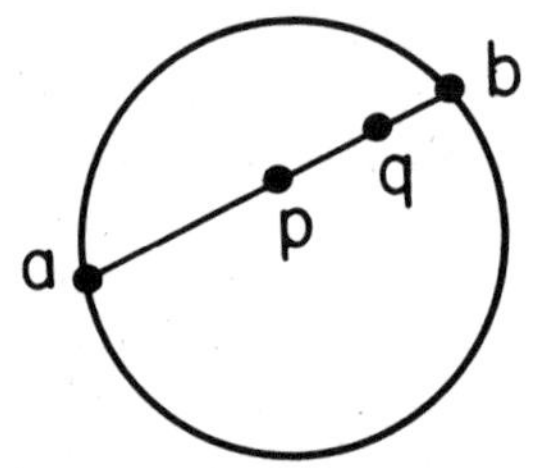


FIGURE 2. The Cayley formula for noneuclidean distance, in the projective disk model

Poincaré [**1882**] re-introduced the Liouville-Beltrami upper half-plane model; and used it to identify the group of orientation preserving isometries of the hyperbolic plane with the group, now usually called $PSL_2\mathbf{R}$, consisting of all fractional linear transformations

$$z \mapsto (az + b)/(cz + d)$$

with real coefficients and with determinant $+1$. As a foundation for the study of automorphic functions, he emphasized the importance of finding discrete groups of hyperbolic isometries. He noted (§XIII) that examples of such discrete groups had been described already by Schwarz [**1873**], and others.

In [**1883**], Poincaré showed that the analogous group of orientation preserving isometries of 3-dimensional hyperbolic space can be identified with the group $PSL_2\mathbf{C}$ consisting of all fractional linear transformations of the plane of points at infinity, with complex coefficients. Again he studied the problem of finding discrete subgroups.

Picard [**1884**] described one particularly interesting example. The ring $\mathbf{Z}[i] \subset \mathbf{C}$ of Gaussian integers gives rise to a discrete group $PSL_2\mathbf{Z}[i] \subset PSL_2\mathbf{C}$, consisting of fractional linear transformations with Gaussian integer coefficients. A fundamental domain for this group, acting on hyperbolic 3-space, is noncompact, but has finite volume.

Bianchi [**1891**] described the analogous construction using the integers of an arbitrary imaginary quadratic number field. The precise volume of a corresponding fundamental domain was computed much later by Humbert [**1919**], in infinitely many cases. (See the Appendix.)

Inspired by examples due to Clifford, Klein [**1890**] posed the following problem:

"... to classify all connectivities which can possibly arise among closed manifolds of constant curvature."

It must be noted in this context that the concept of a (global) manifold was never defined in any satisfactory way during the nineteenth century. For smooth manifolds in euclidean space, the definition given by Betti [**1871**] works only if the normal bundle is trivial. Poincaré [**1895**] used a similar definition and also suggested extending a coordinate patch by analytic continuation. Hadamard [**1898**] clarified the subject by giving a lucid description of surfaces in 3-space in terms of overlapping coordinate patches. Weyl

[**1913**] paved the way for our modern concept of smooth manifold by defining an abstract surface in terms of overlapping coordinate patches. In particular, he gave a clear discussion of the universal covering surface. The generalization to $n$ dimensions was now straightforward. (For a particularly nice presentation, see Cartan [**1928**].) During the same span of time, the study of combinatorial manifolds progressed from the rather vague definitions of Dyck [**1890**] and Poincaré to the precise definition of Brouwer [**1912**]. (See also Tietze [**1908**], Veblen and Alexander [**1913**].) The related concept of topological space is due particularly to Hausdorff [**1914**].

Let us return to the study of manifolds of constant curvature. Killing [**1891**] related Klein's classification problem to the study of discrete groups of isometries, as follows. If $E$ is either $n$-dimensional hyperbolic space, $n$-dimensional euclidean space, or the $n$-dimensional sphere, and if $\Gamma$ is a discrete group of isometries which acts freely on $E$, then he showed that the quotient $E/\Gamma$ is a manifold of constant curvature. He called such a quotient a *Clifford-Klein space form*.

Much later, Hopf [**1925**] clarified this work by introducing a definition of *completeness*, showing that Killing's Clifford-Klein space forms are precisely the *complete Riemannian manifolds of constant curvature*. The concept of completeness was further developed by Cartan [**1928**] and Hopf-Rinow [**1931**].

By definition, a *hyperbolic manifold* is a Clifford-Klein manifold with curvature equal to $-1$. In the years following Killing's paper, a number of examples of hyperbolic manifolds were described. Compare Borel [**1963**]. The study of such manifolds is currently an extremely active area of research, particularly due to the work of Thurston [**1978, 1981**].

Here let me describe two particularly simple examples. The first is due to Gieseking [**1912**]. Start with a regular hyperbolic 3-simplex $\Delta$ having all four vertices on the sphere of points at infinity. Each of the six dihedral angles of $\Delta$ is then equal to $\pi/3$. (Compare the Appendix.) Now identify two faces of $\Delta$ by means of a rotation of $2\pi/3$ about a common vertex. Similarly, identify the opposite two faces by rotating about a common vertex. Then all six edges will be identified. Since the six dihedral angles add up to $2\pi$, they fit smoothly together so that we obtain a non-singular hyperbolic manifold, even in a neighborhood of the common edge. This Gieseking manifold is nonorientable. It is noncompact, but is complete with finite volume.

The second example is due to Riley [**1975**]. Consider the figure eight knot $K$, as shown in Figure 3. The fundamental group $\Pi$ of the complement $S^3 - K$ is generated by two loops $\alpha$, $\beta$, as illustrated, which are subject to a single defining relation $(\alpha\beta^{-1}\alpha^{-1}\beta)\alpha = \beta(\alpha\beta^{-1}\alpha^{-1}\beta)$. Suppose that we look for a faithful representation

$$\Pi \to PSL_2\mathbf{C},$$

and require that the image must be a discrete subgroup. The generators $\alpha$, $\beta \in \Pi$ will correspond to unknown matrices $A$, $B$, which must satisfy a single matrix equation.

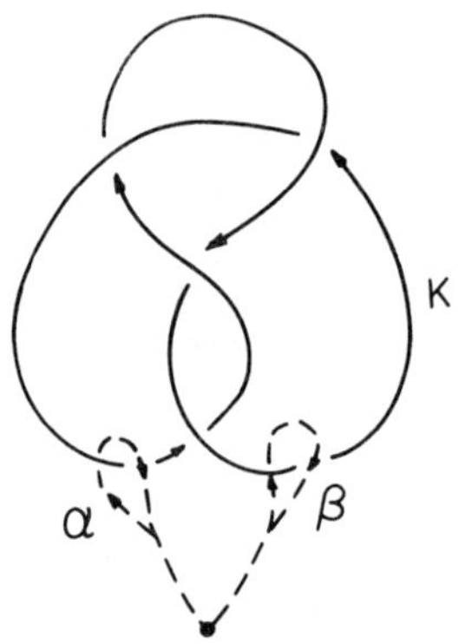


FIGURE 3. The figure eight knot

The possible choices for $A$ and $B$ can be drastically narrowed as follows. Note that every nontrivial element of $PSL_2\mathbf{C}$ is contained in a unique maximal abelian subgroup which is conjugate either to the group consisting of all diagonal matrices $\left[\begin{smallmatrix}\lambda & 0\\ 0 & \lambda^{-1}\end{smallmatrix}\right]$ or to the group of all unipotent matrices of the form $\left[\begin{smallmatrix}1 & \lambda\\ 0 & 1\end{smallmatrix}\right]$, according as the given element does or does not have two linearly independent eigenvectors. Now the generator $\alpha$ of $\Pi$ is contained in a free abelian group $\mathbf{Z} \oplus \mathbf{Z} \subset \Pi$, corresponding to the fundamental group of the boundary of a tubular neighborhood of $K$. It is easy to check that $\mathbf{Z} \oplus \mathbf{Z}$ cannot be embedded as a discrete subgroup of the group of diagonal matrices, which is isomorphic to the multiplicative group $\mathbf{C}^{\cdot}$. Hence $\alpha$, and similarly $\beta$, must map to unipotent elements of $PSL_2\mathbf{C}$, with eigenvalue 1. Using a basis for $\mathbf{C}^2$ consisting of an eigenvector for $\alpha$ and an eigenvector for $\beta$, we may assume that

$$\alpha \mapsto A = \begin{bmatrix} 1 & 1 \\ 0 & 1 \end{bmatrix}, \qquad \beta \mapsto B = \begin{bmatrix} 1 & 0 \\ -\omega & 1 \end{bmatrix}$$

where $\omega$ is some unknown complex constant. Now computation shows that these two matrices satisfy the required relation if and only if $1 + \omega + \omega^2 = 0$, so that $\omega = (-1 \pm \sqrt{-3})/2$. *Thus the requirement that the "peripheral subgroup"* $\mathbf{Z} \oplus \mathbf{Z} \subset \Pi$ *maps isomorphically to a discrete subgroup of* $PSL_2\mathbf{C}$ *leads to a homomorphism*

$$h\colon \Pi \to PSL_2\mathbf{C}$$

*which is uniquely determined, up to inner automorphism and complex conjugation.*

Since the image $h(\Pi)$ is contained in the Bianchi group $PSL_2\mathbf{Z}[\omega]$, it is certainly discrete in $PSL_2\mathbf{C}$. In fact Riley proves that $h$ maps $\Pi$ isomorphically onto a subgroup $h(\Pi) \subset PSL_2\mathbf{Z}[\omega]$ of index twelve. Using the Haken-Waldhausen theory of 3-manifolds, he then proves that *the complement* $S^3 - K$ *is actually homeomorphic to the hyperbolic manifold* $H^3/h(\Pi)$.

In an interesting complement to these two examples, Thurston **[1978]** shows by direct geometric construction that $S^3 - K$ is homeomorphic to the 2-fold orientable covering of the Gieseking manifold. *In particular, the hyperbolic* *manifold* $S^3 - K$ *has a kind of triangulation into two regular ideal hyperbolic 3-simplexes, with their boundaries pasted together by suitable isometries.*

The uniqueness of this Riley representation $\Pi \to PSL_2\mathbf{C}$ illustrates the following fundamental result.

RIGIDITY THEOREM. *If two hyperbolic manifolds of finite volume, with dimension* $n \geqslant 3$, *have isomorphic fundamental groups, then they must necessarily be isometric to each other.*

In the compact case, this was proved by Mostow [**1971, 1973**]. (See also Margulis [**1970**].) For the noncompact case, it was proved by Prasad [**1973**], making essential use of Mostow's work.

It follows that geometric invariants such as volume, the lengths of closed geodesics, and the eigenvalues of the Laplacian operator, are also topological invariants. For an example to show that these particular geometric invariants are not sufficient to determine the topology of a hyperbolic 3-manifold, see Vignéras [**1980**].

In contrast to this Mostow-Prasad rigidity theorem, Thurston has proved a fundamental result which can be stated in a preliminary form as follows.

NONRIGIDITY THEOREM. *Suppose that* $M = H^3/\Gamma$ *is an orientable hyperbolic 3-manifold which is noncompact, but has finite volume. Then there exists an infinite sequence of hyperbolic manifolds*

$$M_j = H^3/h_j(\Gamma)$$

*which have strictly smaller volume, and which approximate the original manifold* $M$ *as* $j \to \infty$ *in the sense that the homomorphisms*

$$h_j\colon \Gamma \to PSL_2\mathbf{C}$$

*tend to the inclusion* $\Gamma \subset PSL_2\mathbf{C}$ *in the topology of pointwise convergence. Furthermore, the volumes of the* $M_j$ *tend to the volume of* $M$.

Of course, according to the Rigidity Theorem, these approximating manifolds $M_j$ cannot be homeomorphic to $M$, or to each other. In particular, the representations $h_j$ cannot be faithful. Their behavior can be described more precisely as follows.

To fix our ideas, suppose that the noncompact manifold $M$ has just one end. Then a neighborhood of infinity in $M$ is smoothly covered by a unique field of half-infinite geodesics, which converge exponentially towards each other as we go out to infinity. Any orthogonal trajectory to this field of geodesics is a flat torus. The fundamental group of such a torus-near-infinity is called a *peripheral subgroup* $\mathbf{Z} \oplus \mathbf{Z}$, embedded in the fundamental group $\Gamma$ of $M$.

As in the discussion of Riley's example, the original representation $\Gamma \subset PSL_2\mathbf{C}$ must map this peripheral subgroup $\mathbf{Z} \oplus \mathbf{Z}$ to a unipotent subgroup, having just one eigenvector with a double eigenvalue of 1. If we perturb this representation slightly within the space of all homomorphisms $\Gamma \to PSL_2\mathbf{C}$, then the single eigenvector will split into two nearby but linearly independent eigenvectors. Hence the subgroup $\mathbf{Z} \oplus \mathbf{Z}$ will map into some subgroup of

$PSL_2\mathbf{C}$ which is isomorphic to the multiplicative group $\mathbf{C}^{\cdot}$. Such a representation of $\mathbf{Z} \oplus \mathbf{Z}$ cannot be faithful, with discretely embedded image. For most choices of perturbation, the image of $\mathbf{Z} \oplus \mathbf{Z}$ will simply be a nondiscrete subgroup of $\mathbf{C}^{\cdot}$. But for certain carefully chosen perturbations $h = h_{pq}$, the images of the two generators $x, y$ of $\mathbf{Z} \oplus \mathbf{Z}$ will be subject to a new relation of the form $h(x^p y^q) = I$; and these images will generate a discretely embedded free cyclic subgroup of $\mathbf{C}^{\cdot}$. In fact Thurston shows that such a perturbation exists whenever $p$, $q$ are relatively prime integers with $|p| + |q|$ sufficiently large. Furthermore, the image $h_{pq}(\Gamma)$ is always a discrete subgroup of $PSL_2\mathbf{C}$; and $h_{pq}$ tends to the inclusion homomorphism as $|p| + |q| \to \infty$.

Topologically, the resulting quotient manifold $M_{pq} = H^3/h_{pq}(\Gamma)$ can be obtained from $M$ by cutting off a neighborhood of infinity bounded by a flat torus $S^1 \times S^1$ and pasting in its place a solid torus $S^1 \times D^2$ which is attached in such a way as to introduce the required relation $x^p y^q = 1$ into the fundamental group. Thurston calls this operation a *Dehn surgery*.

Thus, if $M$ has only one end, the approximating manifolds $M_{pq}$ will always be compact. If $M$ has several ends, then one or more of them can be capped off by analogous Dehn surgeries. In particular, $M$ can always be approximated arbitrarily closely (in a suitable sense) by *compact* hyperbolic manifolds.

Here is a concluding example to test the reader's powers of visualization. Wielenberg [**1978**] has shown that the complement of the *Whitehead link* $W$ of Figure 4 can be given the structure of a hyperbolic manifold $H^3/\Gamma$. In fact $\Gamma$ can be taken as a subgroup of index twelve in the Picard group $PSL_2\mathbf{Z}[i]$. As noted by Whitehead [**1937**], the complement $S^3 - W$ is homeomorphic to the complement of the twisted version on the right. (Proof. Cut along a disk spanning $C$, rotate through any integral multiple of $2\pi$, and then paste back together.) Now by filling in a solid torus neighborhood of $C$ we obtain the manifold $S^3 - K$ from the hyperbolic manifold $S^3 - (K \cup C) \cong S^3 - W$. According to Thurston's theorem, this Dehn surgery must yield a new hyperbolic manifold, in all but a finite number of cases.

The knots $K$ obtained in this way are called *Whitehead doubled knots*. Thus Thurston shows that *the complement of a Whitehead double has a hyperbolic structure whenever the number of twists is sufficiently large*. Furthermore, the volumes of these manifolds tend from below to a finite limit. Note that the figure eight knot occurs as a special case.

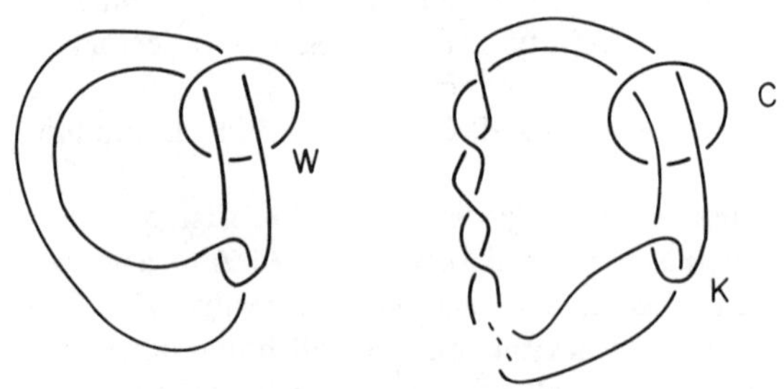


FIGURE 4. The Whitehead Link and a twisted version of it

**Appendix. Volume in hyperbolic 3-space.** Lobachevsky's formulas for the volumes of certain polyhedra in hyperbolic space can best be expressed in terms of the function

$$\text{л}(\theta) = -\int_0^\theta \log|2\sin u|\, du.$$

The Russian л is in honor of Lobachevsky, although he himself never used precisely this expression. Compare Clausen [**1832**], Coxeter [**1935**]. A graph is sketched in Figure 5. The basic properties can be listed as follows.

LEMMA 1. *This function* л$(\theta)$ *is odd, periodic of period* $\pi$, *and satisfies the identity*

$$\text{л}(n\theta) = n \sum_{k \bmod n} \text{л}(\theta + k\pi/n)$$

*for every integer n.*

Evidently, л$(\theta)$ is a continuous function, smooth except at the zeros of $\sin\theta$, taking its maximum at $\theta = \pi/6$.

PROOF OF LEMMA 1. Start with the trigonometric identity

$$2\sin n\theta = \prod_{k=0}^{n-1} 2\sin(\theta + k\pi/n)$$

which can be verified, for small positive values of $\theta$, by substituting $z = e^{-2i\theta}$ in the equation

$$|z^n - 1| = \prod_{k \bmod n} |z - e^{2\pi i k/n}|.$$

The general case follows by analytic continuation.

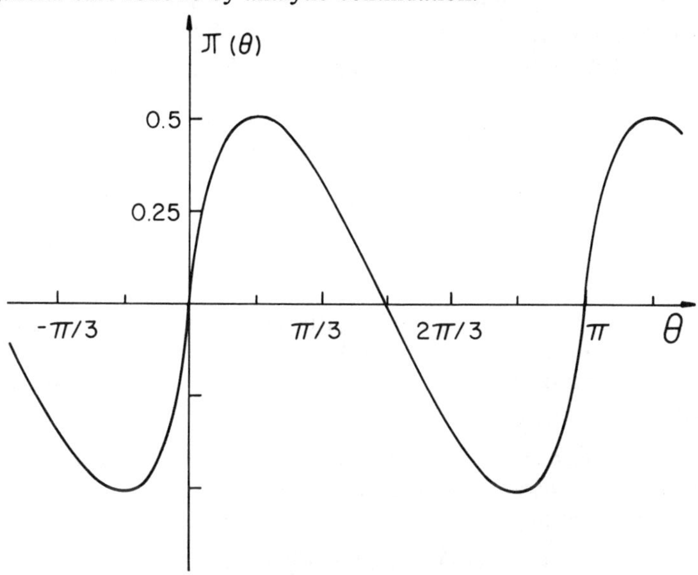


FIGURE 5. Graph of л$(\theta)$

Integrating the negative log absolute value of both sides we obtain

(1) $$Л(n\theta)/n = \sum_{k=0}^{n-1} Л(\theta + \pi k/n) + \text{constant}.$$

For example, if $n = 2$, then

$$Л(2\theta)/2 = Л(\theta) + Л(\theta + \pi/2) + \text{constant}.$$

Substituting $\theta = \pi/2$ or $\theta = 0$ in this last equation and subtracting, we see that $Л(\pi) - Л(0) = 0$. Since the derivative $Л'(\theta) = -\log|2 \sin \theta|$ is periodic of period $\pi$, this proves that $Л$ itself is also periodic of period $\pi$.

Finally, to prove that the constant in equation (1) is zero, we can simply average over all values of $\theta$ mod $\pi$, noting that $Л$ has average zero since it is an odd function. □

This function $Л(\theta)$ is closely related to the *dilogarithm function*

$$L_2(z) = \sum z^n/n^2$$

which has been studied by Euler, Abel, and many others. The precise relationship is described by the identity

$$L_2(e^{2i\theta}) = \pi^2/6 - \theta(\pi - \theta) + 2iЛ(\theta)$$

for $0 \leqslant \theta \leqslant \pi$. (Compare Lewin [**1958**].) This identity can be verified by differentiating both sides for $0 < \theta < \pi$, using the limiting value as $\theta \to 0$ to compute the constant $\pi^2/6$.

Taking the imaginary part of both sides, we obtain the uniformly convergent Fourier series expansion

(2) $$\sum (\sin 2n\theta)/n^2 = 2Л(\theta),$$

which is valid for all values of $\theta$.

For actual computation, it is much better to work with the series

$$Л(\theta) = \theta\left(1 - \log|2\theta| + \sum \frac{B_n(2\theta)^{2n}}{2n(2n+1)!}\right),$$

which converges for $|\theta| \leqslant \pi$ and hence converges quite rapidly for small $\theta$. Here

$$B_1 = \tfrac{1}{6}, \qquad B_2 = \tfrac{1}{30}, \ldots$$

are Bernoulli numbers. This equation can be proved by twice integrating the usual Laurent series expansion for $-\cot \theta$.

Here is an illustration of the utility of this function in computing hyperbolic volumes.

LEMMA 2. *Consider an ideal hyperbolic 3-simplex, that is a simplex $\Delta$ with all four vertices on the sphere of points at infinity. If $\alpha$, $\beta$, $\gamma$ are the dihedral angles along three edges meeting at a common vertex, then $\alpha + \beta + \gamma = \pi$, and*

$$\text{volume}(\Delta) = Л(\alpha) + Л(\beta) + Л(\gamma).$$

It does not matter which particular vertex we choose, since it follows easily that *opposite dihedral angles are equal* so that we obtain the same three dihedral angles $\alpha$, $\beta$, $\gamma$ incident to any vertex.

This computation would follow easily from formulas of Lobachevsky [**1836**], although as far as I know he never worked out this particular example. (Compare Coxeter [**1935**].)

PROOF. We will use the Beltrami upper half-space model, with metric $ds^2 = (dx^2 + dy^2 + dz^2)/z^2$ and associated volume element $dx\,dy\,dz/z^3$, where $z > 0$. Let us position the ideal simplex $\Delta$ so that one of its faces lies in the hemisphere $z = (1 - x^2 - y^2)^{1/2}$, and its opposite vertex lies at the infinite point. Projecting $\Delta$ orthogonally to the unit disk in the $x, y$-plane, we obtain a triangle inscribed in this disk with angles $\alpha$, $\beta$, $\gamma$. These are the angles of a euclidean triangle; so evidently $\alpha + \beta + \gamma = \pi$.

It follows easily from this picture that these angles determine $\Delta$ up to congruence (compare Andreev [**1970**]), and are subject to no other restrictions.

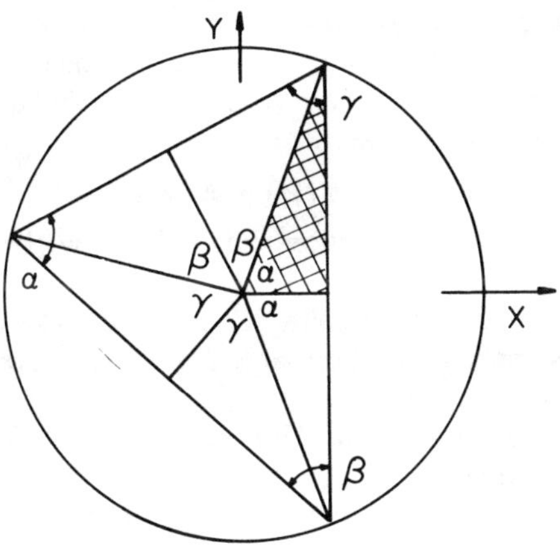


FIGURE 6. A triangle inscribed in the unit disk

If all three angles are acute, then barycentrically subdividing from the origin, we obtain six right triangles as illustrated in Figure 6. (The case of an obtuse angle can be handled by a similar argument, or by analytic continuation.) Take the region in the upper half space lying over just one of these triangles, namely the shaded one. We must integrate the Beltrami volume element $dx\,dy\,dz/z^3$ over the region defined by inequalities

$$z \geqslant \sqrt{1 - x^2 - y^2}, \qquad 0 \leqslant y \leqslant x \tan \alpha, \qquad 0 \leqslant x \leqslant \cos \alpha.$$

Integrating with respect to $z$, we obtain $\frac{1}{2}\,dx\,dy/(1 - x^2 - y^2)$. Integrating with respect to $y$, we obtain

$$\frac{dx}{4A}\left[\log(A + y) - \log(A - y)\right]_0^{x \tan \alpha} = \frac{dx}{4A}\log\frac{A \cos \alpha + x \sin \alpha}{A \cos \alpha - x \sin \alpha}$$

where $A = (1 - x^2)^{1/2}$. Substituting $A = \sin\theta$, $x = \cos\theta$, $dx = -Ad\theta$, and integrating, we get

$$\text{volume} = \frac{1}{4}\int_{\alpha}^{\pi/2} \log\frac{2\sin(\theta+\alpha)}{2\sin(\theta-\alpha)}\,d\theta$$

$$= \left(-Л\left(\frac{\pi}{2}+\alpha\right) + Л(2\alpha) + Л\left(\frac{\pi}{2}-\alpha\right) - Л(0)\right)/4.$$

Now, if we substitute the identity

$$Л(2\alpha) = 2Л(\alpha) + 2Л(\alpha + \pi/2),$$

this reduces easily to volume $= Л(\alpha)/2$. Adding the volumes of the other five regions, we obtain the required formula. □

COROLLARY. *The maximum possible volume of a hyperbolic 3-simplex is* $3Л(\pi/3)$.

This works out as 1.0149416 . . . .

PROOF. We must maximize the continuous function $Л(\alpha) + Л(\beta) + Л(\gamma)$ subject to the constraints $\alpha, \beta, \gamma \geqslant 0$ and $\alpha + \beta + \gamma = \pi$. Since the degenerate cases, where one dihedral angle is zero, all have zero volume, the maximum must occur at an interior point, where the derivative $Л'(\theta) = -\log(2\sin\theta)$ is defined and satisfies

$$Л'(\alpha) = Л'(\beta) = Л'(\gamma),$$

hence $\sin\alpha = \sin\beta = \sin\gamma$. Inspection shows that $\alpha = \beta = \gamma = \pi/3$ is the only interior solution. □

REMARK. Haagerup and Monkholm [**1979**] have proved the analogous theorem in $n$-dimensional hyperbolic space. *The unique n-simplex of maximal volume is the regular ideal n-simplex.*

Now let us compare Lemma 2 with Humbert's formulas for the volumes of certain fundamental domains. The presentation will be based on Borel [**1981**], which contains much further information about arithmetic computations of volume.

Let $\mathbf{Z}[\alpha]$ be the ring of integers in an imaginary quadratic number field $F = \mathbf{Q}[\alpha]$, with discriminant $-D = (\bar{\alpha} - \alpha)^2 < 0$; and let

$$\Gamma = PSL_2\mathbf{Z}[\alpha] \subset PSL_2\mathbf{C}$$

be the corresponding Bianchi group of isometries of hyperbolic 3-space. Borel shows that a fundamental domain for $\Gamma$ has

$$\text{volume} = D^{3/2}\zeta_F(2)/4\pi^2,$$

where $\zeta_F$ is the Dedekind zeta function. Using the standard identity

$$\zeta_F(s) = \zeta_{\mathbf{Q}}(s)\sum_{n=1}^{\infty}\left(\frac{-D}{n}\right)\frac{1}{n^s}$$

(see for example Hecke [**1923**, §49]), where $\zeta_{\mathbf{Q}}(2) = \pi^2/6$, we obtain the formula

$$\text{(3)} \qquad \text{volume} = \frac{D^{3/2}}{24}\sum\left(\frac{-D}{n}\right)\frac{1}{n^2}$$

of Humbert. Here $(-D/n)$ is a generalized quadratic residue symbol, with values $\pm 1$ or 0. In order to express this infinite series in terms of the function $Л(\theta)$, note first that the real-valued function $n \mapsto (-D/n)$ on $\mathbf{Z}/D$ is equal to its own Fourier transform, up to a factor of $\sqrt{-D} = i\sqrt{D}$. More precisely

$$\sum_{k \bmod D} \left(\frac{-D}{k}\right) e^{2\pi i k n/D} = \left(\frac{-D}{n}\right)\sqrt{-D}$$

for every positive integer $n$. (See Hecke §§52, 58.) Substituting this formula in (3), we obtain

$$\text{volume} = \frac{D}{24i} \sum_{k \bmod D} \left(\frac{-D}{k}\right) \sum_{n \geq 1} e^{2\pi i k n/D}/n^2.$$

Evidently the real parts of the exponentials must cancel out. Hence, using the Fourier series expansion (2) we obtain

$$\text{volume} = \frac{D}{12} \sum_{k \bmod D} \left(\frac{-D}{k}\right) Л(k\pi/D).$$

*Thus the volume of Humbert's fundamental domain is a finite linear combination, with rational coefficients, of values of the function* $Л(\theta)$ *at rational multiples of* $\pi$.

This suggests the conjecture that there may be some direct geometrical argument, describing such fundamental domains in terms of cutting and pasting of ideal simplices. (Compare Ash [**1977**].)

As an example, taking $D = 3$ we obtain

$$\text{volume} = \frac{1}{4}\left(Л\left(\frac{\pi}{3}\right) - Л\left(\frac{2\pi}{3}\right)\right) = \frac{1}{2}Л\left(\frac{\pi}{3}\right).$$

According to Riley, the figure eight knot complement must have volume equal to twelve times this number, or $6Л(\pi/3)$. This is precisely twice the volume $3Л(\pi/3)$ of a maximal hyperbolic 3-simplex, thus checking Thurston's observation that the figure eight complement can be triangulated by two copies of a maximal 3-simplex.

Such results suggest the interest of determining number theoretical properties of the real numbers $Л(\theta)$. Here is an explicit guess.

CONJECTURE. *If we consider only angles* $\theta$ *which are rational multiples of* $\pi$, *then every* **Q**-*linear relation*

$$q_1 Л(\theta_1) + \cdots + q_n Л(\theta_n) = 0$$

*is a consequence of the relations*

$$Л(\pi + \theta) = Л(\theta), \qquad Л(-\theta) = -Л(\theta),$$

$$Л(n\theta) = n \sum_{k \bmod n} Л(\theta + k\pi/n)$$

*of Lemma* 1.

For example $л(\pi/3)$ and $л(\pi/6)$ are subject to the $\mathbf{Q}$-linear relation

$$2л(\pi/6) = 3л(\pi/3),$$

as one easily verifies. But the conjecture would imply that $л(\pi/3)$ and $л(\pi/4)$ are rationally independent. In other words, no finite covering manifold of the figure eight complement should have the same volume as some covering of the Whitehead link complement.

A completely equivalent conjecture would be the following. *For any fixed $n > 2$, the $\varphi(n)/2$ real numbers $л(k\pi/n)$, with $0 < k < n/2$ and $k$ relatively prime to $n$, are linearly independent over* $\mathbf{Q}$. Here $\varphi$ is the Euler $\varphi$-function.

The author hopes to discuss this question further in a later paper.[2]

## References


**[1827]** C. F. Gauss, *Disquisitiones generales circa superficies curvas*, Commentationes Gottingensis **6** (Werke **4**, 218–258; German summary, 341–347).

**[1829–30]** N. I. Lobachevsky, *On the foundations of geometry*, Kazan Messenger **25–28**; German transl., 'Zwei geometrische Abhandlungen', Leipzig, 1898.

**[1832]** T. Clausen, *Ueber die Function* sin $\varphi$ + $(1/2^2)$ sin $2\varphi$ + $(1/3^2)$ sin $3\varphi$ + *etc.*, J. Reine Angew. Math. **8**, 298–300.

J. Bolyai, *The absolute science of space, independent of the truth or falsity of Euclid's Axiom* 11 (*which can never be proved a priori*), Maros-Vásárhelyini.

**[1836]** N. I. Lobachevsky, *Imaginary geometry and its application to integration*, Kazan (German transl. Abh. Gesch. Math. **19**, 1904).

**[1837]** ______, *Géometrie imaginaire*, J. Reine Angew. Math. **17**, 295–320.

**[1859]** A. Cayley, *Sixth memoir upon quantics*, Phil. Trans. **149**, 61–91.

**[1868]** B. Riemann, *Ueber die Hypothesen welche der Geometrie zu Grunde liegen*, Abh. K. G. Wiss. Göttingen **13** (from his Inaugural Address of 1854).

E. Beltrami, *Saggio di interpetrazione della geometria non-euclidea*, Gior. Mat. **6**, 248–312 (Also Op. Mat. **1**, 374–405; Ann. École Norm. Sup. **6** (1869), 251–288).

______, *Teoria fondamentale degli spazii di curvatura costante*, Annali di mat. ser. II **2**, 232–255 (Op. Mat. **1**, 406–429; Ann. École Norm. Sup. **6** (1869), 345–375).

**[1871]** F. Klein, *Über die sogenannte Nicht-Euklidische Geometrie*, Math. Ann. **4**, 573–625 (cf. Ges. Math. Abh. **1**, 244–350).

E. Betti, *Sopra gli spazi di un numero qualunque di dimensioni*, Annali di mat. ser. II **4**, 140–158.

**[1873]** H. A. Schwarz, *Ueber diejenigen Fälle in welchen die Gaussische hypergeometrische Reihe eine algebraische Function ihres vierten Elementes darstellt*, J. Reine Angew. Math. **75**, 292–335 (Ges. Math. Abh. **2**, 211–259).

**[1882]** H. Poincaré, *Theorie des groupes fuchsiens*, Acta Math. **1**, 1–62; (Oeuv. **2**, 108–168).

**[1883]** ______, *Mémoire sur les groupes kleinéens*, Acta Math. **3**, 49–92; (Oeuv. **2**, 258–299).

**[1884]** E. Picard, *Sur un groupe de transformations des points de l'espace situés du même côté d'un plan*, Bull. Soc. Math. France **12**, 43–47.

**[1890]** W. Dyck, *Beiträge zur Analysis situs*. II, Math. Ann. **37** 273–316.

F. Klein, *Zur Nicht-Euklidischen Geometrie*, Math. Ann. **37**, 544–572 (Ges. Abh. **1**, 353–383).

**[1891]** L. Bianchi, *Geometrische Darstellung der Gruppen Linearer Substitutionen mit ganzen complexen Coefficienten nebst Anwendungen auf die Zahlentheorie*, Math. Ann. **38**, 313–333.


[2] "On polylogarithms, Hurwitz zeta functions and the Kubert identities" (in preparation).

W. Killing, *Ueber die Clifford-Klein'schen Raumformen*, Math. Ann. **39**, 258–278.

[1895] H. Poincaré, *Analysis situs*, J. École Polyt. **1**, 1–121 (Oeuv. **6**, 193–288).

[1898] J. Hadamard, *Les surfaces à courbures opposées et leur lignes géodésiques*, J. Math. Pures Appl. Ser. 5 **4**, 27–73 (Oeuv. **2**, 729–775).

[1900] C. F. Gauss, Werke **8**, 175–239.

[1908] H. Tietze, *Über die topologischen Invarianten mehrdimensionaler Mannigfaltigkeiten*, Monatshefte für Math. Phys. **19**, 1–118.

[1912] H. Gieseking, *Analytische Untersuchungen ueber topologische Gruppen*, Thesis, Muenster (Compare Magnus (1974), 153).

L. E. J. Brouwer, *Über Abbildung von Mannigfaltigkeiten*, Math. Ann. **71**, 97–115.

[1913] O. Veblen and J. W. Alexander, *Manifolds of n dimensions*, Ann. of Math. (2) **14**, 163–178.

H. Weyl, *Die Idee der riemannschen Fläche*, Teubner, Leipzig.

[1914] F. Hausdorff, *Grundzüge der Mengenlehre*, von Veit, Leipzig.

[1919] G. Humbert, *Sur la mesure des Classes d'Hermite de discriminant donné dans un corps quadratique imaginaire, et sur certains volumes non euclidiens*, Comptes Rendus (Paris) **169**, 448–454.

[1923] E. Hecke, *Vorlesungen über die Theorie der algebraischen Zahlen*, Akad. Verl., Leipzig.

[1925] H. Hopf, *Zum Clifford-Kleinschen Raumproblem*, Math. Ann. **95**, 313–339.

[1928] E. Cartan, *Leçons sur la géométrie des espaces de Riemann*, Gauthier-Villars, Paris.

[1931] H. Hopf and W. Rinow, *Über den Begriff der vollständigen differentialgeometrischen Fläche*, Comment. Math. Helv. **3**, 209–225.

[1935] H. S. M. Coxeter, *The functions of Schläfli and Lobatschefsky*, Quart. J. Math. **6**, 13–29 (pp. 3–20 of twelve geometric essays, S. Ill. Univ. Press, Carbondale, 1968).

[1937] J. H. C. Whitehead, *On doubled knots*, J. London Math. Soc. **12**, 63–71 (Works **2**, 59–67).

[1942] H. S. M. Coxeter, *Non-Euclidean geometry*, Univ. Toronto Press, Toronto.

[1958] L. Lewin, *Dilogarithms and associated functions*, MacDonald, London.

[1963] A. Borel, *Compact Clifford-Klein forms of symmetric spaces*, Topology **2**, 111–112.

[1970] G. A. Margulis, *Isometry of closed manifolds of constant negative curvature with the same fundamental group*, Soviet Math. Dokl. **11**, 722–723.

E. M. Andreev, *On convex polyhedra [of finite volume] in Lobačevskiĭ space*, Mat. Sbornik **81**, 445–478 and **83**, 256–260 (transl. **10**, 413–440 and **12**, 255–259).

[1971] G. D. Mostow, *The rigidity of locally symmetric spaces*, Proc. (1970) Internat. Congr. Math., vol. 2, pp. 187–197, Gauthier-Villars, Paris.

[1973] ______, *Strong rigidity of locally symmetric spaces*, Ann. Math. Studies, No. 78, Princeton Univ. Press, Princeton, N. J.

G. Prasad, *Strong rigidity of* **Q**-*rank* 1 *lattices*, Invent. Math. **21**, 255–286.

[1974] W. Magnus, *Noneuclidean tesselations and their groups*, Academic Press, New York.

[1975] R. Riley, *A quadratic parabolic group*, Math. Proc. Cambridge Philos. Soc. **77**, 281–288.

[1977] A. Ash, *Deformation retracts with lowest possible dimension of arithmetic quotients of self-adjoint homogeneous cones*, Math. Ann. **225**, 69–76.

[1978] W. Thurston, *Geometry and* 3-*manifolds*, Lecture Notes, Princeton University.

N. Wielenberg, *The structure of certain subgroups of the Picard group*, Math. Proc. Cambridge Philos. Soc. **84**, 427–436.

[1979] U. Haagerup and H. Munkholm, *Simplices of maximal volume in hyperbolic n-space*, preprint, Odense University.

[1980] H. J. Munkholm, *Simplices of maximal volume in hyperbolic space, Gromov's norm, and Gromov's proof of Mostow's rigidity theory* (following Thurston), pp. 109–124 of Topology Symposium—Siegen 1979, ed. Koschorke and Neumann, Springer Lecture Notes in Math., no. 788.

R. Rieley, *Applications of a computer implementation of Poincaré's theorem on fundamental polyhedra*, preprint, I.A.S.

______, *Seven excellent knots*, preprint, I.A.S.

M.-F. Vignéras, *Variétés riemanniennes isospectrales et non isométriques*, Ann. of Math. (2) **112**, 21–32.

**[1981]** A. Borel, *Commensurability classes and volumes of hyperbolic* 3-*manifolds*, Ann. Scuola Norm. Sup. Pisa Cl. Sci. (4) **8**, 1–33.

M. Gromov, *Hyperbolic manifolds according to Thurston and Jórgensen*, Sémin. Bourbaki **546**, Springer Lecture Notes in Math., no. 842, pp. 40–53.

D. Sullivan, *Travaux de Thurston sur les groupes quasi-fuchsiens et les variétés hyperboliques de dimension* 3 *fibrées sur* $S^1$, Sémin. Bourbaki **554**, Springer Lecture Notes in Math., no. 842, pp. 196–214.

W. Thurston, *Three dimensional manifolds, Kleinian groups and hyperbolic geometry* (to appear).

______, *Hyperbolic structures on* 3-*manifolds*, I: *Deformation of acylindrical manifolds* (to appear).

SCHOOL OF MATHEMATICS, INSTITUTE FOR ADVANCED STUDY, PRINCETON, NEW JERSEY 08540

# ON THE GEOMETRY OF THE KEPLER PROBLEM

*Reprinted from*

The American Mathematical Monthly
Volume 90 (1983), pp. 353–365

It was a pleasure to write this expository paper, which spans two thousand years of mathematical history. We learn in school that Kepler, in his attempt to understand the orbits of the planets, was led back to the classical Greek theory of conic sections. Also, we learn that Kepler's three laws of planetary motion were incorporated into one universal Law of Gravity by Newton. However, for a proper mathematical understanding of these triumphs of seventeenth century science, we need to make use of geometric ideas which were developed much later, in the nineteenth century. In particular, the non-Euclidean geometry of Lobachevsky and Bolyai, and the inversive geometry of Steiner play an important role.

The first section of this paper describes the classical theory of Kepler orbits. Such orbits are conic sections, whose shape can be described by a dimensionless parameter $\varepsilon \geq 0$, and whose size can be specified by a parameter $\Lambda$ which has the dimensions of length. The associated velocity vector moves around a circle of radius $R = \sqrt{k/\Lambda}$. Here $k$ is the basic constant which specifies the acceleration of a particle at unit distance from the origin. The total energy per unit mass of a particle in such a Kepler orbit is $E = (\varepsilon^2 - 1)R^2/2$. Section 2 gives a more geometric description, interpreting Kepler orbits as geodesics in a space of curvature $K = -2E$, and also studying the limiting behavior as the conic section degenerates into a line segment. There is an appendix describing some classical theory of conic sections, and a second appendix describing well known material on inversive geometry, stereographic projection and non-Euclidean geometry.

The original paper has been reset in TEX, and a number of minor errors have been corrected.

# ON THE GEOMETRY OF THE KEPLER PROBLEM

JOHN MILNOR

*Institute for Advanced Study, Princeton NJ 08540*

**Abstract.** It will be convenient to use the term *Kepler orbit* for any curve $\mathbf{x} = \mathbf{x}(t)$ in 3-space which arises as a solution to the Newtonian two body problem. Hamilton showed that the velocity vector $\mathbf{v} = d\mathbf{x}/dt$, associated with any nondegenerate Kepler orbit, moves along a circle. Following Györgyi, Moser, Osipov and Belbruno, this *velocity circle* can be interpreted as follows. If we fix the total energy $E$, then the manifold $M_E$ consisting of all vectors $\mathbf{v}$ with $\mathbf{v} \cdot \mathbf{v} > 2E$ possesses a natural Riemannian metric of constant curvature $-2E$, whose geodesics are precisely the circles associated in this way with Kepler orbits. In other words, $M_E$ can be identified with one of the three classical geometries, that is with spherical, Euclidean or Lobachevsky space, so that each "straight line" in this geometry corresponds to a unique Kepler orbit.

**1. The Velocity Circle.** In Kepler's first attempts to understand the orbits of the planets, he tried the hypothesis that each orbit is a circle which lies in some plane containing the sun, but is not centered at the sun. This is of course wrong. Yet it does describe the correct answer to a slightly transformed problem.

Consider a particle which is attracted to the origin by a force inversely proportional to the square of the distance, so that its position vector $\mathbf{x} = \mathbf{x}(t)$ satisfies the *Newton differential equation*

$$d^2\mathbf{x}/dt^2 = -k\,\mathbf{x}/|\mathbf{x}|^3. \tag{1}$$

Here $k$ is some fixed positive constant, and $|\mathbf{x}|$ denotes the Euclidean length $\sqrt{\mathbf{x} \cdot \mathbf{x}}$ of the vector $\mathbf{x}$. (This equation is related to the Newtonian two body problem as follows. If two spherical objects with position coordinates $\mathbf{x}_1$ and $\mathbf{x}_2$ move about their common center of gravity in accordance with Newton's laws, then the difference vector $\mathbf{x} = \mathbf{x}_1 - \mathbf{x}_2$ will satisfy (1).) Solutions of this equation, whether elliptical or otherwise, will be called *Kepler orbits*.

*Throughout* §1, *we will assume that the position* $\mathbf{x}$ *and velocity* $\mathbf{v} = d\mathbf{x}/dt$ *are linearly independent vectors at time* $t = t_0$.

THEOREM 1 (Hamilton, 1846). *As $t$ varies, the velocity vector $\mathbf{v} = d\mathbf{x}/dt$ moves along a circle $C$, which lies in some plane $P$ containing the origin, but is not in general centered at the origin. Any such circle can occur, and this "velocity circle" $C$, together with its orientation, determines the orbit $\mathbf{x} = \mathbf{x}(t)$ uniquely.*

---

John Milnor received his Ph.D. at Princeton University in 1954, under the direction of Ralph Fox. As a student he was also strongly influenced by Norman Steenrod, and by a year in Zürich under Heinz Hopf. After teaching for some years at Princeton University, interrupted by stays in Oxford, Berkeley, U.C.L.A., and M.I.T., he moved across town to the Institute for Advanced Study. His mathematical work is centered around the topology of manifolds and related areas of algebra.

The proof will show that the corresponding orbit is either an ellipse, hyperbola, or parabola according as the origin lies inside, outside, or exactly on the velocity circle $C$. In the elliptic case, the velocity vector moves around the entire circle, but in the other two cases only that portion of the circle which is convex towards the origin is actually traversed. (Compare Fig. 1).

To begin the proof, recall that the cross product vector $\mathbf{h} = \mathbf{x} \times \mathbf{v}$ is an invariant of the motion. That is, the derivative $d\mathbf{h}/dt$ is identically zero; as one verifies by an easy calculation. Thinking of $\mathbf{x}(t)$ as the orbit of a particle of unit mass, we will call the length $h = |\mathbf{h}|$ the *angular momentum* of this orbit. Note that $h > 0$ by our linear independence hypothesis.

*It will be convenient to introduce a new system of cartesian coordinates* $\mathbf{x} = (x, y, z)$ *so that the vector* $\mathbf{h}$ *points along the positive* $z$*-axis.* The equation $\mathbf{x}(t) \times \mathbf{v}(t) = \mathbf{h} = (0, 0, h)$ then implies that the vectors $\mathbf{x}(t)$ and $\mathbf{v}(t)$ always lie in the $(x, y)$-plane. Setting $\mathbf{x}(t) = (r\cos\theta, r\sin\theta, 0)$, a straightforward computation shows that

$$h = r^2\, d\theta/dt. \tag{2}$$

The constancy of this expression $r^2\, d\theta/dt$ is just *Kepler's Second Law*, which asserts that a line segment from the origin to $\mathbf{x}(t)$ traverses equal areas in equal times.

[It is interesting to note that Kepler's interpretation of this law was very different from our modern interpretation, since he had no concept of inertia. He believed that the sun exerts a sideways force, inversely proportion to distance, which pushes the planets around their orbits.]

Let us write the Newton equation as $d\mathbf{v}/dt = -k(\cos\theta, \sin\theta, 0)/r^2$. Dividing by $d\theta/dt = h/r^2$, and setting $R = k/h$, we obtain

$$d\mathbf{v}/d\theta = -(R\cos\theta, R\sin\theta, 0).$$

Integrating, we obtain

$$\mathbf{v} = -R\int(\cos\theta, \sin\theta, 0)\, d\theta = R(-\sin\theta, \cos\theta, 0) + \mathbf{c},$$

where $\mathbf{c} = (c_1, c_2, 0)$ is a constant of integration. *This proves that the velocity vector* $\mathbf{v}$ *moves along a circle centered at* $\mathbf{c}$*, with radius* $R = k/h$ *inversely proportional to angular momentum, and lying in the plane through the origin which is orthogonal to the vector* $\mathbf{h}$.

It is interesting to note that the difference vector $\mathbf{v} - \mathbf{c} = R(-\sin\theta, \cos\theta, 0)$ is always orthogonal to the position vector $\mathbf{x} = r(\cos\theta, \sin\theta, 0)$. *Hence the angle between the two velocity vectors, as measured around this circle* $C$*, is equal to the angle between corresponding position vectors as seen from the origin.*

The ratio $\varepsilon = |\mathbf{c}|/R$, that is the distance of the center from the origin divided by the radius, is called the *eccentricity* of the circle $C$ with respect to the origin. *It will*

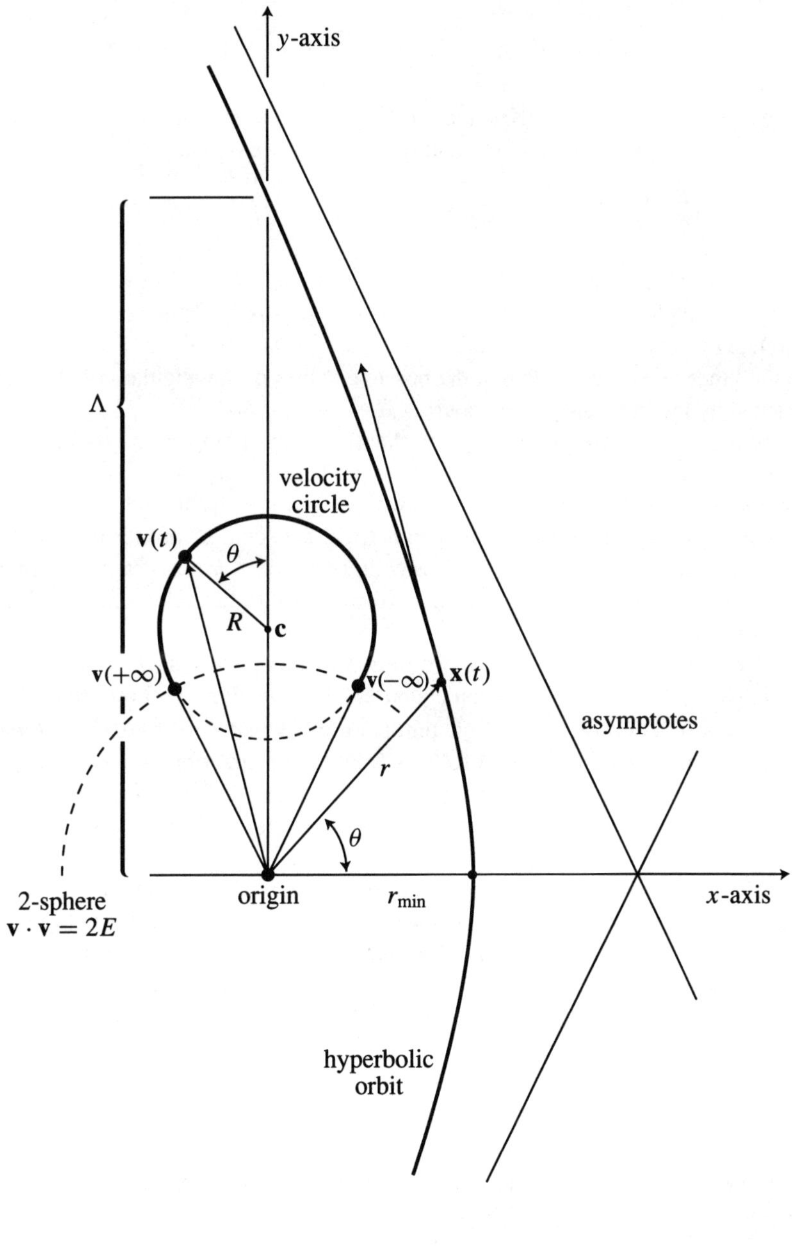
y-axis
Λ
velocity
circle
v(t)
θ
R
c
v(+∞)
v(−∞)
x(t)
asymptotes
r
θ
2-sphere
v · v = 2E
origin
r_min
x-axis
hyperbolic
orbit

Fig. 1.

*be convenient to choose our coordinates $x$, $y$ for the plane so that the center* $\mathbf{c}$ *of the circle $C$ lies on the positive $y$-axis.* With this convention, we can write

$$\mathbf{x} = r(\cos\theta, \sin\theta, 0), \qquad \mathbf{v} = R(-\sin\theta, \varepsilon + \cos\theta, 0).$$

Substituting these expressions in the equation $\mathbf{h} = \mathbf{x} \times \mathbf{v}$ we obtain the formula $h = rR(1 + \varepsilon\cos\theta)$ for angular momentum; and solving for $r$ we obtain

$$r = \Lambda/(1 + \varepsilon\cos\theta), \tag{3}$$

where $\Lambda = h^2/k$. *This is precisely the equation of a conic section of eccentricity $\varepsilon$, with focus at the origin, in polar coordinates.* (See Appendix 1.) This conic section is either a circle, ellipse, parabola, or hyperbola according as $\varepsilon = 0$, $0 < \varepsilon < 1$, $\varepsilon = 1$, or $\varepsilon > 1$. In the latter two cases, note that the possible values of the angular coordinate $\theta$ are constrained by the inequality $1 + \varepsilon\cos\theta > 0$.

Our coordinate system has been chosen so that the point of closest approach, where $r = r_{\min}$, lies on the positive $x$-axis, with $\theta = 0$. This corresponds to the classical convention that the angular coordinate $\theta$, known as the *anomaly*, should be measured from this point of closest approach. The geometrical constant $\Lambda$, proportional to the square of angular momentum, is known classically as the *semi-latus-rectum*. It can be described as the distance from origin to orbit in a direction at right angles to the direction of closest approach.

To complete the proof of Theorem 1, we must show that every such conic section (3) really yields a solution to the Newton equation. But equation (3) shows that $r$ can be expressed as a function of $\theta$, and Equation (2) implies that $t = \int r(\theta)^2\, d\theta/h$ can also be expressed as a function of $\theta$. It follows from the inverse function theorem that $\theta$ and hence $\mathbf{x}$ can be expressed as functions of $t$; and it is not difficult to check that the functions constructed in this way do indeed satisfy Equation (1). Details will be left to the reader. ■

Another classical invariant associated with a solution $\mathbf{x} = \mathbf{x}(t)$ of Equation (1) is the *energy* $E = \mathbf{v}\cdot\mathbf{v}/2 - k/r$ (or more properly the energy per unit mass). A straightforward calculation shows that the derivative $dE/dt$ is zero, so that $E$ is indeed constant along any orbit. Note that the inequality $\mathbf{v}\cdot\mathbf{v} > 2E$ must always be satisfied.

In terms of the velocity circle, with radius $R$ and center $\mathbf{c}$, energy is given by

$$2E = \mathbf{c}\cdot\mathbf{c} - R^2 = -(1 - \varepsilon^2)k^2/h^2, \tag{4}$$

as one can check by evaluating $E$ at any point of the orbit. This computation can be expressed more geometrically by the following statement, which is essentially due to Euclid. (Compare Coxeter [**4**] pp. 8, 81.)

LEMMA 1. *Let $\mathbf{v}_1$ and $\mathbf{v}_2$ be two points of the circle $C$ which lie on a common line through the origin. Then $\mathbf{v}_1\cdot\mathbf{v}_2 = \mathbf{c}\cdot\mathbf{c} - R^2$; hence $\mathbf{v}_1\cdot\mathbf{v}_2$ is equal to $2E$.*

*Proof.* Let $\mathbf{w}$ be the point of the circle which is diametrically opposite to $\mathbf{v}_1$. Setting $\mathbf{v}_1 = \mathbf{c} + \mathbf{e}$ and $\mathbf{w} = \mathbf{c} - \mathbf{e}$, we see that

$$\mathbf{v}_1 \cdot \mathbf{w} = \mathbf{c} \cdot \mathbf{c} - \mathbf{e} \cdot \mathbf{e} = \mathbf{c} \cdot \mathbf{c} - R^2 = 2E.$$

Since $\mathbf{v}_1, \mathbf{v}_2, \mathbf{w}$ is a right triangle, it follows that $\mathbf{v}_1$ is orthogonal to the difference vector $\mathbf{v}_2 - \mathbf{w}$. Hence $\mathbf{v}_1 \cdot \mathbf{v}_2$ is equal to $\mathbf{v}_1 \cdot \mathbf{w} = 2E$, as required. ■

In the case of an elliptical orbit, the precise size and period of the orbit are related to the energy $E$ as follows. First note that the *major axis* $2a$ of the ellipse is given by the computation

$$2a = r_{\min} + r_{\max} = \Lambda/(1+\varepsilon) + \Lambda/(1-\varepsilon) = 2\Lambda/(1-\varepsilon^2).$$

Since $\Lambda = h^2/k$, it follows from (4) that

$$2E = -k/a. \tag{5}$$

*Thus the energy $E$ of an elliptic orbit is negative, and $|E|$ is inversely proportional to the major axis of the ellipse.*

The *minor axis* $2b$ of the ellipse, computed by setting $dy/dt = R(\varepsilon + \cos\theta) = 0$, is given by

$$b = y_{\max} = \Lambda/\sqrt{1-\varepsilon^2}.$$

The *area* of the ellipse is then given by

$$A = \pi ab = \pi\Lambda^2/(1-\varepsilon^2)^{3/2}.$$

Applying Kepler's Second Law (2) in the form $dA/dt = h/2$, where $A(t)$ is the area swept out by the line segment from $\mathbf{0}$ to $\mathbf{x}$, we see that the *period* $T$ of such an elliptical orbit satisfies $A/T = h/2$. Together with (4), (5), and the definition $\Lambda = h^2/k$, this implies that

$$T = 2\pi k/(-2E)^{3/2} = 2\pi(a^3/k)^{1/2}, \tag{6}$$

which is a modern statement of *Kepler's Third Law. Thus the period is proportional to $a^{3/2}$ and is inversely proportional to $|E|^{3/2}$.* For further details of these calculations see, for example, Arnold [**1**] or Synge and Griffith [**21**].

*Remark.* A familiar but very misleading version of this Third Law asserts that the period is proportional to the 3/2 power of the "mean distance" of $\mathbf{x}(t)$ from the origin. For this to be correct, we must interpret the *mean distance* to be simply the average $a = \frac{1}{2}(r_{\min} + r_{\max})$ of minimum and maximum distances. The time average $\oint r\,dt \big/ \oint dt$ is not equal to the major semi-axis $a$ but is rather equal to $(1+\varepsilon^2/2)a$. However it is interesting to note that the time average of $r^{-1}$ is precisely equal to $a^{-1}$. Proofs based on Formula (9) of Appendix 1 are easily supplied.

It is noteworthy that *the period of such a periodic orbit depends only on the energy.* This is true for periodic solutions of Lagrangian or Hamiltonian differential equations under quite general conditions. Compare Herglotz [**9**], Wintner [**22,** §100], and Gordon [**6**].

**2. The Levi-Civita Metric.** Moser [**16**] has given a very pretty picture of the space of all elliptic orbits of fixed energy, or fixed period, in terms of *stereographic projection.* (See also Györgyi [**7**].) To carry out Moser's construction, we must introduce a 3-dimensional sphere, consisting of all unit vectors $\mathbf{u} = (u_1, u_2, u_3, u_4)$ in Euclidean 4-space, and project it stereographically from its "north pole" (0, 0, 0, 1) onto the 3-dimensional Euclidean space consisting of vectors of the form $\mathbf{v} = (v_1, v_2, v_3, 0)$. Thus each unit vector $\mathbf{u}$ maps to a unique vector $\mathbf{v} = (u_1, u_2, u_3, 0)/(1 - u_4)$ which has last coordinate zero, and lies on the same straight line through the north pole. *Then every circle on the 3-sphere projects onto a circle or line in 3-space.* (Compare Appendix 2.)

A brief computation shows that the *antipodal map* $\mathbf{u} \mapsto -\mathbf{u}$ from the 3-sphere to itself corresponds to the *negative inversion map* $\mathbf{v} \mapsto -\mathbf{v}/|\mathbf{v}|^2$ from Euclidean 3-space to itself. In other words, diametrically opposite points of the 3-sphere correspond to points $\mathbf{v}_1$ and $\mathbf{v}_2$ in Euclidean 3-space which lie on a common line through the origin, and satisfy $\mathbf{v}_1 \cdot \mathbf{v}_2 = -1$. Evidently the images of *great circles* on the 3-sphere are just those circles, or lines through the origin, which map into themselves under this negative inversion operation. Using Lemma 1, we see that a circle in 3-space has this invariance property if and only if it is one of our "velocity circles" with energy $E = (\mathbf{c} \cdot \mathbf{c} - R^2)/2$ equal to $-1/2$, or equivalently with period $T$ equal to $2\pi k$. *Thus stereographic projection carries great circles on the 3-sphere precisely onto the velocity circles associated with elliptic orbits of energy $-1/2$.*

There is one apparent defect in this picture. Namely those great circles which pass through the poles correspond to straight lines through the origin in velocity space. These do not seem to fit into our picture. However this apparent defect is actually a virtue in disguise. For it enables us to give a precise description of the limiting behavior of elliptic orbits of fixed period as the angular momentum $h$ tends to zero. If $h \to 0$, keeping $E$ fixed and negative, it is easy to see that the minor axis of the orbit ellipse tends to zero, so that the ellipse flattens out and tends towards a straight line segment of length $-k/E$ with one end at the origin. In order for the orbit to vary continuously as $h \to 0$, we must adopt the following.

REFLECTION CONVENTION. *If an orbit has angular momentum $h = 0$, and if $dr/dt < 0$ for some values of $t$ so that $\mathbf{x}(t)$ falls freely along a straight line towards the origin, then $\mathbf{x}(t)$ is reflected back along this same line when it hits the origin.*

(Compare Devaney [**5**], McGehee [**14**].) In fact it is natural to adopt this same Reflection Convention whether the energy is positive, negative or zero.

By rotating our coordinates, we can put such a singular orbit in the form $t \mapsto$

$(r(t), 0, 0)$, where $r(t) \geq 0$ satisfies the differential equation $(dr/dt)^2/2 = E + k/r$. As an example, if $E = 0$, it follows that $r$ is proportional to $(t - t_0)^{2/3}$. If $E < 0$, then the solution curve $r = r(t)$ is a cycloid, periodic in $t$, and smooth except for a cusp wherever $r = 0$. (See Appendix 1.) Note in particular that the derivative $dr/dt$ tends to minus infinity as $r$ decreases to zero, and then jumps to plus infinity. *If* $\mathbf{x}(t)$ *bounces off the origin, then the velocity vector* $\mathbf{v}(t)$*, moving along a straight line, sweeps through the "point at infinity" in* 3*-space.* This velocity line corresponds, under stereographic projection, to a great circle which sweeps through the north pole of the unit 3-sphere.

*Thus, if we adopt this reflection convention, then there is a precise one-to-one correspondence between orbits of energy* $-1/2$ *and great circles on the unit* 3*-sphere.* The situation for other negative values of $E$ is the same, except for a scale change in 3-space.

According to Osipov **[17]**, **[18]**, and Belbruno **[2]**, **[3]**, there is an analogous description for parabolic or hyperbolic orbits. First consider a parabolic orbit, with energy $E = 0$. Recall that the velocity vector $\mathbf{v}(t)$ associated with such an orbit moves along a circle $C$ which passes through the origin. *Let us transform this problem by applying the inversion operation* $\mathbf{v} \mapsto \mathbf{w} = \mathbf{v}/|\mathbf{v}|^2$. Then a circle $C$ through the origin corresponds to a straight line in the space of vectors $\mathbf{w}$. (See Appendix 2.) *In this way we obtain a precise one-to-one correspondence between orbits of energy zero and straight lines in the Euclidean space consisting of all inverted velocity vectors* $\mathbf{w}$. In this picture, the orbits which bounce off the origin correspond to straight lines through the origin in inverted velocity space.

Finally let us look at the positive energy case, say $E = +1/2$. Since $\mathbf{v} \cdot \mathbf{v} > 2E = 1$, it follows that the inverted velocity vector $\mathbf{w} = \mathbf{v}/|\mathbf{v}|^2$ varies over the unit ball $|\mathbf{w}| < 1$. Using Lemma 1, we see that the corresponding velocity circles $C$ are invariant under inversion, and hence intersect the boundary of the unit ball orthogonally. *Thus each orbit of energy* $+1/2$ *corresponds to a circle arc* $t \mapsto \mathbf{w}(t)$ *which spans the unit ball* $|\mathbf{w}| < 1$*, and intersects its boundary* 2*-sphere orthogonally.* Orbits which bounce off the origin correspond to diameters spanning this unit 3-ball. It is natural to compare this picture with the *conformal unit ball model* for the "hyperbolic" non-Euclidean geometry of Lobachevsky. (See Appendix 2.) In this model, discovered by Beltrami and later by Poincaré, points of hyperbolic space correspond to points in the Euclidean unit ball, and hyperbolic straight lines correspond to circle arcs or diameters which span the unit ball, meeting its boundary orthogonally. *Thus the orbits of energy* $+1/2$ *correspond precisely to "straight lines" in this model for hyperbolic* 3*-space.* There is an analogous picture for any positive value of $E$.

In terms of Riemannian geometry, we can put these three different constructions into one common framework as follows. Levi-Civita pointed out that it is possible to simplify solutions to Equation (1) by introducing a *fictitious time parameter* $s = \int dt/r$ along any orbit, where $r = |\mathbf{x}|$. (Compare Appendix 1.) Using this parameter, we will prove the following.

THEOREM 2 (Osipov and Belbruno). *Fixing some constant energy* $E$*, consider the space* $M_E$ *consisting of all velocity vectors* $\mathbf{v}$ *for which* $\mathbf{v} \cdot \mathbf{v} > 2E$*, together with a* *single improper point* $\mathbf{v} = \infty$. *This space possesses one and only one Riemannian metric* $ds^2$ *so that the arc-length parameter* $\int ds$ *along any velocity circle* $t \mapsto \mathbf{v}(t)$ *is precisely equal to the Levi-Civita parameter* $\int dt/|\mathbf{x}(t)|$. *This metric is smooth and complete, with constant curvature* $-2E$, *and its geodesics are precisely the circles or lines* $t \mapsto \mathbf{v}(t)$ *associated with Kepler orbits.*

In other words, there is a unique way of defining the "length" of a curve $\Gamma$ in this space $M_E$ of all compatible velocity vectors so as to coincide with Levi-Civita's integral $\int_\Gamma dt/r$. If $K = -2E$ is positive, then $M_E$, with this definition of length, is isometric to a 3-sphere of radius $1/\sqrt{K}$ in Euclidean 4-space; and the great circles on this sphere correspond to velocity circles. If $K = 0$, then $M_E$ is isometric to Euclidean space; and if $K < 0$, it is isometric to hyperbolic space, with the unit of distance chosen appropriately. In particular, in all three cases, $M_E$ can be made into a smooth manifold even in a neighborhood of the special point $\mathbf{v} = \infty$. Geodesics which pass through this special point correspond to Kepler orbits which bounce off the origin.

*Proof.* The Newton equation $d\mathbf{v}/dt = -k\mathbf{x}/r^3$ implies that $|d\mathbf{v}/dt| = k/r^2$. Multiplying by $dt/ds = r$ and recalling the definition $E = \frac{1}{2}\mathbf{v}\cdot\mathbf{v} - k/r$, we obtain $|d\mathbf{v}/ds| = k/r = \mathbf{v}\cdot\mathbf{v}/2 - E$. Therefore $|ds| = 2|d\mathbf{v}|/(\mathbf{v}\cdot\mathbf{v} - 2E)$, or in other words

$$ds^2 = \frac{4\,d\mathbf{v}\cdot d\mathbf{v}}{(\mathbf{v}\cdot\mathbf{v} - 2E)^2}. \tag{7}$$

*Thus there is one and only one Riemannian metric on* $M_E$ *which satisfies our condition, and it is given by this formula* (7). To describe what happens in a neighborhood of infinity, we work with the inverted velocity coordinate $\mathbf{w} = \mathbf{v}/|\mathbf{v}|^2$. Note that $2E\mathbf{w}\cdot\mathbf{w} < 1$. Computation shows that $d\mathbf{w}\cdot d\mathbf{w} = d\mathbf{v}\cdot d\mathbf{v}/(\mathbf{v}\cdot\mathbf{v})^2$. (See Formula (10) of Appendix 2.) Hence $ds^2 = 4\,d\mathbf{w}\cdot d\mathbf{w}/(1 - 2E\mathbf{w}\cdot\mathbf{w})^2$, where the denominator is always strictly positive. If we set $K = -2E$, this becomes

$$ds^2 = 4\,d\mathbf{w}\cdot d\mathbf{w}/(1 + K\mathbf{w}\cdot\mathbf{w})^2. \tag{7'}$$

*Except for a scale change, this is just the form given by Riemann, in his inaugural dissertation, for a metric of constant curvature* $K$. In the case $K = 0$, this metric is obviously flat, with straight lines as geodesics. In the positive curvature case, we can compare it with the standard metric on a 3-sphere of radius $1/\sqrt{K}$. It is not difficult to check that these two metrics correspond isometrically, via stereographic projection from the 3-sphere to a 3-plane passing through its center; and that great circles correspond precisely to our (inverted) velocity circles. Further details of the proof, particularly in the case $K < 0$, may be found in Appendix 2. ■

COROLLARY. *The function* $\mathbf{x}(t)$ *and* $\mathbf{v}(t)$ *depend smoothly on the time t and the initial conditions* $\mathbf{x}(0)$ *and* $\mathbf{v}(0)$, *even in the neighborhood of an orbit which bounces off the origin, so long as the vectors* $\mathbf{x}(0)$ *and* $\mathbf{x}(t)$ *themselves are not zero.*

Here is an explicitly worked out example, to illustrate the behavior of these functions. Consider the family of parallel straight lines $s \mapsto \mathbf{w} = (-s, \alpha, \beta)/2$ in inverted velocity space, depending on two real parameters $\alpha$ and $\beta$. A short computation shows that these correspond to the family of parabolic orbits

$$\mathbf{x} = k(\alpha^2 + \beta^2 - s^2, 2\alpha s, 2\beta s)/2.$$

Not only this orbital position $\mathbf{x}$, but also the distance

$$r = |\mathbf{x}| = k(\alpha^2 + \beta^2 + s^2)/2$$

and the time

$$t = \int_0 r\, ds = ks(\alpha^2 + \beta^2 + s^2/3)/2$$

can be expressed as smooth functions of the three parameters $\alpha, \beta, s$. Using the implicit function theorem, we can solve for $s$, and hence $\mathbf{x}$, as functions of $\alpha, \beta, t$. These functions are smooth except precisely at those points where $\partial t/\partial s = |\mathbf{x}|$ is zero.

Similarly, we can look at the six parameter family of orbits depending on the initial position vector $\mathbf{x}(0) \neq \mathbf{0}$ and the initial velocity vector $\mathbf{v}(0)$. Using the fact that a geodesic on a smooth manifold depends smoothly on its initial position and direction, together with the identity

$$r = 2k/(\mathbf{v} \cdot \mathbf{v} - 2E) = 2k\mathbf{w} \cdot \mathbf{w}/(1 - 2E\mathbf{w} \cdot \mathbf{w}),$$

where $2E = \mathbf{v}(0) \cdot \mathbf{v}(0) - 2k/|\mathbf{x}(0)|$, we see that $r$ and $t = \int_0 r\, ds$ can also be expressed as smooth functions of the parameters $\mathbf{x}(0) \neq \mathbf{0}$, $\mathbf{v}(0)$ and $s$. In fact it is not difficult to verify the identity

$$\mathbf{x} = 4k\big(2(\mathbf{w} \cdot \mathbf{w}')\mathbf{w} - (\mathbf{w} \cdot \mathbf{w})\mathbf{w}'\big)/(1 - 2E\mathbf{w} \cdot \mathbf{w}),$$

where $\mathbf{w}' = d\mathbf{w}/ds$, which shows that $\mathbf{x}$ can also be expressed a s a smooth function of these parameters. Again, if we use $t$ in place of $s$ as parameter, then we lose the smoothness of the resulting functions only at those points where $\mathbf{x}$ is precisely equal to the zero vector. ■

## Appendix 1. Conic sections.

This will be a brief review of some classical constructions. An *ellipse*, with foci $\mathbf{f}_1$ and $\mathbf{f}_2$, can be defined as the set of all vectors $\mathbf{x}$ in the Euclidean plane which satisfy the equation

$$|\mathbf{x} - \mathbf{f}_1| + |\mathbf{x} - \mathbf{f}_2| = 2a.$$

Here $2a > |\mathbf{f}_1 - \mathbf{f}_2|$ is a constant equal to the *major axis* of the ellipse. The ratio $\varepsilon = |\mathbf{f}_1 - \mathbf{f}_2|/2a$ of the distance between foci to the major axis is called the *eccentricity* of the ellipse. Note that $0 \leq \varepsilon < 1$, where $\varepsilon = 0$ only for a circle. If we choose cartesian coordinates $x, y$ so that $\mathbf{f}_1$ is the origin and $\mathbf{f}_2$ is the point $(-2\varepsilon a, 0)$, then the distances $r$ and $2a - r$ of $(x, y)$ from $\mathbf{f}_1$ and $\mathbf{f}_2$ are given by the equations

$$x^2 + y^2 = r^2, \quad (x + 2\varepsilon a)^2 + y^2 = (2a - r)^2.$$

(Fig. 2.) Subtracting one equation from the other and dividing by $4a$, we obtain

$$r = \sqrt{x^2 + y^2} = \Lambda - \varepsilon x, \tag{8}$$

where $\Lambda$ denotes the constant $(1 - \varepsilon^2)a$, known as the "semi-latus-rectum."

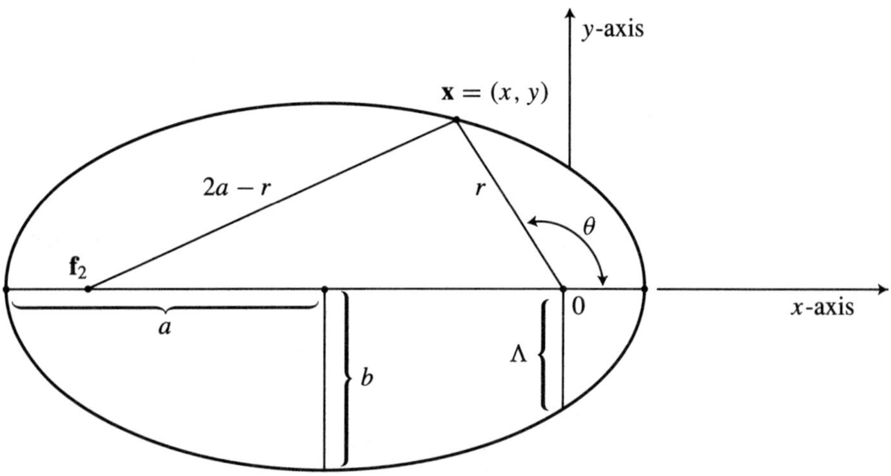


Fig. 2.

More generally, if we fix any constants $\varepsilon \geq 0$ and $\Lambda > 0$, then the locus described by Equation (8) is called a *conic section* with focus at the origin, and with eccentricity $\varepsilon$. Squaring both sides of (8), we obtain

$$(1 - \varepsilon^2)x^2 + y^2 = \Lambda^2 - 2\Lambda\varepsilon x.$$

Evidently this equation describes a *parabola* if $\varepsilon = 1$, or a *hyperbola* if $\varepsilon > 1$. Substituting $x = r\cos\theta$, $y = r\sin\theta$ in (8), we obtain the equation

$$r = \Lambda/(1 + \varepsilon\cos\theta) \tag{3}$$

which defines such a conic section, with focus at the origin, in polar coordinates. In the elliptic case $\varepsilon < 1$, note that the quadratic equation relating $x$ and $y$ can be put in the form $(x + \varepsilon a)^2/a^2 + y^2/b^2 = 1$, where $b = a\sqrt{1 - \varepsilon^2}$. Thus this ellipse is centered at the point $(-\varepsilon a, 0)$, and has principal axes $2a \geq 2b$.

Given the constant $k > 0$, we can introduce the time $t$ into this geometrical picture by means of Kepler's Second Law, $d\theta/dt = h/r^2$, or in other words $t = \int r^2\, d\theta/h$, where $h = \sqrt{k\Lambda}$. Similarly, Levi-Civita's fictitious time can be introduced as the integral $s = \int dt/r = \int r\, d\theta/h$ along an orbit. In practice, it turns out to be easiest to express both time and position as functions of $s$. Substituting (3) in the equations $x = r\cos\theta$, $y = r\sin\theta$ and differentiating, we find that

$$\frac{dx}{ds} = \frac{h}{r}\frac{dx}{d\theta} = -Ry, \qquad \frac{dy}{ds} = \frac{h}{r}\frac{dy}{d\theta} = h\varepsilon + (1-\varepsilon^2)Rx,$$

and therefore $d^2y/ds^2 = 2Ey$, where $R = h/\Lambda$, and $2E = -(1-\varepsilon^2)R^2$. From this, it is not difficult to compute $y, x$, and $t = \int r\, ds = \int(\Lambda - \varepsilon x)\, ds$ as functions of $s$. As an example in the negative energy case $2E = -\omega^2 < 0$, if we normalize so that $s = t = 0$ at the point of closest approach, we find that

$$x = -\varepsilon a + a\cos\omega s, \qquad y = b\sin\omega s,$$

with

$$r = a(1-\varepsilon\cos\omega s), \qquad t = a(s - \varepsilon\omega^{-1}\sin\omega s). \tag{9}$$

(Compare KEPLER.) This form of the equations again makes the elliptical shape of the orbit quite clear. In Kepler's terminology, the angle $\omega s$ is known as the "eccentric anomaly," and the angle $\omega t/a = 2\pi t/T$ is called the "mean anomaly." If the eccentricity $\varepsilon$ tends to 1, keeping $\omega$ and $a = k/\omega^2$ fixed, so that $b$ and $\Lambda$ tend to zero, note that these expressions tend to well behaved limits. The resulting curve in the $t, x$ plane is known as a *cycloid.* In the positive energy case, there are completely analogous formulas involving the hyperbolic sine and cosine functions. In the zero energy case, evidently $y$ is a linear function of $s$, hence $x$ is a quadratic function, and $t$ is a cubic function of $s$.

## Appendix 2. Inversive geometry, stereographic projection, and hyperbolic geometry.

This will be a quick review of well-known material. Further details maybe found for example in Coxeter [4] or in Hilbert and Cohn-Vossen [**10**].

Let $\mathbf{p}$ be a base point in the $n$-dimensional Euclidean space $\boldsymbol{E}$. The operation of *inversion*, in the unit sphere centered at $\mathbf{p}$, is a smooth mapping from $\boldsymbol{E} - \mathbf{p}$ to itself, defined as follows. *Each point* $\mathbf{x} \neq \mathbf{p}$ *of Euclidean space maps to the unique point* $\mathbf{y}$ *which lies on the ray ( = half-line) which starts at* $\mathbf{p}$ *and passes through* $\mathbf{x}$*, such that the Euclidean distance* $|\mathbf{y}-\mathbf{p}|$ *is equal to the reciprocal* $1/|\mathbf{x}-\mathbf{p}|$. Thus each point of the unit sphere $|\mathbf{x}-\mathbf{p}| = 1$ is fixed by this inversion map, but the inside and outside of the sphere are interchanged. More generally, given a constant $\rho > 0$, we can invert in the sphere of radius $\rho$ centered at $\mathbf{p}$. The definition is the same, except that we set $|\mathbf{y}-\mathbf{p}|$ equal to $\rho^2/|\mathbf{x}-\mathbf{p}|$.

It is often convenient to extend this construction by adjoining a formal *point at infinity* to Euclidean space. Then inversion maps $\boldsymbol{E} \cup \infty$ to itself, with the understanding that the base point $\mathbf{p}$ maps to the point $\infty$ and that $\infty$ maps to $\mathbf{p}$.

Here is an example to illustrate the inversion map. *Consider an $(n-1)$-dimensional sphere $S$ in Euclidean space, with two diametrically opposite points* $\mathbf{p}$ *and* $\mathbf{q}$, *and let $P$ be an $(n-1)$-dimensional hyperplane, situated as in Figure 3, so that the half-line from* $\mathbf{p}$ *through* $\mathbf{q}$ *meets $P$ orthogonally at a point* $\mathbf{z} \neq \mathbf{p}$. Then for any $\mathbf{x}$ on $S$ and $\mathbf{y}$ on $P$ which lie on a common line through $\mathbf{p}$, the right triangle $\mathbf{p}, \mathbf{x}, \mathbf{q}$ is similar to the right triangle $\mathbf{p}, \mathbf{z}, \mathbf{y}$. Therefore the ratio $|\mathbf{x}-\mathbf{p}|/|\mathbf{q}-\mathbf{p}|$ is equal to the ratio $|\mathbf{z}-\mathbf{p}|/|\mathbf{y}-\mathbf{p}|$. In other words, the distance from $\mathbf{x}$ to $\mathbf{p}$ is inversely proportional to the distance from $\mathbf{y}$ to $\mathbf{p}$. *This proves that the sphere $S$ maps onto the hyperplane $P \cup \infty$ under inversion from* $\mathbf{p}$, *provided that the constant $\rho$ is chosen correctly; and conversely it proves that $P \cup \infty$ maps onto $S$.*

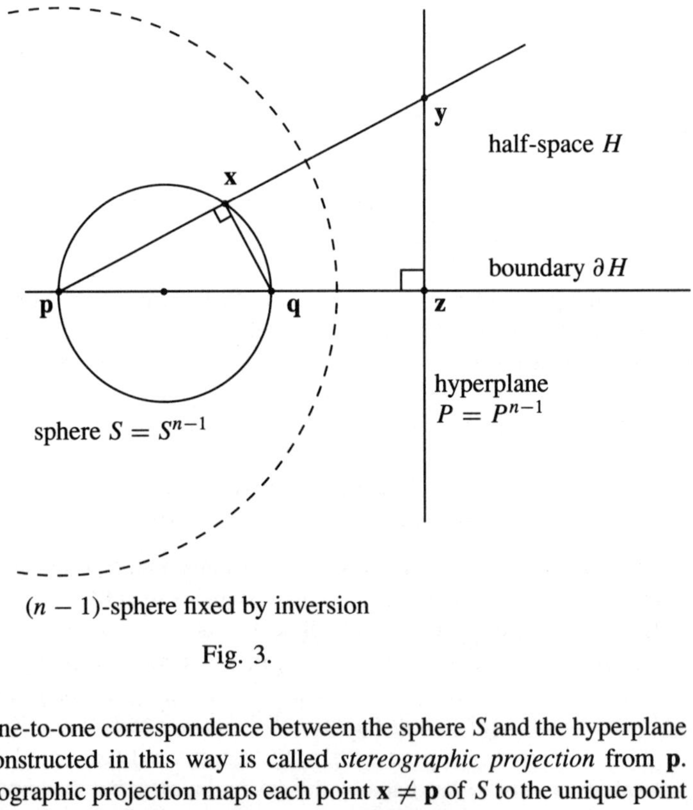


Fig. 3.

*Definition.* The one-to-one correspondence between the sphere $S$ and the hyperplane $P \cup \infty$ which is constructed in this way is called *stereographic projection* from $\mathbf{p}$. Geometrically, stereographic projection maps each point $\mathbf{x} \neq \mathbf{p}$ of $S$ to the unique point $\mathbf{y}$ of $P$ such that $\mathbf{x}$, $\mathbf{y}$, and $\mathbf{p}$ all lie on a common straight line.

If $P^k$ is a plane of dimension $k$ in Euclidean $n$-space, where $k$ can be any number between 1 and $n-1$, then a completely analogous argument shows that the image of

$P^k \cup \infty$ under inversion is either a sphere $S^k$ (if $P^k$ does not contain the base point $\mathbf{p}$), or is equal to $P^k \cup \infty$ itself if $P^k$ does contain $\mathbf{p}$. More generally, we will prove the following.

LEMMA 2 (Steiner, 1824). *The operation of inversion carries every straight line or circle in* $\mathbf{E} \cup \infty$ *to a straight line or circle. Similarly it caries every* $k$*-dimensional sphere or plane to a* $k$*-dimensional sphere or plane. Furthermore it preserves angles. That is, if two lines or circles intersect at angle* $\theta$ *(at a point other than* $\mathbf{p}$*), then their images under inversion also intersect at angle* $\theta$.

To make sense of this statement, we must adopt the convention that the point at infinity belongs to every straight line or hyperplane, but that it does not belong to any circle.

As an immediate corollary of this lemma: *Stereographic projection maps every circle in the sphere* $S$ *to a circle or straight line in the hyperplane* $P \cup \infty$. *Furthermore, stereographic projection also preserves angles.*

*Proof of Lemma 2.* First consider a sphere of dimension $n-1$. By the remarks above, we need only consider the case where this sphere $S^{n-1}$ does not contain the base point $\mathbf{p}$. It will be convenient to choose cartesian coordinates so that $\mathbf{p}$ is the origin. If $\mathbf{x}$ and $\mathbf{x}'$ are two points of $S^{n-1}$ which lie on a common line through the origin $\mathbf{p}$, then, as in the proof of Lemma 1, the scalar product $\mathbf{x} \cdot \mathbf{x}'$ is equal to a nonzero constant, which we can write as $\rho^2/a$. It follows that the image of $\mathbf{x}'$ under inversion is equal to $a\mathbf{x}$. *Therefore, the image of* $S^{n-1}$ *under inversion is precisely equal to the sphere consisting of all points* $a\mathbf{x}$ *with* $\mathbf{x}$ *in* $S^{n-1}$.

The proof for a sphere $S^k$ of arbitrary dimension now follows easily. For $S^k$ can be described as the intersection of the sphere $S^{n-1}$ which has the same center and the same radius with a plane $P^{k+1}$. Since inversion maps both $S^{n-1}$ and $P^{k+1}$ to a sphere or plane of the same dimension, it follows that it maps their intersection to a sphere or plane.

Next consider two smooth curves $\Gamma_1$ and $\Gamma_2$ in Euclidean space which intersect at a point $\mathbf{x} \neq \mathbf{p}$. Then their images $\Gamma_1{}^*$ and $\Gamma_2{}^*$ under inversion intersect at a corresponding point $\mathbf{x}^*$. If $\theta$ is the angle between the tangent vectors of the two curves at $\mathbf{x}$, and $\theta^*$ is the angle between the corresponding tangent vectors at $\mathbf{x}^*$, we must prove that $\theta = \theta^*$. Let $L_i$ be the straight line which is tangent to $\Gamma_i$ at $\mathbf{x}$, let $L_i'$ be the parallel straight line through $\mathbf{p}$, and let $C_i$ be the image of $L_i$ under inversion. Then $L_i'$ is tangent to $C_i$ at $\mathbf{p}$, so the angle between $C_1$ and $C_2$ at $\mathbf{p}$ is equal to $\theta$. Evidently the angle between these two circles at their two points of intersection must be equal. Since $C_i$ is tangent to $\Gamma_i{}^*$, this completes the proof. ■

Now let us describe a very different kind of geometry. *Non-Euclidean geometry*, also called *hyperbolic geometry*, was first introduced by Lobachevsky and Bolyai by means of a collection of geometric axioms. These were identical to the axioms of Euclidean geometry, except for the hypothesis that, within a plane, it is possible to construct more than one parallel to a given line through a given point. There was no proof that these axioms were consistent for many years, until Beltrami showed that hyperbolic geometry could be modeled within Euclidean geometry. (See Milnor **[15]** for references.) One of Beltrami's Euclidean realizations of hyperbolic geometry was the *upper half-space model*, later utilized by Poincaré. It can be described as follows.

Let $H$ be an open half-space, with boundary $\partial H$, in the $n$-dimensional Euclidean space $\boldsymbol{E}$. It will be convenient to make use of Cartesian coordinates $\mathbf{x} = (x_1, \ldots, x_n)$, chosen so that the half-space $H$ is defined by the inequality $x_n > 0$. Thus $\partial H$ is the hyperplane $x_n = 0$. If $\Gamma$ is a smooth curve, described parametrically by the equation $\mathbf{x} = \mathbf{x}(t)$, then we will use the notation $\int_\Gamma |d\mathbf{x}|$ for the Euclidean length $\int |d\mathbf{x}/dt|\, dt$.

*Definition.* By the *hyperbolic length* of a smooth curve $\Gamma$ in the half-space $H$ will be meant the integral $\int_\Gamma |d\mathbf{x}|/x_n$, along $\Gamma$, of the Euclidean length element $|d\mathbf{x}|$ divided by the Euclidean distance from $\partial H$.

A curve $\Gamma$ in $H$ will be called a *hyperbolic line* if it provides the hyperbolically-shortest possible path between any two of its points. In other words, there must be no curve in $H$ which joins two points of $\Gamma$ and has hyperbolic length strictly less than the hyperbolic length of the segment $\Gamma_0$ of $\Gamma$ between these points. (In Riemannian geometry, such a curve $\Gamma$ is called a *minimal geodesic*.)

LEMMA 3 (Beltrami, 1868). *A curve in the half-space $H$ is a hyperbolic line if and only if it is contained in either a Euclidean half-line which meets the boundary of $H$ orthogonally, or a Euclidean semi-circle which meets $\partial H$ orthogonally.*

*Proof.* First suppose that $\Gamma$ is a half-line

$$x_1 = \text{constant}, \ldots, x_{n-1} = \text{constant},\ x_n > 0.$$

If $\Delta$ is any other curve segment joining two points of $\Gamma$, then it easy to check that

$$\int_\Delta |d\mathbf{x}|/x_n \geq \int_\Delta |dx_n|/x_n \geq \int_{\Gamma_0} |dx_n|/x_n.$$

In fact there is a strict inequality unless $\Delta$ is also a vertical line segment. *This proves that $\Gamma$ is a hyperbolic line; and furthermore that it provides the unique hyperbolically-shortest path between any two of its points.*

Now consider an inversion mapping $\mathbf{x} \mapsto \mathbf{y}$, using any base point $\mathbf{p}$ on $\partial H$. To simplify the computation, let us first suppose that $\mathbf{p} = \mathbf{0}$. Differentiating the equation $\mathbf{y} = \rho^2\mathbf{x}/(\mathbf{x}\cdot\mathbf{x})$, we obtain

$$d\mathbf{y} = \rho^2\big((\mathbf{x}\cdot\mathbf{x})\, d\mathbf{x} - 2(\mathbf{x}\cdot d\mathbf{x})\mathbf{x}\big)\big/(\mathbf{x}\cdot\mathbf{x})^2.$$

If we take the dot product of this equation with itself, the terms involving $\mathbf{x}\cdot d\mathbf{x}$ cancel, so that we obtain

$$d\mathbf{y}\cdot d\mathbf{y} = \rho^4 d\mathbf{x}\cdot d\mathbf{x}/(\mathbf{x}\cdot\mathbf{x})^2. \tag{10}$$

More generally, for any choice of base point $\mathbf{p}$, the appropriate formula is

$$d\mathbf{y}\cdot d\mathbf{y} = d\mathbf{x}\cdot d\mathbf{x}(\rho/|\mathbf{x}-\mathbf{p}|)^4.$$

It follows easily that

$$|d\mathbf{y}|/|\mathbf{y}-\mathbf{p}| = |d\mathbf{x}|/|\mathbf{x}-\mathbf{p}|.$$

Since the vector $\mathbf{y}-\mathbf{p}$ is a positive multiple of $\mathbf{x}-\mathbf{p}$, this implies that

$$|d\mathbf{y}|/y_n = |d\mathbf{x}|/x_n,$$

whenever the base point $\mathbf{p}$ belongs to the boundary of $H$. *Thus the inversion mapping* $\mathbf{x}\mapsto\mathbf{y}$, *with any choice of base point in* $\partial H$, *maps the half-space* $H$ *into itself so as to preserve hyperbolic length.* Such a length preserving mapping from $H$ to itself is called a *hyperbolic isometry.*

It follows from Lemma 2 that inversion carries any half-line meeting $\partial H$ orthogonally to a semicircle meeting $\partial H$ orthogonally. Since we have shown that these Euclidean half-lines are hyperbolic lines, it follows that *these semi-circles must also be hyperbolic lines.*

To see that there are no other hyperbolic lines, suppose that we start with two arbitrary points $\mathbf{x}$ and $\mathbf{y}$ of the half-space $H$. Then it is not difficult to check that $\mathbf{x}$ and $\mathbf{y}$ lie on either a Euclidean half-line or a Euclidean semi-circle which meets $\partial H$ orthogonally. Thus we have constructed enough hyperbolic lines to join any two points of $H$; and it follows that we have constructed *all* hyperbolic lines. ■

More generally, a $k$-dimensional subset of $H$ is called a *hyperbolic $k$-plane* if it contains the hyperbolic line joining any two of its points. A similar argument shows that a subset of $H$ is a hyperbolic $k$-plane if and only if it is either a Euclidean half-plane or a Euclidean hemisphere, meeting the boundary of $H$ orthogonally, and having dimension $k$.

Note that the group consisting of all hyperbolic isometries from $H$ to itself is quite large. The proof above shows that any inversion map, with base point in $\partial H$, belongs to this group of isometries. As a further example, it is not difficult to check that the mapping

$$(x_1,\ldots,x_n)\mapsto c_n(x_1,\ldots,x_n)+(c_1,\ldots,c_{n-1},0)$$

is a hyperbolic isometry which carries the point $\mathbf{u}=(0,\ldots,0,1)$ to a completely arbitrary point $\mathbf{c}=(c_1,\ldots,c_n)$ of $H$.

One characteristic property of this geometry is the following. *In any hyperbolic plane, consider a triangle* $\Delta$ *of hyperbolic area* $A$, *bounded by three hyperbolic line segments. The sum of the interior angles of* $\Delta$ *is equal to* $\pi - A$. Here we use the usual Euclidean definition of angle in the upper half-space model, but $A$ must be defined as the integral $\int_\Delta (dA)_{\text{Euclidean}}/(x_n)^2$ so as to be invariant under hyperbolic isometries; where $(dA)_{\text{Euclidean}}$ stands for the usual Euclidean area element.

*Proof.* We may assume that $\Delta$ lies in the plane $x_2 = \cdots = x_{n-1} = 0$. Setting $x = x_1$, $y = x_n$, and integrating $A = \iint_\Delta dx\, dy/y^2$ with respect to the $y$ variable, we obtain $A = \int_{\partial\Delta} dx/y$, to be integrated around the boundary of $\Delta$. Each edge of $\Delta$ is either a vertical line segment, on which $d\mathbf{x}/y$ is zero, or a circle arc

$$x = x_0 + a\cos\theta, \quad y = a\sin\theta,$$

on which $dx/y = -d\theta$. Thus $A = -\int_{\partial\Delta} d\theta$. An elementary argument then shows that this expression is equal to $\pi$ minus the sum of the angles. ■

By way of contrast, for any triangle on the unit 2-sphere the sum of the interior angles is equal to $\pi$ *plus* the area. (Compare Coxeter [**4**], pp. 95, 297.) *On a sphere of radius $r$, the corresponding formula would be*

$$\sum(\text{interior angles}) = \pi + KA, \tag{11}$$

*where* $K = 1/r^2$. Let us take this last formula as a definition of the *curvature* of a space of constant curvature $K$. Then evidently Euclidean space has curvature $K = 0$, and hyperbolic space has curvature $K = -1$. More generally, if we change the definition of "length," multiplying all hyperbolic lengths by $r$ and all areas by $r^2$, then we obtain a space of constant curvature $K = -1/r^2$.

Section 2 makes use of the *conformal unit ball model* for hyperbolic space. This is a slightly different Euclidean model which can be constructed as follows. Let us invert the upper half-space $H$, consisting of all points $\mathbf{x}$ with $x_n > 0$, with respect to a sphere of radius $\rho = \sqrt{2}$ centered at the point $\mathbf{p} = (0, \dots, 0, -1)$. Then it is not difficult to check that the image of $H$ under this inversion is precisely the open unit ball $B$, consisting of all vectors $\mathbf{y}$ with $|\mathbf{y}| < 1$. Let $\mathbf{u}$ be the upward unit vector $-\mathbf{p} = (0, \dots, 0, 1)$. A brief computation, based on the formulas $\mathbf{x}+\mathbf{u} = 2(\mathbf{y}+\mathbf{u})/|\mathbf{y}+\mathbf{u}|^2$ and $|d\mathbf{x}| = 2|d\mathbf{y}|/|\mathbf{y}+\mathbf{u}|^2$, shows that the hyperbolic length element $|d\mathbf{x}|/x_n = |d\mathbf{x}|/(\mathbf{x}\cdot\mathbf{u})$ in the half-space $H$ corresponds to the length element

$$2|d\mathbf{y}|/(1 - \mathbf{y}\cdot\mathbf{y}) \tag{12}$$

in the ball $B$. *This last expression is the Riemann-Beltrami formula for the hyperbolic length element in the unit ball* $\mathbf{y}\cdot\mathbf{y} < 1$.

We can take this ball $B$, with the length element (12), as an alternative model for hyperbolic space. It follows from Lemmas 2 and 3 that the *hyperbolic lines* in this model are just the circle arcs or diameters which span the ball $B$, meeting its boundary sphere orthogonally.

This model gives us a further understanding of the richness of the group of isometries of hyperbolic space, for the subgroup consisting of all hyperbolic isometries of $B$ which fix the origin can evidently be identified with the orthogonal group, consisting of all isometries of Euclidean space fixing the origin. The existence of such a large group of isometries would provide a key step in the proof that hyperbolic geometry, constructed in this way, satisfies all of the axioms of Euclidean geometry with the exception of the parallel axiom.

## References

**1.** V. I. Arnold, Mathematical Methods of Classical Mechanics, Graduate texts in Math, 60, Springer, 1978.

**2.** E. A. Belbruno, Two body motion under the inverse square central force and equivalent geodesic flows, Celest. Mech., 15 (1977) 467–476.

**3.** ———, Regularization and geodesic flows, pp. 1–11 of Classical Mechanics and Dynamical Systems, R. L. Devaney and Z. H. Nitecki, Editors, Dekker, 1981.

**4.** H. S. M. Coxeter, Introduction to Geometry, 2nd ed., Wiley, 1969.

**5.** R. L. Devaney, Blowing up singularities in classical mechanical systems, this MONTHLY, 89 (1982) 535–552.

**6.** W. B. Gordon, On the relation between period and energy in periodic dynamical systems, J. Math. Mech., 19 (1969) 111–114.

**7.** G. Györgyi, Kepler's equation, Fock variables, Bacry's generators and Dirac brackets, Nuovo Cimento 53A, (1968) 717–736.

**8.** W. R. Hamilton, The hodograph or a new method of expressing in symbolic language the Newtonian law of attraction, Proc. Roy. Irish Acad., 3 (1846) 344–353 (Math. Papers v. 2, 287–294, Camb. U. Press, 1940).

**9.** G. Herglotz, Bemerkungen zum dritten Keplerschen Gesetz, Probleme der Astronomie: Festscrift für Hugo v. Seeliger, Springer Berlin, 1924, 197–199.

**10.** D. Hilbert and S. Cohn-Vossen, Geometry and the Imagination, Chelsea, 1952.

**11.** J. Kepler, Astronomia Nova, Prague 1609, §59,60 (cf. "Neue Astronomie," Munich-Berlin 1929, 412–413, or Gesam. Werke 3, Munich, 1937, 480–482).

**12.** M. Kummer, On the regularization of the Kepler problem, Comm. Math. Phys., 84 (1982) 133–152.

**13.** T. Levi-Civita, Fragen der klassischen und relativistichen Mechanik, Springer, 1924.

**14.** R. McGehee, Singularities in classical celestial mechanics, Proc. Int. Congr. Math., Helsinki, 1978, 827–834.

**15.** J. Milnor, Hyperbolic geometry: the first 150 years, Bull. Amer. Math. Soc., 6 (1982), 9–24.

**16.** J. Moser, Regularization of Kepler's problem and the averaging method on a manifold, Comm. Pure App. Math., 23 (1970) 609–636.

**17.** Yu. S. Osipov, Geometrical interpretation of Kepler's problem (Russian). Uspehi Mat. Nauk, 27 #2 (1972) p. 161.

**18.** ———, The Kepler problem and geodesic flows in spaces of constant curvature, Celest. Mech., 16 (1977) 191–208.

**19.** B. Riemann, Ueber die Hypothesen welche der Geometrie zu Grunde liegen, Abh. K. G. Wiss. Göttingen, 13 (1868).

**20.** E. L. Stiefel and G. Scheifele, Linear and Regular Celestial Mechanics, Springer, 1971.

**21.** J. L. Synge and B. A. Griffith. Principles of Mechanics, McGraw Hill, 2nd ed., 1949.

**22.** A. Wintner, The Analytical Foundations of Celestial Mechanics, Princeton U. Press, 1941.

# THE SCHLÄFLI DIFFERENTIAL EQUALITY[1]

*An expository note on Euclidean and non-Euclidean volume*

Let $M^n = M^n_K$ be the spherical, Euclidean or Lobachevsky space of constant curvature $K$, and of dimension $n \geq 2$. By a ***polyhedron*** in $M^n$ we mean a compact subset which can be expressed as a finite intersection of closed half-spaces. Consider a family of polyhedra $P^n \subset M^n$ depending smoothly on one or more parameters. Let $V_n(P^n)$ be the $n$-dimensional volume of $P^n$, and for each $(n-2)$-dimensional face $F$ let $V_{n-2}(F)$ be the $(n-2)$-dimensional volume and $\theta_F$ the dihedral angle between the two $(n-1)$-dimensional faces which meet at $F$. (For the case $n-2=0$, we define the "zero-dimensional volume" $V_0(F)$ of a finite set $F$ to be the number of points in $F$.) *Then*

$$K\, dV_n(P^n) = \frac{1}{n-1} \sum_F V_{n-2}(F)\, d\theta_F, \tag{1}$$

*to be summed over all $(n-2)$-dimensional faces.*

In the case of a family of spherical simplexes, this equation between differential forms in the parameter space was proved by Schläfli in the 1850's. The proof was extended to the case of Lobachevsky space by Sforza in 1907, and the extension to more general polyhedra is easy. Modern proofs may be found in [Kneser] or in [Böhm and Hertel]. Another proof is given in the Appendix.

This note will illustrate the Schläfli equation, with particular emphasis on the transition between Euclidean and non-Euclidean cases for $K$ near zero. Let us first illustrate the meaning of this equation by two very elementary applications.

**Example 1.** For a family of two-dimensional polygons $P^2 \subset M^2$, Equation (1) reads

$$K\, dA(P^2) = \sum_{\mathbf{v}} d\theta_{\mathbf{v}},$$

[1] This note was originally written at the Institute for Advanced Study in 1983. However, it was never published since some proofs were incomplete. The present revision was completed at Stony Brook in April 1993.

to be summed over all vertices of $P^2$, where $A(P^2) = V_2(P^2)$ is the area. Integrating this equation, we obtain

$$K \cdot A(P^2) = \sum_{\mathbf{v}} \theta_{\mathbf{v}} + \text{constant}.$$

The precise value of the constant of integration can easily be determined by looking at the limiting case as the polygon $P^2$ flattens out into a line segment, so that $A \to 0$ and $\sum \theta_{\mathbf{v}} \to (m-2)\pi$, where $m$ is the number of vertices. This yields the classical formula

$$\sum_{\mathbf{v}} \theta_{\mathbf{v}} = K \cdot A(P^2) + (m-2)\pi.$$

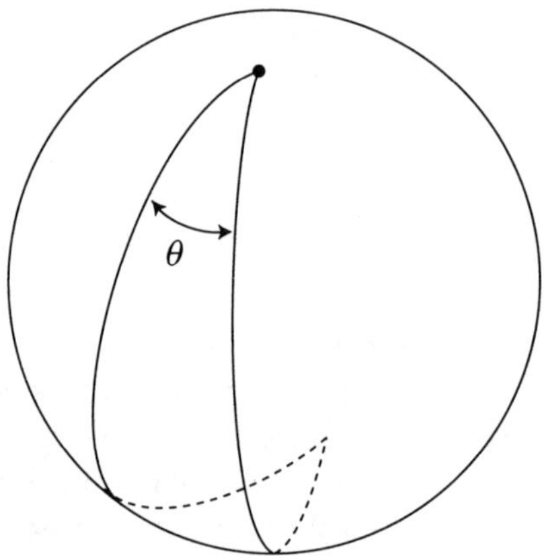


Figure 1. The lune $L^2_\theta \subset S^2$.

**Example 2.** Define an n-dimensional ***lune*** $L^n_\theta$ to be that portion of the unit sphere $S^n \subset \mathbf{R}^{n+1}$ for which the first two coordinates can be written as

$$x_1 = r\cos\varphi, \quad x_2 = r\sin\varphi \qquad \text{with } r \geq 0 \text{ and } 0 \leq \varphi \leq \theta.$$

Then $L^n_\theta$ has just two $(n-1)$-dimensional faces, namely the hemispheres $\varphi = 0$ and $\varphi = \theta$. These intersect in the unique $(n-2)$-dimensional face, namely the sphere $x_1 = x_2 = 0$ which forms their common boundary. The dihedral angle along this face is equal to $\theta$, so the Schläfli equation reads

$$dV_n(L^n_\theta) = \frac{1}{n-1} V_{n-2}(S^{n-2})\, d\theta.$$

Integrating, we obtain

$$V_n(L^n_\theta) = \frac{1}{n-1} V_{n-2}(S^{n-2})\,\theta.$$

In particular, taking $\theta = 2\pi$ so that $L^n_\theta = S^n$, we obtain the familiar recursive formula

$$V_n(S^n) = \frac{2\pi}{n-1} V_{n-2}(S^{n-2}).$$

Starting with $V_0(S^0) = 2$ and $V_1(S^1) = 2\pi$, this formula can be used inductively to compute the volume of the unit sphere of any dimension.

One important special case of the Schläfli equation is the following.

Suppose (as in Example 2) that we are given some family of $n$-dimensional polyhedra $P^n$ parameterized by one or more of the dihedral angles $\theta_F$. We suppose that these preferred dihedral angles can vary independently of each other, with all other dihedral angles fixed. It follows that $V_n(P^n)$ can be expressed (locally at least) as a function of the $\theta_F$, and Equation (1) takes the form

$$K\frac{\partial V_n}{\partial \theta_F} = \frac{1}{n-1} V_{n-2}(F). \tag{2}$$

Here we must assume that $K \neq 0$. For in the Euclidean case $K = 0$, the Schläfli equation gives a linear relation between the differentials $d\theta_F$, so the dihedral angles can never vary independently. Even in the non-Euclidean case, there are many examples where the angles cannot vary independently. For a polyhedron which is combinatorially an $n$-dimensional cube in spherical or Lobachevsky space there are $2n(n-1)$ dihedral angles, but computation shows that there are $n(3n-1)/2$ degrees of freedom up to isometry. Thus for $n = 2$ there are not enough dihedral angles to determine the shape, while for $n \geq 4$ there are too many dihedral angles, so that they cannot vary independently.

On the other hand, according to [Andreev], given any polyhedron in Lobachevsky 3-space with all dihedral angles less than or equal to $\pi/2$, we can vary these dihedral angles independently of each other, and the resulting polyhedron is uniquely determined up to isometry. Hence this formula (2) applies. Similarly, for the family of all $n$-simplexes $\Delta^n$ in the space of curvature $K \neq 0$, it is well known that the $n(n+1)/2$ dihedral angles $\theta_F$ can vary independently, and that these dihedral angles determine $\Delta^n$ up to isometry. (Lemma 1 below.) Thus we can express $V_n(\Delta^n)$ as a function of the $\theta_F$ and Equation (2) will be satisfied.

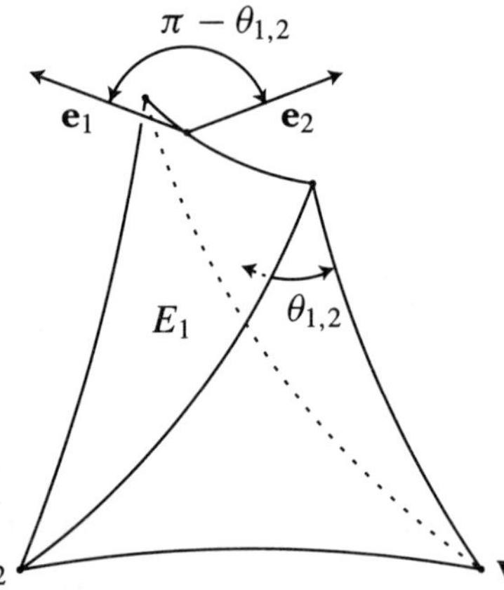


Figure 2. A 3-simplex in Lobachevsky space.

Here is a more precise statement, which includes the Euclidean case. Let $E_0, E_1, \ldots E_n$ be the $(n-1)$-dimensional faces of the simplex $\Delta^n$. Each $(n-2)$-dimensional face can be described uniquely as an intersection $F_{ij} = E_i \cap E_j$. We will identify the collection of all dihedral angles of a simplex with the symmetric matrix $\Theta = [\theta_{ij}]$ where $\theta_{ij}$ is the dihedral angle between $E_i$ and $E_j$ for $i \neq j$, and where the diagonal entries $\theta_{ii}$ are set equal to $\pi$ by definition. It is sometimes more convenient to work with the ***exterior dihedral angles*** $\pi - \theta_{ij}$. If $\mathbf{e}_i$ is the outward unit normal vector to $\Delta^n$ along the face $E_i$, note that $\pi - \theta_{ij}$ is just the angle between $\mathbf{e}_i$ and $\mathbf{e}_j$, measured at any point of $F_{ij} = E_i \cap E_j$. The following closely related construction is even more convenient.

**Definition**. The ***Gram matrix*** $[g_{ij}]$ of the simplex $\Delta^n \subset M^n_K$ is the $(n+1) \times (n+1)$ matrix defined by

$$g_{ij} = \mathbf{e}_i \cdot \mathbf{e}_j = \cos(\pi - \theta_{ij}) = -\cos(\theta_{ij}). \tag{3}$$

All diagonal entries $g_{ii}$ are equal to one. It will be convenient to call a matrix ***unidiagonal*** if it has this property. Note that the affine space consisting of all symmetric unidiagonal $(n+1) \times (n+1)$ matrices has dimension $N = n(n+1)/2$.

Historical note: J. G. Gram (1850–1916) was a Danish mathematician and insurance executive.

It will be convenient to say that two polyhedra $P_1 \subset M^n_{K_1}$ and $P_2 \subset M^n_{K_2}$ are ***similar***, or have the same ***shape***, if they are isometric up to a change of scale. Here we must allow the possibility that $K_1 \neq K_2$; for if we multiply all distances by a constant $r > 0$, then the curvature $K$ will be multiplied by $1/r^2$. Evidently two polyhedra with the same shape must necessarily have the same dihedral angles.

**Lemma 1**. *A simplex must lie either in spherical space, Lobachevsky space, or Euclidean space according as the determinant D of its Gram matrix is positive, negative, or zero. The shape of the simplex is uniquely determined by this matrix* $[g_{ij}]$, *or equivalently by the matrix of angles* $\Theta = [\theta_{ij}]$.

Thus the collection of all possible shapes of simplexes can be identified with the collection of all possible Gram matrices, in the space of dimensional $N = n(n+1)/2$ consisting of all symmetric unidiagonal $(n+1) \times (n+1)$ matrices.

**Lemma 2**. *The collection of all possible Gram matrices associated with finite nondegenerate n-simplexes in any one of the three geometries forms a convex open set* $C_n$ *in this N-dimensional space. Furthermore, the subset* $D = 0$, *corresponding to Euclidean simplexes, forms an* $(N-1)$*-dimensional topological cell* $C_n^0$ *which cuts* $C_n$ *into two open sub-cells* $C_n^+$ *and* $C_n^-$, *corresponding to the spherical and Lobachevsky cases respectively.*

Proofs, as well as a more precise description of the convex set $C_n$ will be given later.

We will be particularly interested in "shape invariants" of a simplex $\Delta^n$, that is quantities such as dihedral angles, or the ratio of two edge lengths, which are invariant under a change of scale. One important example of a shape invariant is the "***volume corrected curvature***"

$$\kappa = K V_n(\Delta^n)^{2/n}. \tag{4}$$

In fact, if all lengths are multiplied by $r > 0$, then the volume will be multiplied by $r^n$ and $K$ will be multiplied by $1/r^2$, so that the expression (4) remains unchanged. In view of Lemma 1, we can express this invariant as a function $\kappa = \kappa(\Theta)$ of the matrix of dihedral angles. Evidently $\kappa(\Theta)$ can be described as the unique value of the curvature $K$ such that there exists a simplex of unit volume in the space of curvature K with this given matrix of dihedral angles.[1] In dimension $n = 2$ we have the simple formula

$$\kappa(\Theta) = \theta_{01} + \theta_{02} + \theta_{12} - \pi.$$

(Compare Example 1 above.) For higher dimensions there is a large liter-

[1] Closely related is the "curvature corrected volume" $|\kappa|^{n/2} = |K|^{n/2} V_n$. For $K \neq 0$ this can be described as the volume of a similar simplex in the space of curvature $\pm 1$. However, this expression is less useful since it is not smooth as a function of $\Theta$ when $n$ is odd, and since it does not determine the sign of $K$.

ature, starting with Lobachevsky in the negative curvature case and Schläfli in the positive curvature case, to understand the computation of volume, and hence of $\kappa$, as a function of $\Theta$. See the list of references at the end.

In terms of this function $\kappa(\Theta)$, the Schläfli equation (2) can be written as follows

$$\frac{n(n-1)}{2}\frac{\partial\kappa}{\partial\theta_{ij}} = \frac{V_{n-2}(F_{ij})}{V_n^{1-2/n}} > 0. \tag{5}$$

(Here we think of the $\theta_{ij}$ with $i < j$ as the independent variables.) In fact we will prove the following.

> **Lemma 3.** *The function $\kappa(\Theta)$ is well defined, smooth, and in fact real analytic as a function of $N = n(n+1)/2$ variables. It is positive, negative or zero according as $K$ is positive, negative, or zero, and satisfies the differential equation* (5).

More precisely, if we use the $g_{ij}$ in place of the $\theta_{ij}$, as independent variables, the function $\kappa$ is defined and real analytic throughout the convex open set $C_n$.

**Remark.** It is conjectured that $\kappa$ extends continuously to the boundary of $C_n$, and that this extended function is non-zero except on the closure of $C_n^0$.

The proofs begin as follows.

**Proof of Lemmas 1 and 2.** We work in the affine space consisting of all symmetric $(n+1)\times(n+1)$ matrices $[g_{ij}]$ with diagonal entries equal to 1. To each such matrix we assign an $(n+1)$-dimensional vector space with basis $\mathbf{e}_0, \ldots, \mathbf{e}_n$ and with the inner product $\mathbf{e}_i \cdot \mathbf{e}_j = g_{ij}$. We first consider the spherical case. Let $C_n^+$ be the convex open set consisting of matrices $[g_{ij}]$ which are positive definite, so that the associated inner product space is Euclidean. Intersecting the unit sphere $S^n$ in this Euclidean space with the half-spaces $\{\mathbf{x} : \mathbf{x}\cdot\mathbf{e}_i \le 0\}$, we obtain a spherical simplex $\Delta^n$ with the specified dihedral angles. In terms of the dual basis $\mathbf{e}_i^*$ for the inner product space[1], the vertices of the simplex are the points $\mathbf{v}_i = -\mathbf{e}_i^*/\|\mathbf{e}_i^*\|$. It is not difficult to check that we obtain all nondegenerate spherical simplexes in this way.

For the Euclidean and Lobachevsky cases, we consider a larger convex open set $B_n$, consisting of matrices $[g_{ij}]$ for which every principal minor is positive definite, so that every $n$ of our $n+1$ basis vectors span a Euclidean

[1] The dual basis is defined by the requirement that $\mathbf{e}_i \cdot \mathbf{e}_j^* = \delta_{ij}$, where $[\delta_{ij}]$ is the identity matrix.

inner product space. We will see that the set-theoretic difference $B_n \setminus C_n^+$ has $2^n$ connected components. The required set $C_n$ of matrices can be described as the union of $C_n^+$ with just one of these $2^n$ complementary components. It will follow easily from this description that $C_n$ is also convex and open.

First consider a matrix $[g_{ij}]$ in $B_n$ with determinant $D = 0$. Then there is an essentially unique linear relation $\sum \beta_i\, g_{ij} = 0$ between the rows. We can set $g_{ij} = \mathbf{e}_i \cdot \mathbf{e}_j$ where the $n+1$ vectors $\mathbf{e}_i$ span an $n$-dimensional Euclidean vector space, and satisfy the corresponding relation $\sum \beta_i \mathbf{e}_i = \mathbf{0}$. These coefficients $\beta_i$ must all be non-zero. If the coefficients $\beta_i$ all have the same sign, then it is not difficult to see that the inequalities $\sum \mathbf{x} \cdot \mathbf{e}_i \le 1$ describe a Euclidean simplex with the required dihedral angles. On the other hand, if the $\beta_i$ do not have the same sign, then there is no (bounded!) simplex with these dihedral angles. An equivalent condition is that all of the entries in the rank one matrix $[c_{ij}] = [\partial D/\partial g_{ji}]$ of cofactors[1] must be strictly positive. (Since the determinant is zero, the matrix product $\sum g_{ij} c_{jk}$ is the zero matrix, so the rows and columns of $[c_{ij}]$ must be proportional to $(\beta_0, \ldots, \beta_n)$. In fact, if we normalize by setting $\beta_i$ equal to the volume of the parallelepiped spanned by the $\mathbf{e}_h$ with $h \ne i$, then we can write $c_{ij} = \beta_i \beta_j$.) To see that the set of all such matrices forms an open topological cell, note that a Gram matrix of this form is uniquely determined by the positive definite $n \times n$ submatrix formed by the $g_{ij}$ with $1 \le i, j \le n$, together with the positive ratios $(\beta_1 : \beta_2 : \cdots : \beta_n)$. The corresponding $\beta_0$ is then uniquely determined by the requirement that $g_{00} = 1$. Any positive definite unidiagonal symmetric matrix, and any positive ratios can occur.

If $D < 0$, so that we have a Minkowski inner product space of signature $(++\cdots+-)$, then either sheet $S$ of the two-sheeted hyperboloid $\mathbf{x} \cdot \mathbf{x} = -1$ provides a model for Lobachevsky space.[2] Two points $\mathbf{x}$ and $\mathbf{y}$ belong to the same sheet if and only if $\mathbf{x} \cdot \mathbf{y} < 0$. We can try to construct the simplex $\Delta^n \subset S$, defined by the inequalities $\mathbf{x} \cdot \mathbf{e}_i \le 0$, with vertices $\mathbf{v}_i = -e_i^*/\|e_i^*\|$, just as in the case $D > 0$. (Here we set $\|\mathbf{e}\| = \sqrt{|\mathbf{e} \cdot \mathbf{e}|}$.) The problem is that these vertices may not all belong to the same sheet of the hyperboloid. A necessary and sufficient condition that we obtain an honest finite simplex is that the numbers $\mathbf{e}_i^* \cdot \mathbf{e}_j^*$, which can never be zero, must be all negative. An equivalent condition is that all entries in the matrix of cofactors $[c_{ij}]$ must be strictly positive, since $\mathbf{e}_i^* \cdot \mathbf{e}_j^* = c_{ij}/D$. These conditions will be satisfied

---

[1] We take all of the $g_{ij}$ to be independent variables in this equation for $c_{ij}$, even though in the actual application the matrix $[g_{ij}]$ is symmetric and unidiagonal.

[2] We will make use of several different models for Lobachevsky space. For introductions to this subject see [Thurston], [Hilbert and Cohn-Vossen], [Coxeter 1969], or [Milnor 1982].

precisely on one connected component of $B_n \setminus C_n^+$, and the union of $C_n^+$ with this component is evidently convex. Further details will be left to the reader. □

**Remark.** Along the Euclidean locus $D = 0$, since the determinant function $D$ has the same zero manifold as the function $\kappa$, we can conclude that the gradient matrix

$$\left[\frac{\partial D}{\partial \theta_{ij}}\right] = \left[\frac{\partial D}{\partial g_{ij}}\frac{dg_{ij}}{d\theta_{ij}}\right] = [c_{ij}\sin\theta_{ij}],$$

with all diagonal entries set equal to zero, is a scalar multiple of $[\partial\kappa/\partial\theta_{ij}]$, or in other words of $[V_{n-2}(F_{ij})]$ (again with all diagonal entries zero).

**Proof of Lemma 3.** Recall the definition $\kappa = K V_n^{2/n}$. For $K \neq 0$, if we keep $K$ fixed, then differentiating with respect to $\theta_{ij}$ and using the Schläfli equation (2) we obtain

$$\frac{\partial\kappa}{\partial\theta_{ij}} = K\cdot\frac{2}{n}V_n^{2/n-1}\frac{\partial V_n}{\partial\theta_{ij}} = \frac{2}{n(n-1)}V_n^{(2/n)-1}V_{n-2}(F_{ij}),$$

which is just the required differential equation (5). (Note that the exponent on $V_n$ is uniquely determined by the requirement that both sides of the equation must be scale invariant.) Intuitively, we can obtain the same equation for the case $K = 0$ simply by passing to the limit as $K \to 0$. However, some care is needed since it is not completely obvious that both sides have well defined limits as $K \to 0$. We must prove that the "shape" of a simplex can be described as a function of the $\theta_{ij}$ or the $g_{ij}$ which remains smooth and real analytic, even as we vary through the Euclidean case $D = 0$. In particular, we claim that any reasonable invariant of shape, such as the ratio of two edge lengths, or the expression $\kappa = K V_n^{2/n}$, can be expressed as a smooth function of the $\theta_{ij}$.

To prove smoothness and real analyticity, we make use of the family of conformal metrics

$$ds^2 = \frac{4\sum dx_i{}^2}{\left(1 + K\sum x_i{}^2\right)^2} \tag{6}$$

which depend smoothly on the curvature $K$. Here, for $K < 0$ we must restrict to the ball $\sum x_i{}^2 < 1/|K|$ where the denominator is positive. For $K \neq 0$ the hyperplanes in this metric can be described either as Euclidean hyperplanes through the origin, or as Euclidean spheres which map into themselves under the inversion

$$\mathbf{x} \mapsto -\frac{1}{K} \frac{\mathbf{x}}{\|\mathbf{x}\|^2}. \tag{7}$$

(This inversion corresponds to the antipodal map on spherical space, when $K > 0$.)

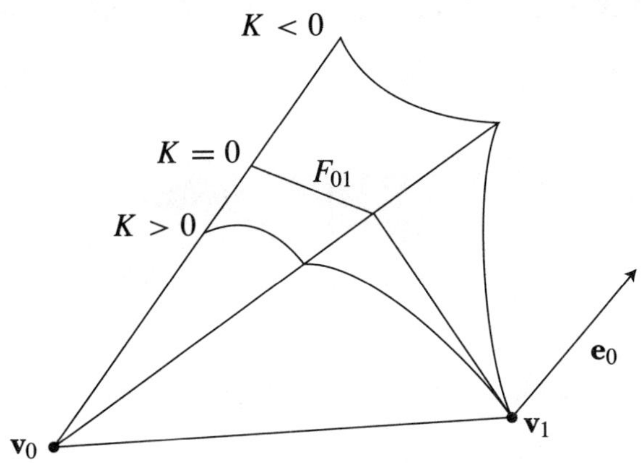


Figure 3. A one-parameter family of simplexes with all dihedral angles but $\theta_{01}$ fixed, parameterized by the curvature $K$, using the conformal metric (6).

Let us start with the Euclidean case $K = 0$. To fix our ideas, since there is just one relation $\det(g_{ij}) = 0$, we can choose a family of Euclidean polyhedra parameterized by the collection of all dihedral angles $\theta_{ij}$ but one. More explicitly, we can express $\theta_{01}$ as a smooth function of the remaining $\theta_{ij}$. Furthermore, let us fix the vertices $\mathbf{v}_0$ and $\mathbf{v}_1$, with $\mathbf{v}_0$ at the origin. Let $\lambda_{ij} = \|\mathbf{v}_i - \mathbf{v}_j\|$ be the Euclidean edge length and let $a_i = (\mathbf{v}_j - \mathbf{v}_i) \cdot \mathbf{e}_i$ be the $i^{\text{th}}$ altitude. Now let us pass to non-Euclidean simplexes by varying the parameter $K$ in (6). Since the faces $E_1, \ldots, E_n$ through the origin remain hyperplanes for all values of $K$, we can leave these hyperplanes unchanged. Further, we leave the vertex $\mathbf{v}_1$, and the direction of the outward normal vector $\mathbf{e}_0$ to $E_0$ at $\mathbf{v}_1$ unchanged, so that all of the dihedral angles $\theta_{ij}$ at $\mathbf{v}_0$ and $\mathbf{v}_1$ will remain unchanged. However, the face $E_0$ will now be replaced by a Euclidean sphere, whose curvature varies monotonically with $K$. More precisely, let us express the Euclidean curvature vector for this sphere at $\mathbf{v}_1$ as a multiple $-k\mathbf{e}_0$, where $\mathbf{e}_0$ is the outward unit normal to $E_0$. By definition, this means that the sphere is centered at $\mathbf{v}_1 - \mathbf{e}_0/k$. Applying the inversion (7) to $\mathbf{v}_1$, we see that the point $-\mathbf{v}_1/(K\|\mathbf{v}_1\|^2) = -\mathbf{v}_1/(K\lambda_{01}^2)$ also belongs to the sphere, and hence has distance $1/|k|$ from $\mathbf{v}_1 - \mathbf{e}_0/k$. A brief computation then shows that

$$k = \frac{2Ka_0}{1 + K\lambda_{01}^2},$$

where the constant $a_0 = \mathbf{e}_0 \cdot \mathbf{v}_1$ is the altitude in the Euclidean case. In particular, $k$ can be expressed as a smooth function of $K$, throughout a neighborhood of $K = 0$, and the derivative $\partial k/\partial K$ is strictly positive.

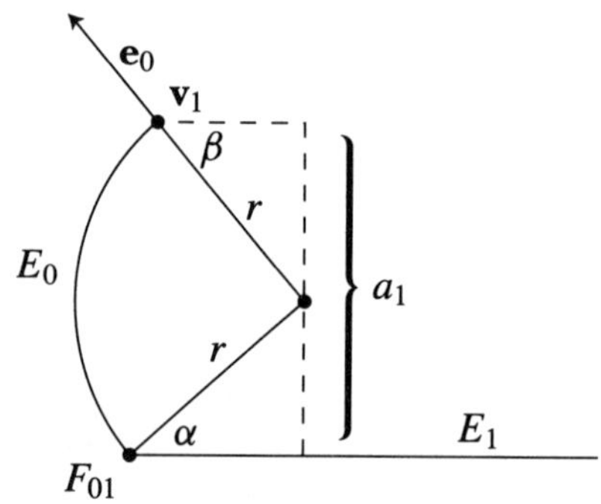


Figure 4. Diagram in the plane through vertex $\mathbf{v}_1$ spanned by the vectors $\mathbf{e}_0$ and $\mathbf{e}_1$. (Here $r = 1/k$ and $\theta_{01} = \alpha + \pi/2$. The vector $\mathbf{e}_1$ points straight down.) The case $K > 0$ is illustrated, but the case $K < 0$ would be similar.

The remaining dihedral angle $\theta_{01}$ can now be expressed as a function of $k$, and hence of $K$. In fact we will show that

$$\frac{\partial\theta_{01}}{\partial k} = \frac{a_1}{\sin\theta_{01}} > 0, \qquad \text{hence} \qquad \frac{\partial\theta_{ij}}{\partial K} > 0.$$

As $k$ varies, the point $\mathbf{v}_1$, the vectors $\mathbf{e}_0$ and $\mathbf{e}_1$, the altitude $a_1$, and the angle $\beta$ all remain constant. However, the radius $r = 1/k$ and the angle $\theta_{01} = \alpha + \pi/2$ vary as functions of $k$. We see from Figure 4 that the sum $r\sin\alpha + r\sin\beta$ takes the constant value $a_1$, equal to the Euclidean distance from $\mathbf{v}_1$ to the opposite face $E_1$. Multiplying by $k$ and differentiating the resulting equation $\sin\alpha + \sin\beta = ka_1$ with respect to $k$, we see that $(\cos\alpha)\partial\alpha/\partial k = a_1$, and the stated equation follows. Thus $\partial\theta_{01}/\partial K > 0$. Now it follows from the implicit function theorem that we can solve for $K$ as a smooth function of the $\theta_{ij}$. Hence the non-Euclidean volume $V_n(P_n)$ is a smooth function of the $\theta_{ij}$, and therefore the product $\kappa = KV_n^{2/n}$ also depends smoothly on the $\theta_{ij}$. $\square$

## Appendix. A proof of the Schläfli equality

**Step 1.** We first consider a very special one-parameter family of polyhedra in Lobachevsky space, with $K = -1$. We will use the upper half space model, with coordinates $x_1, \ldots, x_{n-1}$ and $y > 0$, and with metric

$$ds^2 = \left(dx_1^2 + \cdots + dx_{n-1}^2 + dy^2\right)/y^2.$$

Let $P_t^n$ be the polyhedron, depending on a single parameter $t$, which is obtained by intersecting some fixed polyhedron $P_*^n$ with the half-space $x_{n-1} \leq t$. (A 2-dimensional slice is shown in Figure 5.) If $E_t$ is the "right hand

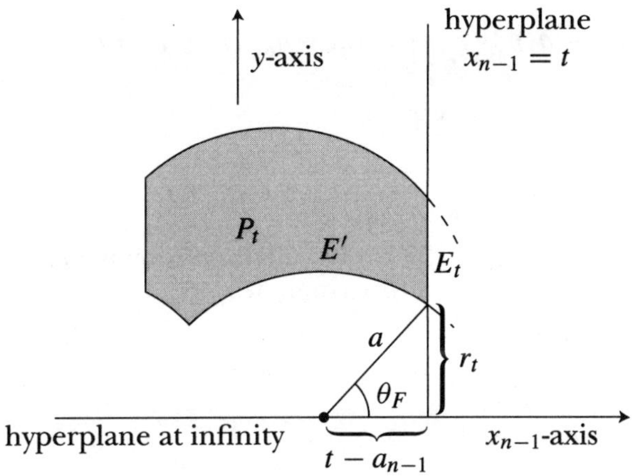


Figure 5. A family of polyhedra in the upper half-space model.

face", obtained by intersecting $P_t^n$ with the plane $x_{n-1} = t$, then clearly the derivative of

$$V_n(P_t) = \int_{P_t} dx_1 \cdots dx_{n-1} dy/y^n$$

with respect to the parameter $t$ is given by

$$\frac{dV_n(P_t)}{dt} = \int_{E_t} dx_1 \cdots dx_{n-2} dy/y^n.$$

Integrating with respect to $y$ and multiplying by $K = -1$, we obtain the formula

$$K\frac{dV_n}{dt} = \frac{1}{n-1}\int_{\partial E_t} \pm dx_1 \cdots dx_{n-2}/y^{n-1}, \tag{8}$$

to be integrated over the boundary of $E_t$, where the top $(n-2)$-dimensional faces $F$ of $E_t$ are counted positively but the bottom faces are counted negatively. (Any "side faces", on which the form $dx_1 \cdots dx_{n-2}$ is zero, can be ignored.)

If $F$ is any $(n-2)$-dimensional face of $E_t$, then we will show that the contribution of $F$ to the integral in (8) is precisely equal to $V_{n-2}(F)d\theta_F/dt$, as required by (1) or (2). Let $F = E_t \cap E'$, where $E'$ is the adjoining $(n-1)$-dimensional face. If $E'$ is a vertical plane, passing through the infinite point of the half-space model, then $F$ makes no contribution to (8) and $d\theta_F/dt = 0$. Thus it suffices to consider the case where $E'$ is a Euclidean hemisphere, with equation

$$(x_1 - a_1)^2 + \cdots + (x_{n-1} - a_{n-1})^2 + y^2 = a^2,$$

so that the intersection $F$ has equation

$$(x_1 - a_1)^2 + \cdots + (x_{n-2} - a_{n-2})^2 + y^2 = {r_t}^2,$$

with ${r_t}^2 = a^2 - (t - a_{n-1})^2$. The $(n-2)$-dimensional *Euclidean* volume element on this hemisphere can be written as $r_t\, dx_1 \cdots dx_{n-2}/y$. Hence the Lobachevsky volume $V_{n-2}(F)$ is the integral over $F$ of $r_t\, dx_1 \cdots dx_{n-2}/y^{n-1}$. On the other hand, the dihedral angle $\theta_F$ satisfies

$$-\cos\theta_F = \pm(t - a_{n-1})/a, \quad \sin\theta_F = r_t/a,$$

taking the plus sign or the minus sign according as the face $E'$ lies above or below $P_t$. Differentiating the left hand equation, we get $\sin\theta_F\, d\theta_F/dt = \pm 1/a$ hence $d\theta_F/dt = \pm 1/r_t$. Thus the product $V_{n-2}(F)\, d\theta_F/dt$ is precisely the required integral over $F$.

**Step 2.** Next let us consider an $n$-parameter family of polyhedra $P_\Pi$ obtained by intersecting some fixed polyhedron $\hat{P}$ in Lobachevsky space with a half-space bounded by the variable hyperplane $\Pi$. It will be convenient to switch to the projective disk model for Lobachevsky space. Consider a unit sphere $S^{n-1}$ in the Euclidean space $\mathbf{R}^n$, which in turn is embedded in the real projective space $\mathbf{RP}^n$. Then points of Lobachevsky space correspond to points of the open disk $D^n$ bounded by $S^{n-1}$, while hyperplanes $\Pi$ in Lobachevsky space extend to hyperplanes in $\mathbf{RP}^n$ which intersect $D^n$. These correspond in a one-to-one fashion with points $p$ in the complementary region $\mathbf{RP}^n \setminus \overline{D}^n$. The correspondence is such that every line from $\Pi \cap S^{n-1}$ to $p$ is tangent to the sphere. (Compare Figure 6.) If we allow $p$

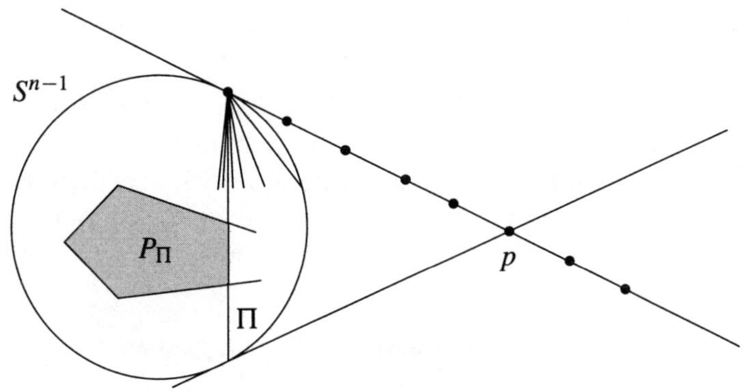


Figure 6. Family of polyhedra in the projective unit disk model.

to vary along a line tangent to the sphere, then we obtain a one parameter family of hyperplanes $\Pi_t$ which sweep out all of Lobachevsky space. If we then switch back to the upper half-space model, letting the fixed point of tangency correspond to the point at infinity, then these hyperplanes $\Pi_t$ correspond precisely to a family of mutually parallel hyperplanes of the form $x_{n-1} =$ constant, as considered in Step 1.

Let us start with any point $p$ whose associated hyperplane $\Pi$ intersects the fixed polyhedron $\hat{P}$ transversally. Then the same will be true for any point $p'$ close to $p$. In order to prove Equation (1) for this $n$-parameter family, it suffices to find $n$ linearly independent vectors at $p$ for which the equation is true. But we have already proved the equation for all directions at $p$ which point along lines tangent to $S^{n-1}$. Since the cone tangent to $S^{n-1}$ from $p$ contains $n$ linearly independent vectors, the conclusion follows.

**Step 3.** Now let us start with a convex polyhedron $P$ in Lobachevsky $n$-space such that precisely $n$ of its $(n-1)$-dimensional faces meet transversally at any vertex. (For example this condition is satisfied for simplex or cube, but not for an icosahedron.) Let $E_1, \ldots, E_k$ be the $(n-1)$-dimensional faces of $P$. Then $P$ can be described as the intersection of $k$ closed half-spaces, and fits into a smooth $nk$-parameter family which is obtained by varying these half-spaces independently. Clearly Equation (1) is true for this $nk$-parameter family, since it suffices to vary the faces one at a time.

**Step 4.** We have proved Equation (1) for simplexes in the case $K = -1$. It follows for all negative values of $K$, since it is formulated so as to be invariant under a change in scale in which all lengths are multiplied by some positive constant. Proceeding as the proof of Lemma 3, we see by analytic continuation that it is true for simplexes in the case $K \geq 0$ also. (Thus we have reproved Schläfli's original statement.)

**Step 5.** A completely arbitrary polyhedron $P$ in a space of constant curvature can be cut up into simplexes by some rectilinear subdivision. If we vary $P$ within a small neighborhood in the parameter space, then it is not difficult to vary the subdivision smoothly. Since both sides of Equation (1) are additive with respect to such a smoothly varying subdivision, this completes the proof. □

**Miscellaneous Concluding Remarks.** In the Lobachevsky case, it is also of interest to consider a family of polyhedra with one or more vertices on the sphere of points at infinity. If $n \neq 3$, then Equation (1) remains true as stated. If $n = 3$, then some of the edge lengths $V_1(F)$ become infinite. However, if we remove a small horospherical neighborhood of each infinite vertex $\hat{\mathbf{v}}$ before measuring edge lengths, then (1) remains valid. Here it is important to note that the various dihedral angles $\theta_F$ along edges extending to $\hat{\mathbf{v}}$ form the angles of a Euclidean triangle. Hence the sum of the corresponding differentials $d\theta_F$ is zero.

Here is another elementary application which may help to illustrate Schläfli's equation. Every convex polyhedron $P$ in the $n$-dimensional spherical space, with $K = 1$, has a dual polyhedron $P^*$ consisting of all points of distance at least $\pi/2$ from $P$. (Compare p. 223 of *Euler characteristic and finitely additive Steiner measures,* in this volume.) Each exterior dihedral angle $\pi - \theta_F$ of $P$ is equal to a corresponding edge length $V_1(F^*)$ of $P^*$, and conversely each edge length of $P$ is equal to a corresponding exterior dihedral angle of $P^*$. In the 3-dimensional case, summing over all edges $F$, we assert that

$$V_3(P) + V_3(P^*) + \frac{1}{2}\sum V_1(F)V_1(F^*) = \pi^2.$$

This equation can be verified by checking that the total differential of each side is zero, and that the equation is true in the limiting case as $P$ shrinks to a point and $P^*$ expands to a hemisphere.

There is a curious formal analogy between this differential form

$$\sum V_{n-2}(F)\, d\theta_F$$

which occurs in the Schäfli equation and the ***Dehn invariant***

$$\sum V_{n-2}(F) \otimes \theta_F \in \mathbf{R} \otimes_{\mathbf{Z}} (\mathbf{R}/2\pi\mathbf{Z}).$$

In particular, both expressions are constituted so as to be additive when two polyhedra are pasted together along a common face. Compare [Dupont and Sah], as well as [Milnor 1983].

## Bibliography


***General Background***

D. Hilbert and S. Cohn-Vossen, "Geometry and the Imagination," Chelsea, 1952.

***Non-Euclidean Geometry and Volume Computations***

E. M. Andreev, *On convex polyhedra in Lobachevskii spaces,* Math. USSR Sb. **10** (1970), 413–440.

K. Aomoto, *Analytic structure of Schläfli function,* Nagoya J. Math **68** (1977), 1–16.

A. Bloch and G. Guillaumin, *Sur la volume des polyèdres non euclideans,* C. R. Acad. Sci. Paris **224** (1947), 1690–1692.

J. Böhm, *Zu Coxeters Integrationsmethode in gekrümmten Räumen,* Math. Nachr. **27** (1964), 179–214.

——— and E. Hertel, "Polyedergeometrie in $n$-dimensionalen Räumen konstanter Krümmung," Birkhauser, 1981.

H. S. M. Coxeter, *The functions of Schläfli and Lobatschefsky,* Quart. J. Math. **6** (1935), 13–29.

———, "Non-Euclidean Geometry," 5th ed. Toronto, 1965.

H. Kneser, *Der Simplexinhalt in der nichteuklidische Geometrie,* Deutsche Math **1** (1936), 337–340.

B. L. Laptev, *The volume of a pyramid in Lobacevskii space,* Kazan. Gos. Univ. Uc. Zap. **114,** no. 2 (1954), 53–77. (Russian)

N. I. Lobachevsky, *On the foundations of geometry,* Kazan Messenger **25–28** (1829–30).

W. Maier, *Inhaltsmessung im $R_3$ fester Krümmung,* Arch. Math. **5** (1954), 226–273.

J. Milnor, *Hyperbolic geometry: the first 150 years,* Bull. A. M. S. **6** (1982), 9–24. (This volume, pp. 245–260.)

H. W. Richmond, *The volume of a tetrahedron in Elliptic space,* Quart. J. Math. **34** (1903), 175–177.

L. Schläfli, *On the multiple integral $\int^n dx\,dy \cdots dz$, whose limits are $p_1 = a_1 x + b_1 y + \cdots + h_1 z > 0$, $p_2 > 0, \ldots, p_n > 0$, and $x^2 + y^2 + \cdots + z^2 < 1$,* Quart. J. Math **2**, 269–301.

G. Sforza, *Sul volume dei poliedri nell'ipotesi non euclidea,* Soc. Nat. Math. Modena (4) **9** (1907).

E. B. Vinberg, *Discrete groups generated by reflections in Lobacevskii spaces,* Math. USSR Sb. **1** (1967), 429–444.

***Miscellaneous Papers***

J .L. Dupont and C.-H. Sah, *Scissors congruences,* II, J. Pure Appl. Alg. **25** (1982), 159–195.

J. Milnor, *Polylogarithms, Hurwitz zeta functions, and the Kubert identities,* Enseign. Math. (2) **29** (1983), 281–322. (See Appendix 3.)